International Series in Operations Research & Management Science

Volume 343

The book series **International Series in Operations Research and Management Science** encompasses the various areas of operations research and management science. Both theoretical and applied books are included. It describes current advances anywhere in the world that are at the cutting edge of the field. The series is aimed especially at researchers, advanced graduate students, and sophisticated practitioners.

The series features three types of books:

• Advanced expository books that extend and unify our understanding of particular areas.

• Research monographs that make substantial contributions to knowledge.

• Handbooks that define the new state of the art in particular areas. Each handbook will be edited by a leading authority in the area who will organize a team of experts on various aspects of the topic to write individual chapters. A handbook may emphasize expository surveys or completely new advances (either research or applications) or a combination of both.

The series emphasizes the following four areas:

Mathematical Programming: Including linear programming, integer programming, nonlinear programming, interior point methods, game theory, network optimization models, combinatorics, equilibrium programming, complementarity theory, multiobjective optimization, dynamic programming, stochastic programming, complexity theory, etc.

Applied Probability: Including queuing theory, simulation, renewal theory, Brownian motion and diffusion processes, decision analysis, Markov decision processes, reliability theory, forecasting, other stochastic processes motivated by applications, etc.

Production and Operations Management: Including inventory theory, production scheduling, capacity planning, facility location, supply chain management, distribution systems, materials requirements planning, just-in-time systems, flexible manufacturing systems, design of production lines, logistical planning, strategic issues, etc.

Applications of Operations Research and Management Science: Including telecommunications, health care, capital budgeting and finance, economics, marketing, public policy, military operations research, humanitarian relief and disaster mitigation, service operations, transportation systems, etc.

This book series is indexed in Scopus.

Matthias Seifert
Editor

Judgment in Predictive Analytics

Springer

Editor
Matthias Seifert
IE Business School
Madrid, Madrid, Spain

ISSN 0884-8289 ISSN 2214-7934 (electronic)
International Series in Operations Research & Management Science
ISBN 978-3-031-30084-4 ISBN 978-3-031-30085-1 (eBook)
https://doi.org/10.1007/978-3-031-30085-1

This Springer imprint is published by the registered company Springer Nature Switzerland AG
The registered company address is: Gewerbestrasse 11, 6330 Cham, Switzerland

To Sara and Hana

Preface

One of the critical challenges in predictive analytics relates to the question of how to effectively design and utilize algorithmic models in a world of continuous change. In recent years, we have witnessed technological milestone achievements in the fields of artificial intelligence, neural networks, machine learning, cloud computing, and others, which allow decision-makers to have better and more sophisticated predictive tools at their disposal. In fact, a recent survey suggested that the global AI market will be valued at $407 billion by the year 2027 with predictive analytics holding the potential to significantly reduce operational costs, for instance related to inventory management and delivery optimization (Beasley, 2021).

Despite its potential to significantly improve business decisions, one of the biggest restraints in the rise of predictive analytics lies in what is often referred to as the human factor. In particular, given the increasing complexity of algorithmic models, companies face a need for employing skilled workers to both design such models and manage their output. This is important, because effectively using predictive analytics requires the joint effort of multiple human actors at various stages of the modelling process. For example, in the FMCG sector, the tasks of (1) recording and pre-processing demand data, (2) building the forecasting model, and (3) interpreting model outputs are typically carried out by different managers at different points of time. Furthermore, the process of making judgmental adjustments to machine-generated outputs often involves the collective opinions of *several* experts, who may be conflicting in their subjective assessment of the decision or forecasting task at hand.

Taking this into account, past research has suggested that the role of judgment in human-machine collaboration represents a double-edged sword. In some contexts, human experts have been shown to be highly proficient in their diagnostic ability to detect abnormal deviations from historical data patterns. This can prove to be extremely valuable in situations where disruptive changes in the task environment make static models become obsolete. In other situations, however, human intervention can also exert a detrimental effect on predictive performance as judgments are likely to suffer from various cognitive biases.

Hence, the effectiveness of predictive analytics critically depends on firms' ability to recognize *when* and *how* human intervention can add value as well as to understand in what manner expert judgments and model outcomes can be aggregated to create crowd wisdom effects.

Addressing this challenge from a scholarly perspective lies at the heart of this edited book. Defining *predictive analytics* as the class of analytical methods that makes predictions about future outcomes using historical data combined with statistical modeling, data mining techniques, and machine learning, the chapters in this book aim at providing a comprehensive overview of extant research on various key issues of human–model interactions.

To achieve this purpose, the book is divided into three main parts, each shedding light on one critical aspect related to role of judgment in predictive analytics. The chapters included in this book have been contributed by key scholars and together intend to showcase the state-of-the-art of research in this field.

Part I focuses on research regarding the role of judgment in human–machine interactions. In particular, in the context of predictive analytics, human intervention can occur either during the various stages of the model building process or when it comes to utilizing machine-generated outputs. Past research has highlighted a number of ways in which human judgment can make a valuable contribution, but also raised concerns as to when it may be systematically flawed. Taking this into account, Chap. 1 draws on the literature of augmented decision-making to discuss how *algorithm aversion*—a behavioral phenomenon accounting for the human tendency to discount algorithmic model outputs—may impact the role of judgment in predictive analytics. Specifically, the authors offer a systematic contextualization of this behavioral phenomenon and propose methods that could help mitigating it.

Chapter 2 then shifts the focus to the role of human judgment in the context of augmented reality, where the objective is to develop a machine learning-based object detection system. Adopting the lens of design thinking, the authors discuss various types of subjective decisions that need to be taken during the development process and showcase how they critically influence the functionality of the final outcome.

Chapter 3 represents a reprint of a seminal article on the role of human judgment in the selection of forecasting models. The authors compare the judgmental performance of forecasting practitioners when assessing the appropriateness of demand forecasting models against a commonly used algorithm.

Chapter 4 (reprint) includes one of the rare studies examining the complementarity of human judgment and statistical models by systematically decomposing judgments into their linear and nonlinear forecasting achievements. Specifically, the authors rely on a Brunswikian lens model to study how forecasters' cognitive ability to interpret linearities and nonlinearities in the task environment may add predictive value over and beyond linear models.

Chapter 5 then concludes Part I of the book by studying behavioral changes in the adjustment of machine-generated forecasts. In particular, the authors rely on a laboratory experiment to examine the extent to which experts modify forecasts when adjustments may occur at various stages of the forecasting process.

Part II of this edited series highlights the role of judgment in collective forecasting. Taking this into account, Chap. 6 discusses the issue of skill identification in forecasting crowds. In the context of probability estimation tasks, the authors categorize forecasting skills into various measures and use two focal applications to study which of the measures are likely to be most strongly correlated with forecasting accuracy.

Chapter 7 subsequently provides a comprehensive overview of performance-weighted aggregation methods that are commonly used in collective forecasting settings. The authors distinguish between several different approaches to weigh the input of individual experts and discuss the benefits and drawbacks in their practical implementation.

In Chap. 8, the authors direct attention to the role of time horizon in extracting the wisdom of crowds. The temporal distance to the resolution of the forecast event has a direct impact on the judgmental accuracy of forecasters and, thus, changes the way in which individual judgments should be optimally aggregated. In this chapter, the authors describe various aggregation challenges resulting from this time dependence and use real-world data to derive several recommendations for forecasting practice.

The final part of this edited volume (Part III) showcases how the characteristics of the task environment influence the role of human judgment in predictive analytics. In particular, as the configuration of the forecasting environment critically influences the appropriateness of algorithmic models, Chap. 9 sheds light on the value of scenario analysis as a means to improve predictive performance. The authors provide a comprehensive summary of the extant literature on the use of scenario-based forecasting and report the results of a behavioral experiment, which identifies both scenario tone and extremity as important drivers of judgmental predictions when a model-based forecast is available.

Chapter 10 provides a general taxonomy of external events in time series forecasting, which can be broadly classified according to their event characteristics and impact. The authors continue by discussing various judgmental and modelling-based approaches to effectively incorporate event characteristics into time series forecasts.

Finally, Chap. 11 discusses the challenges of using predictive analytics from the perspective of the organizational context. The author adopts the lens of mindful organizing to propose that an excessive emphasis on forecast accuracy may prevent organizations from collective learning going forward.

In sum, this edited volume seeks to provide a comprehensive overview of research addressing the role of human judgment in predictive analytics. We hope that readers will find the ideas included in the chapters thought-provoking and stimulating for carrying out future studies on this important topic.

Madrid, Spain Matthias Seifert

Reference

Beasley, K. (2021). *Unlocking the power of predictive analytics with AI.* Forbes Technology Council. Accessed from https://www.forbes.com/sites/forbestechcouncil/2021/08/11/unlocking-the-power-of-predictive-analytics-with-ai/?sh=17fc929b6b2a

Acknowledgments

This work has been generously supported by the Ministerio de Ciencia e Innovación in Spain (agency ID: 10.13039/501100011033) with the grant numbers PID2019-111512RB-I00-HMDM and ECO2014-52925-P.

Contents

Part I
Judgment in Human-Machine Interactions

Chapter 1
Beyond Algorithm Aversion in Human-Machine Decision-Making

Jason W. Burton, Mari-Klara Stein, and Tina Blegind Jensen

Keywords Algorithm aversion · Human-machine · Decision-making · Hybrid intelligence

1 Introduction

In the 1950s, several influential papers dispelled with the idea of the utility maximizing "economic man" portrayed in classical rational choice theory (e.g., Brunswik, 1955; Edwards, 1954; Hammond, 1955; Meehl, 1954; Miller, 1956; Simon, 1955). Instead, it was argued that human decision-makers are *boundedly rational*, meaning they are constrained by the limited computational capacities of the mind and by limited access to information in the environment (Simon, 1955). In contrast to "economic man," real human decision-makers usually cannot calculate the expected utility of all possible alternatives due to either uncertainty about relevant possibilities, inability to perform such mental calculations, or both. Thus, if one is striving for optimality in decision-making, an attractive approach is to augment human (i.e., clinical, expert, intuitive) judgment with a machine (i.e., statistical, actuarial, mechanical, algorithmic) aid.

The case for augmented, machine-aided decision-making—hereinafter referred to as *human-machine decision-making*—stems from the longstanding observation that human judgment tends to be less accurate than judgment reached via a formal decision rule, formula, or algorithm (Ægisdóttir et al., 2006; Dawes, 1979; Dawes et al., 1989; Goldberg, 1965; Grove et al., 2000; Grove & Meehl, 1996; Kuncel et al., 2013; Meehl, 1954). This early observation clearly and controversially suggested

J. W. Burton (✉) · T. B. Jensen
Department of Digitalization, Copenhagen Business School, Frederiksberg, Denmark
e-mail: jb.digi@cbs.dk

M.-K. Stein
Department of Digitalization, Copenhagen Business School, Frederiksberg, Denmark

Department of Business Administration, Tallinn University of Technology, Tallinn, Estonia

M. Seifert (ed.), *Judgment in Predictive Analytics*, International Series in Operations Research & Management Science 343,
https://doi.org/10.1007/978-3-031-30085-1_1

that people would benefit from more heavily relying on machine-generated judgments, as opposed to those made by human clinicians, managers, teachers, or judicial officials alone. Yet, alongside those demonstrating the superiority of machine decision-making over human decision-making were others who emphasized the valuable human role in providing the inputs that machines needed to mechanically combine, thereby giving rise to jointly made, human-machine decisions (Dana & Thomas, 2006; Edwards, 1962; Einhorn, 1972; Sawyer, 1966). Thanks to recent advances in computing capacity, data availability, and analytics, opportunities for the design and application of human-machine decision-making have become unignorable. However, the success of any human-machine decision system rests on the assumption that its human users will effectively accept and utilize its mechanical outputs—an assumption that recent research has questioned (e.g., Dietvorst et al., 2015; Logg et al., 2019).

In this chapter, we explore the concept of *algorithm aversion* in human-machine decision-making. Beginning with a brief overview of the human versus machine debate in judgment and decision-making, we then outline the business case for augmented, human-machine decision-making before defining algorithm aversion, its implications, and antecedents. Finally, we conclude by highlighting conceptual and methodological limitations of existing research on algorithm aversion and point towards future directions for improving its study to both enhance academic understanding and inform real-world practice.

Given the focus of the chapter is on the relationship between human and machine decision-making in general, we use the term *decision-making* interchangeably with terms like *judgment*, *forecasting*, and *prediction*. Likewise, it is also worth noting that what we describe as *human-machine decision-making* has also been referred to as *augmented decision-making* (e.g., Baudel et al., 2021; Burton et al., 2020), *decision/forecasting support systems* (e.g., Alavi & Henderson, 1981; Prahl & Van Swol, 2017; Promberger & Baron, 2006), and *decision aids* (e.g., Dietvorst et al., 2016; Kuncel et al., 2013) in the existing literature.

2 The Human vs. Machine Debate in Judgment and Decision-Making

While contemporary discussions of machine decision-making often center on the imagery of automated algorithms and artificial intelligence, the origins of the human versus machine debate in judgment and decision-making pre-date "machines" as we know them today. In 1954, Paul Meehl published his book, *Clinical versus Statistical Prediction: A Theoretical Analysis and a Review of the Evidence*, in which he surveyed 20 empirical studies to compare the performance of two methods for prediction when given a set of data. In what he labelled the *clinical* (or *nonmechanical*) method, such decision-making is achieved via a human judge's informal, holistic inference without the application of any straightforward equation or rule. Whereas in the *statistical* (or *mechanical*) method, the data are mechanically combined with a formal equation or rule, and then an actuarial table is consulted with the resulting output. In other words, statistical decision-making can be easily carried out and reproduced by a clerk, whereas clinical decision-making relies on a trained professional's evaluation and weighing of the data. Crucially, Meehl (1954) viewed these two methods as mutually exclusive. Every decision was either reached through the human's clinical method or the machine's statistical method, and upon tallying up the "wins" for each method, Meehl (1954) reported that the clinical outperformed the statistical only once.[1] This observation and the ensuing debate about the merits of each method is what has been known as the clinical-statistical controversy, and is what we discuss today as the human versus machine debate in judgment and decision-making.

The message received from Meehl's (1954) book was that more decisions should be delegated away from human experts and left to machines—that is, formal rules, models, or algorithms. Indeed, this conclusion was reified in his subsequent article bluntly posing the question, "When shall we use our heads instead of a formula?" (Meehl, 1957). The answer, he argued, is that when there is disagreement between the prediction of a clinician's intuition and the prediction of a mathematical formula, going with our heads is "very, *very* seldom" Meehl, 1957, p. 273) the correct choice. But why is this the case? Beyond the general inconsistency in human evaluations of complex data (i.e., the same person evaluating the same data twice may give different answers), Meehl (1954) was rather sympathetic to what he called "the special powers of the clinician" (p. 24). He described these special powers with a simple example case of trying to predict whether an individual will go to the cinema on a certain night. In the example, a mechanical analysis of all the relevant, available data (e.g., the demographics of the individual and it being a Friday) suggests there is a 90% probability that the individual will go to the cinema. However, if a clinician is aware that, in addition to the available data, the individual in question has a broken leg, then the clinician can outperform the mechanical prediction because this single, extra datum reduces the probability from 90% to near zero. While this so-called "broken leg problem" is easily dismissed by machine advocates as being so improbable that it bears little to no empirical consequence, Meehl (1954) took it seriously and explained that "there may be *many (different) rare kinds of factors*," and "improbable factors as a class, each of which considered singly will not appear in a statistical analysis as significant, may contribute heavily to the 'misses'" (p. 25). Nevertheless, Meehl (1954) and his analysis cautioned against relying on human intuition as a sensor for broken leg cases.

Objections to Meehl's conclusions are well-chronicled. As itemized by Grove and Meehl (1996), these ranged in perspective from the conceptual (e.g., "the goal is to understand, not to predict") to the ethical (e.g., "it is inhumane and degrading") to

[1] McNemar (1955) later identified an error in the one study that suggested superiority of the clinical method (Hovey & Stauffacher, 1953), meaning it should have been recorded as a tie between the clinical and statistical methods.

the logistical (e.g., "we do not have a regression equation or actuarial table") (also see Dawes, 1979; Dawes et al., 1989). While reviewing each recognized criticism is out of scope for this chapter, the general narrative is that expert human practitioners were highly skeptical that a machine could outperform their judgmental accuracy, arguing that Meehl's (1954) conclusion could not be generalized to the types of decisions and services they provided in the real world. Despite straightforward rebuttals to such objections, and several follow-up studies and meta-analyses demonstrating the superiority of machine decision-making across hundreds of studies in wide ranging contexts (e.g., Ægisdóttir et al., 2006; Dawes, 1979; Dawes et al., 1989; Goldberg, 1965; Grove et al., 2000; Grove & Meehl, 1996; Kuncel et al., 2013; Meehl, 1954), the narrative of the Luddite-like clinician and layperson carried on [e.g., "the hostility to algorithms" in Kahneman's (2013, p. 227)].

Although there has always been a clear empirical question as to whether people object to or discount judgments made by machines, the recorded "rage against the machine" in judgment and decision-making was largely anecdotal in nature when the narrative was first taking hold. For instance, one of the primary (and few)[2] challenges to Meehl (1954) in the academic literature came from Robert Holt, who did not object to the merits of the statistical method. For Holt the central concern was with Meehl's "formulation of the issues rather than his arguments, which are sound" (Holt, 1958, p. 1). In fact, he agreed that "when the necessary conditions for setting up a pure actuarial system exist, the odds are heavy that it can outperform clinicians in predicting almost anything in the long run" (Holt, 1970, p. 348). However, Holt (1958, 1970, 1986) argued that Meehl and his contemporaries focused too narrowly on a false dichotomy between the clinical, human method and the statistical, machine method. Instead of keeping score of the "wins" for one method or the other, the more important task for Holt (1958) was "to find the optimal combination of actuarially controlled methods and sensitive clinical judgment" (p. 12).

Holt's (1958, 1970, 1986) compatibilist view—whereby decision-making is conceptualized as a joint, human-machine process—was emphatically denounced by Meehl. As Meehl (1986) explained, neither he nor anyone else disputed the value of consulting a clinician in setting up a statistical model, but practically-speaking, the relevant data must be combined through *either* human judgment or a formal equation, two methods which will naturally disagree in many cases as his analyses implicitly showed. While this re-assertation of Meehl's (1954) original distinction between the clinical and statistical method is unambiguous, others continued to distinguish different components of the decision-making process—namely, data collection versus data combination—and approached the question of humans vs. machines as a matter for humans *and* machines (e.g., Dana & Thomas,

[2] The other primary opposition came decades later in Gary Klein's (1993, 1997, 2008) studies on naturalistic decision-making. These studies promote reliance on expert intuition by focusing on real world contexts marked by time pressure and high-stake consequences, rather than artificial experiments. While Kahneman acted as his contemporary adversary, they reconciled their positions by agreeing that the comparative performance of human versus machine judgment depends on the environment in which it takes place (Kahneman & Klein, 2009).

2006; Edwards, 1962; Einhorn, 1972; Sawyer, 1966; for a review see Kleinmuntz, 1990). Here, the basic observation was that where machines can reliably apply mathematical rigor to combine data, human judgment serves as a vital measuring device to provide the data itself. In doing so, such human-machine decision-making integrates the machine's superiority in data combination with the so-called "special powers of the clinician."

3 Human-Machine Decision-Making

In what is perhaps the earliest concrete proposal for human-machine decision-making, Edwards (1962) introduced a prospective design for Probabilistic Information Processing (PIP) systems. The basic idea of the PIP system starts with the premise that the raw data decisions are based on are often fallible or incomplete, and so a type of Bayesian processing is needed to account for not only the conditional probability that a certain hypothesis is true given the data, but also the conditional probability of the data given the hypothesis. While in some cases such conditional probabilities can be calculated directly from the raw data, in other cases imperfections in the data (e.g., sampling error; missing observations) will result in those calculations being misleading. In the latter cases, Edwards (1962) proposed that human judges could serve as intermediaries by interpreting raw data and estimating the conditional probabilities needed to enable a mechanical combination via Bayes' theorem.[3] Although Edwards' (1962) PIP system was hypothetical at the time, his speculation was later followed up by Einhorn (1972) in an empirical study examining the accuracy of pathologists' predictions for cancer patients' survival. What Einhorn (1972) found was that both the informal, holistic predictions of the pathologists and the formal, mechanical predictions of a pre-specified formula could be outperformed by a hybrid method in which the expert pathologists identified and weighed the importance of cues that were then mechanically combined. This finding, along with similar results in other contexts—e.g., the mechanical combination of human judgments (Camerer, 1981; Dawes, 1971; Goldberg, 1970), the aggregation of human and machine judgments (Blattberg & Hoch, 1990; Lawrence et al., 1986; Pankoff & Roberts, 1968), and the human adjustment of machine outputs (Wolfe & Flores, 1990)—provided the initial evidence base from which interest in human-machine decision-making was born.

Fast forward to the present day and the topic of human-machine decision-making is subject to much discussion both in and outside of academia. With the digitalization of society naturally leading to the production of massive amounts of data (Holst, 2021), advances in computing allowing for increased storage and accessibility of said data (Hilbert & Lopez, 2011), and rapid developments in machine learning

[3] Bayes' theorem is a mathematical formula for calculating a conditional probability that a hypothesis is true given some evidence (for an in depth review see Joyce, 2003).

enabling large-scale analytics (Hindman, 2015), businesses and society at large are motivated to find human-machine configurations that can efficiently leverage data for decision-making and value creation. By delegating specific components of decision-making tasks (e.g., data collection, evaluation, combination, and implementation) to humans or machines, such configurations can take shape in a variety of ways (for taxonomies see Parasuraman et al., 2000; Pescetelli et al., 2021; Zellner et al., 2021). These configurations include relatively simple systems that aggregate human and machine judgments (Fildes & Petropoulos, 2015; Yaniv & Hogarth, 1993), and systems that allow for human decision-makers to adjust the data fed into a machine model or the model outputs themselves (Grønsund & Aanestad, 2020; Sanders & Manrodt, 2003; Wolfe & Flores, 1990). They also include more complex systems that use machines to mediate collective decision-making processes (Burton et al., 2021a, b; Rosenberg et al., 2017). While each of these configurations aims to increase decision accuracy, no single configuration suits all contexts. For example, an aggregation of human and machine judgments is most likely to perform well in situations where the humans and machines draw on different sets of information (Blattberg & Hoch, 1990; Yaniv & Hogarth, 1993). Human adjustments to machine outputs are most likely to bring about benefits for decision tasks in highly-variable environments (Wolfe & Flores, 1990), and machine-mediated collective decision-making processes can only be used in digital contexts where there is ready access to data on all individuals' judgments and their interactions (Pescetelli et al., 2021).

Notwithstanding the challenge of identifying the appropriate configuration for any given context, the benefits of human-machine decision-making are twofold. First and foremost is the potential to enhance decision accuracy, which, in principle, is achieved because humans and machines have complementary strengths and weaknesses (Blattberg & Hoch, 1990; Jarrahi, 2018). Where machines' computational power allows for precise, reliable analyses of data that would otherwise be intractable for humans, they can be unhelpfully rigid in the face of uncertainty as they are limited to whatever data has been pre-specified. In contrast, humans may possess domain knowledge and flexibility that allows them to recognize unexpected "broken leg" cues and adapt under uncertainty, but might also display motivational biases and inconsistencies in judgment. Thus, when used in combination, the hope is that humans' and machines' errors will cancel out.

A secondary benefit of human-machine decision-making is the potential to widen the range of decision-making problems that can be addressed. Since machines allow for the ingestion of larger volumes of data, there is the possibility of increasing the granularity of predictive decisions when guided by human domain expertise. Along these same lines, the speed with which machines can take in data coupled with the human ability to account for short-term variability means raw data can be translated into actionable insights at a faster pace (e.g., *nowcasting*, Sills et al., 2009).

Evidently, these opportunities of human-machine decision-making are widely recognized. Surveys of forecasting practitioners show that reliance on pure human judgment has decreased while the use of human-machine decision-making has increased (Fildes & Petropoulos, 2015). And relatedly, major state-sponsored funding programs have been introduced to stimulate research in this space, such as the Hybrid Forecasting Competition supported by the Intelligence Advanced Research Projects Activity (IARPA) in the United States.[4]

However, the adoption of human-machine decision-making in practice also brings new challenges. Academics and policymakers alike have written extensively about how, at a societal scale, reliance on increasingly complex, opaque decision-making systems may introduce new forms of discrimination, undermine privacy, and generally reduce the human experience (e.g., Mittelstadt et al., 2016; Newell & Marabelli, 2015; Partnership on AI, 2019; The Parliamentary Office of Science and Technology, 2020; Tutt, 2017; Zerilli et al., 2019). At its core, much of the challenge for any given system can be traced back to the interaction between the machines and human users. That is to say, the success of any human-machine decision-making in practice relies on the ability of the human users to understand and discriminately utilize machines' outputs. A complacent over-reliance on machine decision-making tools can lead to unfounded conclusions because such tools are built to find associations rather than meaning in data—for example, the failure of Google Flu Trends (Ginsberg et al., 2009; Lazer & Kennedy, 2015) and the failure of Nike's i2 forecasting system (Worthen, 2003). On the other hand, a misguided under-utilization of mechanical outputs will necessarily undo the potential benefits of any human-machine decision-making system. With these concerns in mind, recent research efforts have sought to empirically test human perceptions and acceptance of machine-generated judgments. Here, the finding of so-called *algorithm aversion* (Dietvorst et al., 2015) has prompted many to believe that human-machine decision-making is undermined by a human tendency to discount machine outputs.

4 Beyond Algorithm Aversion: What Is Algorithm Misuse?

Since the anecdotal accounts of a resistance to algorithms (or the statistical method) in the wake of Meehl's (1954) book, an array of empirical studies have presented evidence of a general algorithm aversion—"a biased assessment of [algorithmic outputs] which manifests in negative behaviors and attitudes towards the algorithm compared to a human agent" (Jussupow et al., 2020, p. 4). For example, laypeople rate decisions made by professionals (i.e., doctors, lawyers, scientists) more positively when they are told the decision has been made with human judgment rather than a statistical formula (Eastwood et al., 2012). Illusory knowledge and overconfidence lead to a neglect of algorithmic outputs that results in inferior decision accuracy (Arkes et al., 1986; Cadario et al., 2021; Sieck & Arkes, 2005), and judgmental errors made by algorithmic decision aids lead to a greater decrease in utilization as compared to errors made by human advisors (Dietvorst et al., 2015; Prahl & Van Swol, 2017). However, current understandings of algorithm aversion

[4] https://www.dni.gov/index.php/newsroom/press-releases/item/1785-iarpa-launches-hybrid-forecasting-competition-to-improve-predictions-through-human-machine-integration

are not as clear-cut as is often portrayed, and a number of parallel studies suggest that people prefer algorithmic judgment to human judgment under certain conditions. For instance, people can be more easily persuaded by advice from an "expert system" than a human because they view it as more objective and rational, even when the expert system provides erroneous advice (Dijkstra, 1999; Dijkstra et al., 1998). Further, people prefer algorithmic news curation over human editorial curation (Thurman et al., 2019) and weigh advice more when it is framed as coming from an algorithm than a group of people (Logg et al., 2019). Taken together, these studies demonstrate the difficulty of conceptualizing algorithm aversion in a way that is both coherent and inclusive of all relevant past findings. For this reason, a precise, agreed upon definition that can produce accurate predictions for when and to what degree algorithm aversion (or appreciation) will arise remains elusive.

One contributing factor to this confusion is the fact that studies of algorithm aversion have appeared in the existing literature under a variety of guises across several decades. For instance, Alavi and Henderson (1981) examined design considerations to "increase both *utilization of and satisfaction with [a decision support system]*" (p. 1310). Arkes et al. (1986) examined the "factors influencing the *use of a decision rule* in a probabilistic task" (p. 93). Whitecotton (1996) studied "the effects of experience and confidence on *decision aid reliance*" (p. 194), and Önkal et al. (2009) examined "the *relative influence of advice from human experts and statistical methods* on forecast adjustments" (p. 390) (italics added to emphasize differences in terminology).

Moreover, experiments have also operationalized algorithm aversion in different ways. Study participants may have faced a binary task of taking advice from either a human or an algorithm in a simulated decision-making scenario, a quantitative task of updating a judgmental estimate after being provided with the estimate of a human or algorithmic advisor, or an evaluative task of rating their trust in advice from a human versus an algorithm (for a review see Jussupow et al., 2020). As a result, making comparisons across the literature requires a careful eye, as reconcilable results can easily be misunderstood to be in opposition to one another. For example, in the paper where the term "algorithm aversion" was originally coined, Dietvorst et al. (2015) specifically demonstrated that people preferred their own judgment or the judgment of another person over that of a superior but imperfect algorithm after seeing the algorithm err. In other words, the algorithm aversion identified by Dietvorst et al. (2015) refers specifically to an asymmetry whereby human decision-makers are more tolerant of human error than algorithmic error. Often overlooked in that study, however, is that participants displayed either indifference to or a preference for algorithmic judgment over human judgment *before* seeing the algorithm perform (i.e., in the control condition) (Dietvorst et al., 2015). Thus, while it is easy to understand why a casual reader might expect a subsequent paper's presentation of an "algorithm appreciation" effect—whereby people asymmetrically rely on algorithmic advice *more than* advice from a human advisor or their own judgment (Logg et al., 2019)—to be contradictory evidence, this is not actually the case. As Logg et al. (2019) highlight, their "results are consistent with those of Dietvorst et al. (2015)" (p. 99) because they consider only scenarios where the human decision-maker has no knowledge of the algorithm's past performance. Indeed, a recent review of 29 published experimental studies between 2006 and 2019 identified ten papers with conclusive evidence of algorithm aversion, four papers with conclusive evidence of algorithm appreciation, and 15 with mixed or inconclusive results (Jussupow et al., 2020).

Fundamentally, the varied terminologies, operationalizations, and experimental results highlight the importance of recognizing the label of algorithm aversion as a misnomer. This is because the crucial implication for human-machine decision-making at large is not an *aversion* to algorithmic outputs per se, but a broader class of algorithm misuse. As mentioned in the previous section, human-machine decision-making requires the discriminate use of said outputs, meaning that both under-utilization and over-utilization can lead to undesirable consequences. Notwithstanding this confusion in the literature, however, the accumulated findings of past research have pointed out a range of factors that may increase the likelihood of human decision-makers to misuse algorithmic outputs.

5 Causes of Algorithm Aversion and Algorithm Misuse

Algorithm aversion and the general misuse of machine outputs in human-machine decision-making can occur due to features of the individual human user, the machine aid, the task environment, and the interactions between them. As identified in a systematic review of the literature from 1950 to 2018 conducted by the present authors, these include (1) prior knowledge accrued by the human user, (2) a real or perceived lack of decision control afforded by the machine aid's design, (3) lacking or misguided incentives imposed by the task environment, (4) the alignment of the human's and machine's decision-making processes, and (5) the alignment of the human's and machine's decision-making objectives (Burton et al., 2020).[5] In the remainder of this section, we provide a brief overview of each of these five factors as antecedents to algorithm aversion.

5.1 Prior Knowledge

When placed in a human-machine decision-making system, human users' prior knowledge is likely to shape their subjective perceptions of a machine's performance, functionality, and intentions. Be it gained through direct past experience or through the testimony of others, the nature of this prior knowledge has been shown to influence how machine outputs are utilized. Here, two important dimensions have

[5]These five factors fall under the fives themes presented in Burton et al. (2020): expectations and expertise, decision autonomy, incentivization, cognitive compatibility, and divergent rationalities.

been differentiated in existing literature: the degree of general, computing or statistics knowledge and the degree of specialized, task-specific expertise.

If a human decision-maker has been trained to work with statistical models and is familiar with how machine decision aids function, then it is more likely that the outputs of a machine aid will be properly interpreted and utilized (Mackay & Elam, 1992; Whitecotton, 1996; Green & Hughes, 1986). While this observation is straightforward, different factors have been highlighted in explanations of this effect. For instance, one factor is the importance of understanding statistical concepts like uncertainty so that errors observed in small samples do not unduly undermine the acceptance of high but imperfect accuracy in the long run (Arkes et al., 1986; Ashton et al., 1994; Dietvorst & Bharti, 2020; Einhorn, 1986; Rebitschek et al., 2021). In addition, some have pointed out the need to undo the folk belief that human advisors learn from their errors whereas machines cannot, which may cause human users to abandon machine aids upon seeing them err (Berger et al., 2021; Dietvorst et al., 2015, 2016; Highhouse, 2008; Renier et al., 2021). Through either explanation it seems clear that for human users to appropriately respond to machine-generated outputs they must have a sufficient level of "algorithmic literacy."

On the other hand, if a human decision-maker possesses expertise in a specific task domain, it may be *less* likely for the outputs of a machine aid to be properly utilized (Arkes et al., 1986; Ashton et al., 1994; Logg et al., 2019; Montazemi, 1991; also see Lawrence et al., 2006 for a discussion how domain knowledge affects forecasting in general). This is arguably because domain expertise can lead a human decision-maker to believe that a machine aid is unnecessary or inferior with respect to their own judgment. In principle, this faith in one's own expertise may be warranted in certain circumstances when the so-called "special powers of the clinician" are used to recognize cues in the environment that have not been accounted for by a machine aid. However, empirical evidence suggests that this is often not the case: reported expertise associates with overconfidence in one's judgment, in turn leading to the neglect of machine-generated outputs to the detriment of ultimate judgmental accuracy (Cadario et al., 2021; Logg et al., 2019; Whitecotton, 1996).

5.2 Decision Control

Another factor shown to influence the utilization of a machine decision aid is whether the human users are granted an opportunity to exert control over the machine aid's functionality. For example, in a follow-up to their initial study coining the term algorithm aversion, Dietvorst et al. (2016) demonstrated how participants in a mock forecasting task reported higher satisfaction with and greater confidence in their algorithmic aid when they were able to modify its outputs. Relatedly, it also seems plausible that an ability to modify a machine aid's *inputs* may associate with a greater uptake of its outputs. In a recent analysis of large-scale behavioral data, Lin et al. (2022) found that diabetes patients displayed no change in adherence to machine-generated insulin dosage advice after seeing the machine err—thereby not supporting Dietvorst et al.'s (2015) original presentation of algorithm aversion. However, the machine aid in question for Lin et al. (2022) specifically asks the users to provide inputs to be analyzed, which potentially allows for the users to perceive a sense of control and confidence that the machine will not be undermined by uniqueness neglect (cf. Jung & Seiter, 2021). Further underscoring the importance of decision control, Jussupow et al.'s (2020) review noted that algorithm aversion tends to be observed in experiments where participants are presented with fully-automated machine outputs (i.e., outputs that could only be accepted or rejected in a binary fashion), whereas algorithm appreciation and inconclusive results appeared more frequently in experiments that present participants with "advisory" machine outputs (i.e., outputs that can be accepted or rejected in a continuous fashion).

The desirable effect of decision control on human users' experience seems duly recognized, with survey findings indicating that practitioners' most used forecasting method is judgmentally adjusting the statistical outputs of model (Fildes & Petropoulos, 2015). Yet it must also be recognized that granting human users too much control could undermine the benefits of a machine aid. In Dietvorst et al. (2016), for example, participants given the opportunity to freely adjust the outputs of an algorithm produced less accurate decisions than those whose adjustments were restricted to a pre-specified range. However, reported satisfaction with and confidence in the algorithmic aid did not differ between participants granted complete control and those granted restricted control (Dietvorst et al., 2016). This observation thereby suggests that designers of machine aids should target a middle ground to accommodate human users' need for autonomy without permitting the neglect of machine outputs altogether.

5.3 *Incentive Structures*

Looking beyond characteristics of the human user or the machine aid, incentive structures in the task environment also influence the degree to which machine outputs are utilized. As highlighted by Brown (2015) and Hafenbrädl et al. (2016), human-machine decision-making inherently imposes costs of time and effort because it requires the integration of multiple judgments. Thus, extra motivation is needed to balance these costs with the potential benefits of decision performance, otherwise machine aids may simply be ignored or worked around in practice (Christin, 2017).

While for some tasks the incentives are explicit, such as monetary rewards for judgmental accuracy, other tasks may impose more implicit social incentives. In cases of the former, a seemingly common-sense prediction would be that algorithm aversion could be lessened by explicitly rewarding decision-makers for accuracy, since it has long been shown that statistical, machine methods often outperform human judgment alone (e.g., Meehl, 1954). But counterintuitively, early experimental work has demonstrated how introducing monetary incentives for decision accuracy may *decrease* reliance on an algorithmic aid (Arkes et al., 1986). The effect of such explicit incentives seems to depend on their competitive, game theoretic structure. As explained in Burton et al. (2020),

> if a decision maker is incentivized to make the best decision (relative to peers on a case-by-case competition basis) rather than a good decision (relative to one's own performance in the long run), then he or she would need to find a way to gain a unique advantage over competitors. If all competitors have access to the same or similar algorithmic aids, then the decision maker would put him or herself at a disadvantage by utilizing the algorithmic judgment since this would mean simply mirroring, rather than surpassing, the performance of other decision makers (p. 5).

Still, the role of explicit incentivization on the utilization of machine outputs remains poorly understood. A reader of the literature quickly comes across seemingly contradictory results. For example, Dietvorst et al.'s (2015) observation of algorithm aversion came in the presence of monetary rewards for accuracy with no competition among participants; Prahl and Van Swol (2017) did not observe algorithm aversion despite competitive monetary rewards; and Önkal et al. (2009) observed algorithm aversion in the absence of monetary incentives.

The variable findings on the influence of explicit accuracy incentives point toward the need to consider how implicit, social incentives might feature in human users' utilization of machine outputs. This is perhaps most conspicuous in professional environments (e.g., medicine, law, consultancy) that have long promoted the image of the "expert intuiter," where reliance on a machine aid may be stigmatized as a signal of incompetence (Alexander et al., 2018; Brown, 2015; Eastwood et al., 2012; Highhouse, 2008; Klimoski & Jones, 2008; Kuncel, 2008; Önkal et al., 2009). In such environments, a human user may feel discouraged from utilizing machine outputs out of a desire to maintain an "aura of omniscience" in front of clients, patients, or managers (Arkes et al., 2007; Eastwood et al., 2012; Sanders & Courtney, 1985). Going a step further, there may also be decision tasks for which "accuracy" is ill-defined, and the perceptions of peers and stakeholders are the most explicit incentives available. Indeed, empirical results suggest that algorithm aversion may specifically arise in subjective and moral decision tasks (Bigman & Gray, 2018; Castelo et al., 2019). In cases such as these, a general conclusion is that human users' utilization of machine outputs can be incentivized through social influence. For example, Alexander et al. (2018) demonstrate how informing a human user that a machine aid has been utilized by others can increase adherence to the machine's outputs. While this seems to suggest that algorithm aversion may gradually dissipate as human-machine decision-making naturally becomes more prevalent and socially acceptable, it also suggests that, in the meantime, interventions could target perceptions of social norms to reduce algorithm aversion.

5.4 Alignment of Decision-Making Processes

Human-machine decision-making requires human users to incorporate interaction with machine aids into their decision-making processes. For this interaction to be successful, the human user must be both metacognitively aware of their own abilities and sufficiently understand the machine aid to recognize the conditions under which its outputs should be accepted, adjusted, or ignored. With interpersonal human-to-human interactions, individuals can pick up on cues in their peer's behavior to infer that peer's mental state and enable efficient communication (i.e., *theory of mind*, Premack & Woodruff, 1978). Yet when interacting with a machine, such cues may be perceived differently or be entirely unavailable. For instance, Scherer et al. (2015) show that people expect high-stakes decisions to be preceded by slow, careful deliberation, presumably indicating an effortful consideration of evidence. And while a slower response time for a machine aid may indeed be interpreted as a signal of effortful exertion, this is typically perceived negatively by human users and associates with greater distrust of the machine's outputs (Efendić et al., 2020). For a human user to calibrate their trust and confidence in a machine aid there must be an alignment between the decision processes of a human user and machine aid so that the user can readily recognize when to delegate tasks to the machine, and how to recognize situations where it requires adjustment (Brown, 2015; Muir, 1987; Sieck & Arkes, 2005).

Crucially, however, this need for alignment of decision processes goes both ways. Not only does the human user need to be able to recognize when the machine aid is most likely to produce (in)valid outputs, but the machine aid must also be designed in a way that caters to cognitive and situational limitations faced by the user. For example, research shows that people display overconfidence (Arkes et al., 1986; Brown & Jones, 1998; Eining et al., 1997; Sieck & Arkes, 2005) and conservatism in response to machine-generated outputs (Lim & Connor, 1996). Given that human decision-makers often lack the metacognitive control to correct for these limitations on their own, designers of machine aids should expect these cognitive tendencies to affect how the outputs of the aid are utilized. Relatedly, human decision-makers in the real world often face constraints on things such as the information they can access or the time they have to make a decision. With this in mind, outputs generated by a machine aid may be more likely to be utilized if that aid can fit into heuristic-based processes the user typically follows, rather than demanding the user learn an entirely new procedure (e.g., *fast-and-frugal trees*, Hafenbrädl et al., 2016).

5.5 Alignment of Decision-Making Objectives

In addition to aligned decision processes, human-machine decision-making also requires an alignment of the human's and machine's decision objectives. That is, for a human user to value and accept the outputs of a machine aid, the user must first trust that the machine will help them reach what they believe to be a "rational" or "good" decision. Typically, algorithms and machine aids aim for optimization. Given a set of data, a machine might calculate the expected utility across a set of possible alternatives so that the "optimal" choice is identified (as the alternative with the maximum expected utility). While this objective of maximizing expected utility may be appropriate in a world of *risk*, where all relevant probabilities and alternative outcomes are known, such optimization is unfeasible in a world of *uncertainty*, where alternatives and probabilities are unknowable or cannot be reliably calculated with available data (Gigerenzer & Gaissmaier, 2015; Knight, 1921). Under uncertainty, the decision-making objective is thus not to optimize, but to *satisfice*; to satisfy minimal needs so that a sufficient, "good enough" decision is reached (Simon, 1956). Whereas many of the most typical machine decision aids (e.g., linear models) assume their applications take place under risk in order to function, a vast majority of real-world decision tasks arguably take place under uncertainty. This suggests that for certain decision tasks, a human user and a machine aid may be targeting fundamentally different outcomes, which may in turn affect how the machine's outputs are utilized.

Indeed, this distinction between risk and uncertainty in the context of human-machine decision-making has been hinted at in early literature mentioning the effects of task structure on decision aid utilization (Benbasat & Taylor, 1978; Carey & Kacmar, 2003; Er, 1988; Green & Hughes, 1986; Kahn & Baron, 1995; Sage, 1981; Sanders & Courtney, 1985). Here, the basic result is that while most decision aids are more suited to structured tasks (i.e., decisions under risk), human decision-makers are more likely to utilize decision aids in unstructured tasks (i.e., decisions under uncertainty). This presents a problem because it suggests that human users turn to machine aids specifically in the instances where those aids are most likely to err, which may lead to an unwarranted decrease in trust in the machine (as in Dietvorst et al., 2015). However, counter to this point, Dietvorst and Bharti (2020) find that people prefer human over algorithmic judgment when faced with uncertainty. They show that this is because human judges in their experiments pursued decision outcomes according to a concave loss function and thus preferred decision-making methods with more variance and less bias. In other words, people have diminishing sensitivity to error (i.e., a small error is viewed as nearly as bad as a large error), and therefore prefer decision-making methods that have a chance of producing a near-perfect decision, even if that method (e.g., human judgment) performs worse in the long run (Dietvorst & Bharti, 2020). Whereas this finding does not fit neatly into the presumption that human decision-makers seek to satisfice under uncertainty, it emphasizes the need for an alignment of human and machine decision objectives and calls for continued research in this space.

6 Towards Improved Methods and Metrics for Understanding and Resolving Algorithm Misuse

The prior knowledge of human users, opportunities for decision control, incentive structures in the environment, and the alignment of decision processes and objectives have all been shown to factor into how machine outputs are utilized. But do these past findings provide us with a firm enough understanding of algorithm aversion to resolve it in the real world? Despite the substantial and growing body of literature, this seems unlikely. As with many topics in the social and behavioral sciences, the issue of algorithm misuse faces a large, high-dimensional parameter space. Even if one is to accept just the five "known" factors outlined in the previous section, the amount of data needed to predict interactions with more than mere speculation is vast. While the many multi-disciplinary research efforts on algorithm aversion referenced throughout this chapter have indeed provided a wide array of relevant data and results, their usefulness is limited by a lack of systematicity. Studies of algorithm aversion have largely been conducted in singular, stand-alone instances whereby each individual study may provide a valid result, yet little attention has been paid to how the collection of results fit together for real-world solutions (Almaatouq et al., 2022; Watts, 2017). This can be noted in the adoption of several varying methods and measures. Past research has applied vignette-based experiments (e.g., Eastwood et al., 2012), incentivized forecasting tasks (e.g., Prahl & Van Swol, 2017), and ethnographic field studies (e.g., Christin, 2017), and relied on ranging operationalizations such as weight on advice (e.g., Logg et al., 2019), binary utilization (e.g., Dietvorst et al., 2015), and subjective evaluations (e.g., Cadario et al., 2021). As a result, comparing and generalizing findings from one study to another is exceedingly difficult. What might be a focal variable of interest in one study (e.g., incentivization schemes, Arkes et al., 1986) is likely to be treated elsewhere as a nuisance variable and specified seemingly arbitrarily (e.g., incentivizating participants to compete with one another, Prahl & Van Swol, 2017), and it is unclear how one is to know whether that (arbitrary) handling of the variable affects results. While employing a variety of methods can be informative and ensure that a result is not an artefact of any particular study design, the value of a "many methods" approach requires consensus as to what the phenomenon in question is and how to operationalize it. Such a consensus seems underdeveloped in the algorithm aversion literature. For instance, is algorithm aversion a response to seeing a machine aid err (Dietvorst et al., 2015) or a general distrust of computers (Promberger and Baron, 2006)? This lack of systematicity in research not only limits academic understanding, but practically speaking, it also means that any intervention geared towards resolving algorithm misuse may be misguided depending on which antecedents and which "algorithm misuse" it targets.

To introduce systematicity it seems necessary to develop standardized methods and metrics. But to make strides towards achieving this, we first need to answer the normative question: how *should* human users respond to machine-generated judgments? An obvious answer here may be that human decision-makers should utilize outputs in whatever way results in decisions that achieve maximum accuracy. Indeed, new quantitative metrics in this respect have recently been proposed. With business contexts in mind, for example, Baudel et al. (2021) quantify the success of human-machine decision-making, M_1, with a ratio:

$$M_1 = \frac{H_{aided}}{\max\{H_{unaided}, A\}}$$

where H_{aided} is the accuracy achieved by the human decision-maker after being aided by a machine, $H_{unaided}$ is the initial accuracy achieved by the human decision-maker in the absence of the machine aid, and A is the accuracy of the machine aid on its own. As they explain, $M_1 > 1$ may be observed when the human decision-maker and the machine aid draw on different information or weigh information differently, in which case a successful "collaboration effect" arises where the human-machine judgment outperforms both the human and machine alone (Baudel et al., 2021). A reader familiar with the advice taking literature might recognize that this measure bears some resemblance to the "weight on advice" (WOA) measure (for a review see Bonaccio & Dalal, 2006), which has previously been adapted for studies of algorithm aversion (most typically without using absolute values, e.g., Logg et al., 2019; Önkal et al., 2009; Prahl & Van Swol, 2017). Calculating WOA proceeds as follows:

$$\boldsymbol{WOA} = \frac{|\ \boldsymbol{aidedEstimate} - \boldsymbol{initialEstimate}\ |}{|\ \boldsymbol{machineEstimate} - \boldsymbol{initialEstimate}\ |}$$

where, assuming a numeric scale, *aidedEstimate* is a decision-maker's ultimate judgment, *initialEstimate* is the decision-maker's initial, unaided judgment, and *machineEstimate* is the machine aid's judgment (or advice). But crucially, M_1 accounts for decision accuracy whereas WOA does not. Put simply, a high WOA indicating adherence to a machine aid is practically meaningless if it does not benefit accuracy. Notwithstanding potential ethical concerns (e.g., Mittelstadt et al., 2016), there is nothing inherently good or bad about preferring human or machine judgment from a decision-making perspective, and the only anti- or pro-machine "bias" of consequence is one that incurs a systematic accuracy cost (for a review bias in the context of human rationality, see Hahn & Harris, 2014).[6] An informative measure of algorithm aversion, misuse, or whatever term is used should take this into account.

Answering the normative question is more difficult, but arguably most important, in decision scenarios where accuracy is ill-defined or incalculable. This is perhaps most obvious for markedly subjective decision tasks where "accuracy" depends on

[6]We note that in experimental set ups using variants of the WOA measure (e.g., Logg et al., 2019; Önkal et al., 2009), decision accuracy is often artificially accounted for because the machine aid's judgment (or advice) is guaranteed to be highly accurate by the experimenter. However, this is not necessarily guaranteed in the real world, where achieving a high WOA may be undesirable in some circumstances (i.e, *complacency bias* or *automation bias*, Baudel et al., 2021; Parasuraman & Manzey, 2010; Zerilli et al., 2019).

the tastes of the individual user. However, the problem of defining accuracy also applies to contrived experimental studies where, despite potentially having a pre-specified truth against which decisions can be assessed, it is difficult to distinguish a systematic bias that incurs an accuracy cost *on average* from sporadic judgmental errors. For this reason, an appropriate normative model of "optimal" responding to machine judgments is needed so that real-world accuracy costs can be identified as systematic deviations from said model. Unfortunately, it is unclear what the appropriate normative model is with respect to algorithm misuse and human-machine decision-making, particularly given their context-dependent nature. One reasonable proposal for a normative model is provided by Logg et al. (2019), who argue that users should respond to machine-generated outputs the same way they respond to human-generated outputs. While this proposal is sensible if all other things are held precisely equal, it seems insufficient in general. Presumably, a user should weigh machine judgment differently than human judgment depending on characteristics of the machine aid, human judge, and decision task. In cases where, for example, a machine aid is being used outside of its intended domain of application, it would likely be inadvisable to weigh its outputs equally to advice provided by a knowledgeable human judge. Showing that users respond differently to machine-generated outputs versus human-generated outputs does not necessarily imply "bias" or "irrationality" to be corrected.

7 Conclusion

While advances in computing technologies promise human-machine decision-making systems capable of achieving high judgmental accuracy in evermore domains of application, this can only be realized if algorithm aversion—and the misuse of machine outputs in general—is properly understood and resolved. In this chapter, we contextualized recent findings of algorithm aversion within the longstanding human versus machine debate in judgment and decision-making, emphasizing that such "aversion" may be less prevalent than is often assumed. Making note of early proposals for human-machine decision-making (e.g., Edwards, 1962), we explained how humans and machines can be configured so that their strengths and weaknesses complement one another. Yet crucially, the potential benefits of this human-machine decision-making can be undone by *both* under- and over-utilization of machine outputs on behalf of the human user. With this in mind, we then outlined the contemporary notion of algorithm aversion and the variety of related, sometimes contradictory results before summarizing five key influencing factors. These factors—prior knowledge, decision control, incentive structures, and the alignment of decision processes and objectives—should be considered by anyone with an interest in understanding or predicting the success of human-machine decision-making. However, academics and practitioners alike should be wary of the limitations on current understandings. Without more standardized methods and measures, and an appropriate normative model against which findings can be calibrated, linking supposed under- or over-utilization of machine outputs to real-world consequences may not be straightforward.

References

Ægisdóttir, S., White, M. J., Spengler, P. M., Maugherman, A. S., Anderson, L. A., Cook, R. S., Nichols, C. N., Lampropoulos, G. K., Walker, B. S., Cohen, G., & Rush, J. D. (2006). The meta-analysis of clinical judgment project: Fifty-six years of accumulated research on clinical versus statistical prediction. *The Counseling Psychologist, 34*(3), 341–382. https://doi.org/10.1177/0011000005285875

Alavi, M., & Henderson, J. C. (1981). An evolutionary strategy for implementing a decision support system. *Management Science, 27*(11), 1309–1323.

Alexander, V., Blinder, C., & Zak, P. J. (2018). Why trust an algorithm? Performance, cognition, and neurophysiology. *Computers in Human Behavior*. https://doi.org/10.1016/j.chb.2018.07.026

Almaatouq, A., Griffiths, T. L., Suchow, J. W., Whiting, M. E., Evans, J., & Watts, D. J. (2022). Beyond playing 20 questions with nature: Integrative experiment design in the social and behavioral sciences. *Behavioral and Brain Sciences, 2022*, 1–55. https://doi.org/10.1017/s0140525x22002874

Arkes, H. R., Dawes, R. M., & Christensen, C. (1986). Factors influencing the use of a decision rule in a probabilistic task. *Organizational Behavior and Human Decision Processes, 37*, 93–110.

Arkes, H. R., Shaffer, V. A., & Medow, M. A. (2007). Patients derogate physicians who use a computer-assisted diagnostic aid. *Med Decis Making, 27*(2), 189–202. https://doi.org/10.1177/0272989X06297391

Ashton, A. H., Ashton, R. H., & Davis, M. N. (1994). White-collar robotics: Levering managerial decision making. *California Management Review, 37*, 83–109.

Baudel, T., Verbockhaven, M., Cousergue, V., Roy, G., & Laarach, R. (2021). ObjectivAIze: Measuring performance and biases in augmented business decision systems. In C. Ardito, R. Lanzilotti, A. Malizia, H. Petrie, A. Piccinno, G. Desolda, & K. Inkpen (Eds.), *Human-computer interaction – INTERACT 2021* (Vol. 12934, pp. 300–320). Springer. https://doi.org/10.1007/978-3-030-85613-7_22

Benbasat, I., & Taylor, R. N. (1978). The impact of cognitive styles on information system design. *MIS Quarterly, 2*(2), 43–54.

Berger, B., Adam, M., Rühr, A., & Benlian, A. (2021). Watch me improve—Algorithm aversion and demonstrating the ability to learn. *Business & Information Systems Engineering, 63*(1), 55–68. https://doi.org/10.1007/s12599-020-00678-5

Bigman, Y. E., & Gray, K. (2018). People are averse to machines making moral decisions. *Cognition, 181*, 21–34. https://doi.org/10.1016/j.cognition.2018.08.003

Blattberg, R. C., & Hoch, S. J. (1990). Database models and managerial intuition: 50% model + 50% Manager. *Management Science, 36*(8), 887–899. https://doi.org/10.1287/mnsc.36.8.887

Bonaccio, S., & Dalal, R. S. (2006). Advice taking and decision-making: An integrative literature review, and implications for the organizational sciences. *Organizational Behavior and Human Decision Processes, 101*(2), 127–151. https://doi.org/10.1016/j.obhdp.2006.07.001

Brown, R. V. (2015). Decision science as a by-product of decision-aiding: A practitioner's perspective. *Journal of Applied Research in Memory and Cognition, 4*, 212–220. https://doi.org/10.1016/j.jarmac.2015.07.005

Brown, D. L., & Jones, D. R. (1998). Factors that influence reliance on decision aids: A model and an experiment. *Journal of Information Systems, 12*(2), 75–94.

Brunswik, E. (1955). Representative design and probabilistic theory in a functional psychology. *Psychological Review, 62*(3), 193–217. https://doi.org/10.1037/h0047470

Burton, J. W., Stein, M., & Jensen, T. B. (2020). A systematic review of algorithm aversion in augmented decision making. *Journal of Behavioral Decision Making, 33*(2), 220–239. https://doi.org/10.1002/bdm.2155

Burton, J. W., Almaatouq, A., Rahimian, M. A., & Hahn, U. (2021a). Rewiring the wisdom of the crowd. *Proceedings of the Annual Meeting of the Cognitive Science Society, 43*, 1802–1808. Retrieved from https://escholarship.org/uc/item/7tj34969

Burton, J. W., Hahn, U., Almaatouq, A., & Rahimian, M. A. (2021b). Algorithmically mediating communication to enhance collective decision-making in online social networks. *ACM Collective Intelligence Conference, 2021*(9), 1–3. Retrieved from https://www.acm-ci2021.com/program

Cadario, R., Longoni, C., & Morewedge, C. K. (2021). Understanding, explaining, and utilizing medical artificial intelligence. *Nature Human Behaviour*. https://doi.org/10.1038/s41562-021-01146-0

Camerer, C. (1981). General conditions for the success of bootstrapping models. *Organizational Behavior and Human Performance, 27*(3), 411–422. https://doi.org/10.1016/0030-5073(81)90031-3

Carey, J. M., & Kacmar, C. J. (2003). Toward a general theoretical model of Computerbased factors that affect managerial decision making. *Journal of Managerial Issues, 15*(4), 430–449.

Castelo, N., Bos, M. W., & Lehmann, D. R. (2019). Task-dependent algorithm aversion. *Journal of Marketing Research, 56*(5), 809–825. https://doi.org/10.1177/0022243719851788

Christin, A. (2017). Algorithms in practice: Comparing web journalism and criminal justice. *Big Data & Society, 1–14*. https://doi.org/10.1177/2053951717718855

Dana, J., & Thomas, R. (2006). In defense of clinical judgment … and mechanical prediction. *Journal of Behavioral Decision Making, 19*(5), 413–428. https://doi.org/10.1002/bdm.537

Dawes, R. M. (1971). A case study of graduate admissions: Application of three principles of human decision making. *American Psychologist, 26*(2), 180–188. https://doi.org/10.1037/h0030868

Dawes, R. M. (1979). The robust beauty of improper linear models in decision making. *American Psychologist, 34*(7), 571–582.

Dawes, R. M., Faust, D., & Meehl, P. E. (1989). Clinical versus actuarial judgment. *Science, 243*(4899), 1668–1674. https://doi.org/10.1126/science.2648573

Dietvorst, B. J., & Bharti, S. (2020). People reject algorithms in uncertain decision domains because they have diminishing sensitivity to forecasting error. *Psychological Science, 31*(10), 1302–1314. https://doi.org/10.1177/0956797620948841

Dietvorst, B. J., Simmons, J. P., & Massey, C. (2015). Algorithm aversion: People erroneously avoid algorithms after seeing them err. *Journal of Experimental Psychology: General, 144*(1), 114–126. https://doi.org/10.1037/xge0000033

Dietvorst, B. J., Simmons, J. P., & Massey, C. (2016). Overcoming algorithm aversion: People will use imperfect algorithms if they can (even slightly) modify them. *Management Science, 64*(3), 1155–1170. https://doi.org/10.1287/mnsc.2016.2643

Dijkstra, J. J. (1999). User agreement with incorrect expert system advice. *Behaviour & Information Technology, 18*(6), 399–411. https://doi.org/10.1080/014492999118832

Dijkstra, J. J., Liebrand, W. B. G., & Timminga, E. (1998). Persuasiveness of expert systems. *Behaviour & Information Technology, 17*(3), 155–163. https://doi.org/10.1080/014492998119526

Eastwood, J., Snook, B., & Luther, K. (2012). What people want from their professionals: Attitudes toward decision-making strategies. *Journal of Behavioral Decision Making, 25*, 458–468. https://doi.org/10.1002/bdm.741

Edwards, W. (1954). The theory of decision making. *Psychological Bulletin, 51*(4), 380–417.

Edwards, W. (1962). Dynamic decision theory and probabilistic information processings. *Human Factors, 4*(2), 59–74. https://doi.org/10.1177/001872086200400201

Efendić, E., Van de Calseyde, P. P. F. M., & Evans, A. M. (2020). Slow response times undermine trust in algorithmic (but not human) predictions. *Organizational Behavior and Human Decision Processes, 157*, 103–114. https://doi.org/10.1016/j.obhdp.2020.01.008

Einhorn, H. J. (1972). Expert measurement and mechanical combination. *Organizational Behavior and Human Performance, 7*(1), 86–106. https://doi.org/10.1016/0030-5073(72)90009-8

Einhorn, H. J. (1986). Accepting error to make less error. *Journal of Personality Assessment, 50*(3), 387–395. https://doi.org/10.1207/s15327752jpa5003_8

Eining, M. M., Jones, D. R., & Loebbecke, J. K. (1997). Reliance on decision aids: An examination of auditors' assessment of management fraud. *Auditing: A Journal of Practice & Theory, 16*(2), 1–19.

Er, M. C. (1988). Decision support systems: A summary, problems, and future trends. *Decision Support Systems, 4*, 355–363.

Fildes, R., & Petropoulos, F. (2015). Improving forecast quality in practice. *Foresight: The International Journal of Applied Forecasting, 36*, 5–12.

Gigerenzer, G., & Gaissmaier, W. (2015). Decision making: Nonrational theories. In *International Encyclopedia of the Social & Behavioral Sciences* (pp. 911–916). Elsevier. https://doi.org/10.1016/B978-0-08-097086-8.26017-0

Ginsberg, J., Mohebbi, M. H., Patel, R. S., Brammer, L., Smolinski, M. S., & Brilliant, L. (2009). Detecting influenza epidemics using search engine query data. *Nature, 457*(7232), 1012–1014. https://doi.org/10.1038/nature07634

Goldberg, L. R. (1965). Diagnosticians vs. diagnostic signs: The diagnosis of psychosis vs. neurosis from the MMPI. *Psychological Monographs: General and Applied, 79*(9), 1–28. https://doi.org/10.1037/h0093885

Goldberg, L. R. (1970). Man versus model of man: A rationale, plus some evidence, for a method of improving on clinical inferences. *Psychological Bulletin, 73*(6), 422–432. https://doi.org/10.1037/h0029230

Green, G. I., & Hughes, C. T. (1986). Effects of decision support systems training and cognitive style on decision process attributes. *Journal of Management Information Systems, 3*(2), 83–93. https://doi.org/10.1080/07421222.1986.11517764

Grønsund, T., & Aanestad, M. (2020). Augmenting the algorithm: Emerging human-in-the-loop work configurations. *The Journal of Strategic Information Systems, 29*(2), 101614. https://doi.org/10.1016/j.jsis.2020.101614

Grove, W. M., & Meehl, P. E. (1996). Comparative efficiency of informal (subjective, impressionistic) and formal (mechanical, algorithmic) prediction procedures: The clinical–statistical controversy. *Psychology, Public Policy, and Law, 2*(2), 293–323. https://doi.org/10.1037/1076-8971.2.2.293

Grove, W. M., Zald, D. H., Lebow, B. S., Snitz, B. E., & Nelson, C. (2000). Clinical versus mechanical prediction: A meta-analysis. *Psychological Assessment, 12*(1), 19–30. https://doi.org/10.1037/1040-3590.12.1.19

Hafenbrädl, S., Waeger, D., Marewski, J. N., & Gigerenzer, G. (2016). Applied decision making with fast-and-frugal heuristics. *Journal of Applied Research in Memory and Cognition, 5*, 215–231. https://doi.org/10.1016/j.jarmac.2016.04.011

Hahn, U., & Harris, A. J. L. (2014). What does it mean to be biased: Motivated reasoning and rationality. In *Psychology of learning and motivation* (Vol. 61, pp. 41–102). Elsevier. https://doi.org/10.1016/B978-0-12-800283-4.00002-2

Hammond, K. R. (1955). Probabilistic functioning and the clinical method. *Psychological Review, 62*(4), 255–262.

Highhouse, S. (2008). Stubborn reliance on intuition and subjectivity in employee selection. *Industrial and Organizational Psychology, 1*(3), 333–342. https://doi.org/10.1111/j.1754-9434.2008.00058.x

Hilbert, M., & Lopez, P. (2011). The World's technological capacity to store, communicate, and compute information. *Science, 332*(6025), 60–65. https://doi.org/10.1126/science.1200970

Hindman, M. (2015). Building better models: Prediction, replication, and machine learning in the social sciences. *The Annals of the American Academy of Political and Social Science, 659*(1), 48–62. https://doi.org/10.1177/0002716215570279

Holst, A. (2021). Volume of data/information created, captured, copied, and consumed worldwide from 2010 to 2025. *Statista*. Retrieved from https://www.statista.com/statistics/871513/worldwide-data-created/

Holt, R. R. (1958). Clinical and statistical prediction: A reformulation and some new data. *The Journal of Abnormal and Social Psychology, 56*(1), 1–12. https://doi.org/10.1037/h0041045

Holt, R. R. (1970). Yet another look at clinical and statistical prediction: Or, is clinical psychology worthwhile? *American Psychologist, 25*(4), 337–349. https://doi.org/10.1037/h0029481

Holt, R. R. (1986). Clinical and statistical prediction: A retrospective and would-be integrative perspective. *Journal of Personality Assessment, 50*(3), 376–386. https://doi.org/10.1207/s15327752jpa5003_7

Hovey, H. B., & Stauffacher, J. C. (1953). Intuitive versus objective prediction from a test. *Journal of Clinical Psychology, 9*(4), 349–351.

Jarrahi, M. H. (2018). Artificial intelligence and the future of work: Human-AI symbiosis in organizational decision making. *Business Horizons, 61*(4), 577–586. https://doi.org/10.1016/j.bushor.2018.03.007

Joyce, J. (2003). Bayes' Theorem. In *The Stanford Encyclopedia of philosophy* (Fall 2021). Retrieved from https://plato.stanford.edu/archives/fall2021/entries/bayes-theorem/.

Jung, M., & Seiter, M. (2021). Towards a better understanding on mitigating algorithm aversion in forecasting: An experimental study. *Journal of Management Control*. https://doi.org/10.1007/s00187-021-00326-3

Jussupow, E., Benbasat, I., & Heinzl, A. (2020). Why are we averse towards algorithms? A comprehensive literature review on algorithm aversion. *ECIS 2020 Proceedings, 2020*, 1–18.

Kahn, B. E., & Baron, J. (1995). An exploratory study of choice rules favored for high-stakes decisions. *Journal of Consumer Psychology, 4*(4), 305–328.

Kahneman, D. (2013). *Thinking, fast and slow* (1st ed.). Farrar, Straus & Giroux.

Kahneman, D., & Klein, G. (2009). Conditions for intuitive expertise: A failure to disagree. *American Psychologist, 64*(6), 515–526. https://doi.org/10.1037/a0016755

Klein, G. (1993). A recognition-primed decision (RPD) model of rapid decision making. In *Decision making in action: Models and methods* (pp. 138–147). Ablex Publishing.

Klein, G. (1997). Developing expertise in decision making. *Thinking & Reasoning, 3*(4), 337–352. https://doi.org/10.1080/135467897394329

Klein, G. (2008). Naturalistic decision making. *Human Factors: The Journal of the Human Factors and Ergonomics Society, 50*(3), 456–460. https://doi.org/10.1518/001872008X288385

Kleinmuntz, B. (1990). Why we still use our heads instead of formulas: Toward an integrative approach. *Psychological Bulletin, 107*(3), 296.

Klimoski, R., & Jones, R. G. (2008). Intuiting the selection context. *Industrial and Organizational Psychology, 1*(3), 352–354. https://doi.org/10.1111/j.1754-9434.2008.00061.x

Knight, F. H. (1921). *Risk, uncertainty, and profit*. Houghton Mifflin.

Kuncel, N. R. (2008). Some new (and old) suggestions for improving personnel selection. *Industrial and Organizational Psychology, 1*(3), 343–346. https://doi.org/10.1111/j.1754-9434.2008.00059.x

Kuncel, N. R., Klieger, D. M., Connelly, B. S., & Ones, D. S. (2013). Mechanical versus clinical data combination in selection and admissions decisions: A meta-analysis. *Journal of Applied Psychology, 98*(6), 1060–1072. https://doi.org/10.1037/a0034156

Lawrence, M., Edmundson, R. H., & O'Connor, M. J. (1986). The accuracy of combining judgemental and statistical forecasts. *Management Science, 32*(12), 1521–1532. https://doi.org/10.1287/mnsc.32.12.1521

Lawrence, M., Goodwin, P., O'Connor, M., & Önkal, D. (2006). Judgmental forecasting: A review of progress over the last 25 years. *International Journal of Forecasting, 22*(3), 493–518. https://doi.org/10.1016/j.ijforecast.2006.03.007

Lazer, D., & Kennedy, R. (2015). What we can learn from the epic failure of Google flu trends. *Wired*. Retrieved from https://www.wired.com/2015/10/can-learn-epic-failure-google-flu-trends/

Lim, J. S., & Connor, M. O. (1996). Judgmental forecasting with interactive forecasting support systems. *Decision Support Systems, 16*, 339–357.

Lin, W., Kim, S. H., & Tong, J. (2022). *What drives algorithm use? An empirical analysis of algorithm use in type 1 diabetes self-management*. https://doi.org/10.2139/ssrn.3891832

Logg, J. M., Minson, J. A., & Moore, D. A. (2019). Algorithm appreciation: People prefer algorithmic to human judgment. *Organizational Behavior and Human Decision Processes, 151*, 90–103. https://doi.org/10.1016/j.obhdp.2018.12.005

Mackay, J. M., & Elam, J. J. (1992). A comparative study of how experts and novices use a decision aid to solve problems in complex knowledge domains. *Information Systems Research, 3*(2), 150–172. https://doi.org/10.1287/isre.3.2.150

McNemar, Q. (1955). Review of the book clinical versus actuarial prediction. *American Journal of Psychology, 68*, 510.

Meehl, P. E. (1954). *Clinical versus statistical prediction: A theoretical analysis and a review of the evidence*. University of Minnesota Press.

Meehl, P. E. (1957). When shall we use our heads instead of the formula? *Journal of Counseling Psychology, 4*(4), 268–273. https://doi.org/10.1037/h0047554

Meehl, P. E. (1986). Causes and effects of my disturbing little book. *Journal of Personality Assessment, 50*(3), 370–375. https://doi.org/10.1207/s15327752jpa5003_6

Miller, G. A. (1956). The magical number seven, plus or minus two: Some limits on our capacity for processing information. *Psychological Review, 101*(2), 343–352. https://doi.org/10.1037/0033-295X.101.2.343

Mittelstadt, B. D., Allo, P., Taddeo, M., Wachter, S., & Floridi, L. (2016). The ethics of algorithms: Mapping the debate. *Big Data & Society, 3*(2), 205395171667967. https://doi.org/10.1177/2053951716679679

Montazemi, A. L. I. R. (1991). The impact of experience on the design of user interface. *International Journal of Man-Machine Studies, 34*(5), 731–749.

Muir, B. M. (1987). Trust between humans and machines, and the design of decision aids. *International Journal of Man-Machine Studies, 27*(5–6), 527–539. https://doi.org/10.1016/S0020-7373(87)80013-5

Newell, S., & Marabelli, M. (2015). Strategic opportunities (and challenges) of algorithmic decision-making: A call for action on the long-term societal effects of 'datification'. *The Journal of Strategic Information Systems, 24*(1), 3–14. https://doi.org/10.1016/j.jsis.2015.02.001

Önkal, D., Goodwin, P., Thomson, M., Gonul, S., & Pollock, A. (2009). The relative influence of advice from human experts and statistical methods on forecast adjustments. *Journal of Behavioral Decision Making, 22*, 390–409. https://doi.org/10.1002/bdm.637

Pankoff, L. D., & Roberts, H. V. (1968). Bayesian synthesis of clinical and statistical prediction. *Psychological Bulletin, 70*(6), 762–773. https://doi.org/10.1037/h0026831

Parasuraman, R., & Manzey, D. H. (2010). Complacency and bias in human use of automation: An attentional integration. *Human Factors: The Journal of the Human Factors and Ergonomics Society, 52*(3), 381–410. https://doi.org/10.1177/0018720810376055

Parasuraman, R., Sheridan, T. B., & Wickens, C. D. (2000). A model for types and levels of human interaction with automation. *IEEE Transactions on Systems, Man, and Cybernetics - Part A: Systems and Humans, 30*(3), 286–297. https://doi.org/10.1109/3468.844354

Partnership on AI. (2019). *Report on algorithmic risk assessment tools in the U.S. Criminal Justice System*. Partnership on AI. Retrieved from https://partnershiponai.org/paper/report-on-machine-learning-in-risk-assessment-tools-in-the-u-s-criminal-justice-system/

Pescetelli, N., Rutherford, A., & Rahwan, I. (2021). Modularity and composite diversity affect the collective gathering of information online. *Nature Communications, 12*(1), 3195. https://doi.org/10.1038/s41467-021-23424-1

Prahl, A., & Van Swol, L. (2017). Understanding algorithm aversion: When is advice from automation discounted? *Journal of Forecasting, 36*, 691–702. https://doi.org/10.1002/for.2464
Premack, D., & Woodruff, G. (1978). Does the chimpanzee have a theory of mind? *Behavioral and Brain Sciences, 1*(4), 515–526. https://doi.org/10.1017/S0140525X00076512
Promberger, M., & Baron, J. (2006). Do patients trust computers? *Journal of Behavioral Decision Making, 19*(5), 455–468. https://doi.org/10.1002/bdm.542
Rebitschek, F. G., Gigerenzer, G., & Wagner, G. G. (2021). People underestimate the errors made by algorithms for credit scoring and recidivism prediction but accept even fewer errors. *Scientific Reports, 11*(1), 20171. https://doi.org/10.1038/s41598-021-99802-y
Renier, L. A., Schmid Mast, M., & Bekbergenova, A. (2021). To err is human, not algorithmic – Robust reactions to erring algorithms. *Computers in Human Behavior, 124*, 106879. https://doi.org/10.1016/j.chb.2021.106879
Rosenberg, L., Pescetelli, N., & Willcox, G. (2017). Artificial Swarm Intelligence amplifies accuracy when predicting financial markets. In *2017 IEEE 8th Annual Ubiquitous Computing, Electronics and Mobile Communication Conference (UEMCON)*, pp. 58–62. https://doi.org/10.1109/UEMCON.2017.8248984.
Sage, A. P. (1981). Behavioral and organizational considerations in the design of information systems and processes for planning and decision support. *IEEE Transactions on Systems, Man, and Cybernetics, 11*(6), 640–678.
Sanders, G. L., & Courtney, J. F. (1985). A field study of organizational factors influencing DSS success. *MIS Quarterly, 9*(1), 77–93.
Sanders, N. R., & Manrodt, K. B. (2003). The efficacy of using judgmental versus quantitative forecasting methods in practice. *Omega, 31*(6), 511–522. https://doi.org/10.1016/j.omega.2003.08.007
Sawyer, J. (1966). Measurement and prediction, clinical and statistical. *Psychological Bulletin, 66*(3), 178–200. https://doi.org/10.1037/h0023624
Scherer, L. D., de Vries, M., Zikmund-Fisher, B. J., Witteman, H. O., & Fagerlin, A. (2015). Trust in deliberation: The consequences of deliberative decision strategies for medical decisions. *Health Psychology, 34*(11), 1090–1099. https://doi.org/10.1037/hea0000203
Sieck, W. R., & Arkes, H. A. L. R. (2005). The recalcitrance of overconfidence and its contribution to decision aid neglect. *Journal of Behavioral Decision Making, 53*, 29–53.
Sills, D., Driedger, N., Greaves, B., Hung, E., & Paterson, R. (2009). ICAST: A prototype thunderstorm nowcasting system focused on optimization of the human-machine mix. In *Proceedings of the World Weather Research Programme Symposium on Nowcasting and Very Short Range Forecasting*, pp. 2, 16.
Simon, H. A. (1955). A behavioral model of rational choice. *The Quarterly Journal of Economics, 69*(1), 99. https://doi.org/10.2307/1884852
Simon, H. A. (1956). Rational choice and the structure of the environment. *Psychological Review, 63*(2), 129–138. https://doi.org/10.1037/h0042769
The Parliamentary Office of Science and Technology. (2020). Interpretable machine learning. *UK Parliament POST*. Retrieved from https://researchbriefings.files.parliament.uk/documents/POST-PN-0633/POST-PN-0633.pdf.
Thurman, N., Moeller, J., Helberger, N., & Trilling, D. (2019). My friends, editors, algorithms, and I: Examining audience attitudes to news selection. *Digital Journalism, 7*(4), 447–469. https://doi.org/10.1080/21670811.2018.1493936
Tutt, A. (2017). An FDA for algorithms. *Administrative Law Review, 69*(1), 83–123. http://www.jstor.org/stable/44648608
Watts, D. J. (2017). Should social science be more solution-oriented? *Nature Human Behaviour, 1*(1), 0015. https://doi.org/10.1038/s41562-016-0015
Whitecotton, S. M. (1996). The effects of experience and confidence on decision aid reliance: A causal model. *Behavioral Research in Accounting, 8*, 194–216.

Wolfe, C., & Flores, B. (1990). Judgmental adjustment of earnings forecasts. *Journal of Forecasting, 9*(4), 389–405. https://doi.org/10.1002/for.3980090407

Worthen, B. (2003). *Future results not guaranteed; contrary to what vendors tell you, computer systems alone are incapable of producing accurate forecasts*. Retrieved from http://www2.cio.com.au/article/168757/future_results_guaranteed/.

Yaniv, I., & Hogarth, R. M. (1993). Judgmental versus statistical prediction: Information asymmetry and combination rules. *Psychological Science, 4*(1), 58–62. https://doi.org/10.1111/j.1467-9280.1993.tb00558.x

Zellner, M., Abbas, A. E., Budescu, D. V., & Galstyan, A. (2021). A survey of human judgement and quantitative forecasting methods. *Royal Society Open Science, 8*(2), 201187.

Zerilli, J., Knott, A., Maclaurin, J., & Gavaghan, C. (2019). Algorithmic decision-making and the control problem. *Minds and Machines, 29*(4), 555–578. https://doi.org/10.1007/s11023-019-09513-7

Chapter 2
Subjective Decisions in Developing Augmented Intelligence

Thomas Bohné, Lennert Till Brokop, Jan Niklas Engel, and Luisa Pumplun

Keywords Augmented intelligence · Augmented reality · Design science · Decisions

1 Introduction

Augmented intelligence is the use of machines to augment humans in conscious intellectual activity such as thinking, reasoning, remembering or decision-making. It is motivated by evolving visions from completely automated work to more nuanced visions balancing human-centered considerations with advancements in Artificial Intelligence (AI) and other technologies. In contrast to autonomous or function-based visions, the idea of augmentation is based on new reciprocal and interdependent relationships between humans and machines. This vision has been described as Industry 5.0 by the European Union or Society 5.0 in Japan. At its core is the belief that it is socially desirable to imagine a future of work in which human abilities and technical capabilities are used in a balanced way.

The idea of augmented intelligence can be traced back to early notions of "augmenting human intellect" (Engelbart, 1962, p. 1), which is recontextualized by rapidly evolving digital technologies of today. In our contribution, we describe the development of such a modern application, a machine learning-based (ML) object detection system for an augmented reality (AR) application. Our development is an attempt to successfully integrate the idea of augmented

T. Bohné (✉) · L. T. Brokop
Cyber-Human Lab, University of Cambridge, Cambridge, UK
e-mail: tmb35@cam.ac.uk

J. N. Engel
Cyber-Human Lab, University of Cambridge, Cambridge, UK

Technische Universität Darmstadt, Darmstadt, Germany

L. Pumplun
Technische Universität Darmstadt, Darmstadt, Germany

M. Seifert (ed.), *Judgment in Predictive Analytics*, International Series in Operations Research & Management Science 343,
https://doi.org/10.1007/978-3-031-30085-1_2

intelligence into human-operated industrial inspection settings that cannot be fully automated. Incidentally, our development also makes for an interesting case study on the role humans play during the development process and how they influence it with their decisions. As it turns out, even a building process and prototype that seems driven by functional criteria is far from objective and influenced by many subjective decisions.

In what follows below, we will offer some details on the design science approach and technical implementation of our development. Our main focus will be on the processes that led to certain decisions during the development. We will draw on concepts from decision-making, especially heuristics, and cluster our decisions using the concept of decision pyramids. We will analyze our decisions and decision paths and how they influenced each other. We will discuss decision types we encountered, their importance and how they influenced the overall development outcome. In so doing, we aim to contribute by highlighting the subjective nature of many decisions underpinning the development of intelligent augmentation and its wider ramifications for the idea of predictive analytics.

2 Theoretical Framework

2.1 *Machine Learning-Based Augmented Reality*

AR describes a situation where the virtual world blends into the real world. We speak of AR as soon as virtual information such as holograms is superimposed on top of the physical environment that is being observed (Szajna et al., 2019). This way it is possible to enable collaboration between humans and machines using various devices. We mainly distinguish between handheld devices (HHDs) such as mobile phones and head-mounted devices (HMDs) such as smart glasses. Modern HMDs have many sensors and are thus able to generate large amounts of data. This is one reason that makes them a very interesting case for ML applications. ML methods allow to generate knowledge from structured and unstructured data that these HMDs deliver. For example, Roth et al. (2020) propose an approach that uses data about the body's movement and angular rate to classify human activities with ML. This data can also be delivered by an HMD and be used to not only tell where the person wearing the HMD is located but also help to identify what they are currently doing.

AR has wide-ranging application fields and is, for example, employed in the aviation industry, plant and mechanical maintenance, consumer technology, nuclear industry, and remote (expert) systems (Palmarini et al., 2018). It also plays a role in health-care as a means to improve surgery (Vávra et al., 2017), is of growing importance in industrial quality inspection (Angrisani et al., 2020), and for infrastructure inspection (Mascareñas et al., 2020). Both the wide range of industrial AR applications and the growing data produced by HMDs makes the development of systems that make joint use of the capabilities of AR and ML a promising approach.

Table 2.1 Classification framework for ML-based AR systems

	Task	Hardware
AR	Which purpose does AR serve in the system (e.g. visualisation)?	On which (AR) device is the application deployed (e.g. mobile-iOS)?
ML	Which purpose does ML serve in the system (e.g. classification)?	Where is the ML model executed (e.g. server/cloud)?

To better understand how ML and AR can interoperate appropriately we reviewed work proposing systems for joint application of ML and AR and came up with a framework to classify these systems based on how the underlying technologies can be combined. Specifically, we ask four questions to structure the framework:

1. What is the purpose of AR in the system?
2. What is the purpose of ML in the system?
3. On which AR-capable device is the application running?
4. On which hardware is the ML model executed?

The answers to the first and second questions describe the immediate purpose of the integration of AR ("AR task") and ML ("ML task"), respectively. Usually, the former is used for visualization of the output that the latter produces in classification or regression tasks. Answering the third question determines on which AR device the system is deployed ("AR Hardware"), thus considering the different maturity levels and platform dependencies of such devices. Given the high computational cost of running a ML model, the answer to the fourth question determines whether the ML model is run on the same device as the AR task or is offloaded to another infrastructure ("ML Hardware"). The framework is illustrated in Table 2.1.

Our literature review indicates that there is a tendency for the ML task to focus on object detection. Object detection is a task in which objects are not only recognized as present in the scene. They also need to be located, telling where an object is. Most researchers have researched object detection tasks with the execution of the models equally divided between on- and off-device computations. While AR is primarily used for visualization, the device usage is balanced between HHDs and HMDs. An overview of the results is provided in Table 2.2.

The examined ML-based AR systems tackle a broad variety of problem domains like visual recognition, quality inspection and assembly. For example, Eckert et al. (2018) and Shen et al. (2020) each created an application that aids the visually impaired by detecting objects in the vicinity of the users and—based on this information—provide them with audio guidance for navigation. The difference between the two applications is that Eckert et al. (2018) use cloud computing for their neural network execution, while Shen et al. (2020) let this step be performed directly on the device. Another example is Freeman (2020), who provides ad-hoc translation and projection of a mobile weather forecast application from English to Chinese. Trestioreanu et al. (2020) tackle a segmentation problem in radiology and von Atzigen et al. (2021) use object detection to improve spinal surgery. However,

Table 2.2 Literature on ML-based AR systems categorized systematically

	Machine learning		Augmented reality	
Authors	Tasks	Hardware[a]	Tasks	Hardware
Bahri et al. (2019)	Object detection	Cloud	Visualisation	HoloLens
Dasgupta et al. (2020)	Object detection	Edge	Visualisation	HoloLens
Eckert et al. (2018)	Object detection, Text2Speech	Cloud	Speech	HoloLens
Farasin et al. (2020)	Object detection	Cloud	Visualisation, tracking	HoloLens
Freeman (2020)	Translation	Cloud	Visualisation	Mobile
Krenzer et al. (2019)	Classification	Device	Feedback	Raspberry Pi
Li et al. (2020)	Object detection	Device	Visualisation	Mobile (Android)
Liu et al. (2019)	Object detection	Edge	Visualisation	Magic Leap One[b]
Pullan et al. (2019)	Classification, anomaly detection	Edge	Visualisation	Mobile (browser)
Shen et al. (2020)	Object detection, Text2Speech	Device	Speech	Mobile (iOS)
Su et al. (2019)	Pose estimation	Edge	Projection	Mobile
Svensson & Atles (2018)	Edge and Object Detection, Segmentation	Device	Projection	Mobile (iOS)
Trestioreanu et al. (2020)	Segmentation	Cloud	Projection	HoloLens
von Atzigen et al. (2021)	Object Detection	Device	Projection	HoloLens
Wortmann (2020)	Object Detection	Device	Projection	Mobile (iOS)

[a]In case it was not clearly stated or ambiguous on which hardware the ML model is executed the most likely scenario was inferred
[b]Liu et al. (2019) did not use the device itself but a computer with the same specifications

the majority of authors that target a specific problem, focus their research on industrial applications. They improve quality control during (Krenzer et al., 2019) and after (Wortmann, 2020) assembly, help engineers make better design choices through generative algorithms (Pullan et al., 2019), and support assembly processes by blending in text instructions (Svensson & Atles, 2018) or superimposing 3D models on the product (Su et al., 2019).

In addition to the papers and application domains already mentioned, there are other contributions that deal exclusively with research on object detection for AR devices, such as Bahri et al. (2019) and Dasgupta et al. (2020). Both use a server to perform object detection and visualize the results with a HoloLens. Farasin et al. (2020) go a different way and implement an algorithm for tracking detected objects. As long as the position of an object is known, there is no need for subsequent neural network executions. Another interesting approach is the combination of ordinary object detection with additional spatial information such as the distance from the camera to the object in order to improve the detection results (Li et al., 2020). Among 10 object detection tasks in the literature review, six chose HMDs as the AR device, while four targeted mobile devices. Out of these six detection tasks that targeted HMDs, only von Atzigen et al. (2021) executed the neural network on the device itself, in this case a HoloLens 1. This is primarily due to the low computing capacity of HMDs and thus a lot of research deals with the issue by off-loading the neural network execution to cloud or edge computers. However, these solutions introduce other latencies like network traffic as well as making the system reliant on additional infrastructure, such as a stable network connection and the availability of servers. Therefore, having one device that successfully runs all necessary ML and AR operations should mostly be the preferred approach, especially during critical tasks such as surgery.

2.2 Design Science

The research approach that was applied during the development of the ML-based object detection system for the HoloLens follows the Design Science Paradigm, which puts forward the creation of a "viable artefact in the form of a construct, a model, a method, or an instantiation" (Hevner et al., 2004, p. 83). Gregor and Hevner (2013, p. 341) define the term artefact as "a thing that has, or can be transformed into, a material existence as an artificially made object or process". Peffers et al. (2008) propose a Design Science Research Methodology (DSRM) Process Model that provides guidance for conducting and evaluating Design Science Research in information systems. The nominal process sequence in this model includes six steps, also referred to as activities. These are (1) problem identification and motivation, (2) definition of the objectives of a solution, (3) design and development, (4) demonstration, (5) evaluation, and (6) communication. A summary of the DSRM Process Model applied in the context of the research presented here, can be found in Table 2.3.

2.3 Decision Making

Design Science aims for a highly objective decision-making process to decide on upcoming issues during the development process. Yet, at many points in this process developers face a situation in which they do not have access to a sufficient amount of information to decide in a purely objective manner.

In general, decisions are made in three different ways by applying logic, statistics or heuristics (Gigerenzer & Gaissmaier, 2011). DSRM tries to build the foundation to decide rationally, thus by applying logic or statistics. In contrast, heuristics differ from the former decision-making approaches and can be defined as follows:

Table 2.3 Summary of our project using the DSRM framework by Peffers et al. (2008)

Problem identification and motivation	Processes that cannot be automated and must rely on human expertise are prone to error. In these settings digital transformation is not about automation, but rather about assisting and increasing human performance.
Objectives of a solution	A system that combines human and machine capabilities through AR and ML can help to reduce the error rate. Performing object detection on an HMD would be a promising base case for this and could be built upon for specific problem domains.
Design and development	A ML-based object detection system for the HoloLens is designed to assess the feasibility of neural network execution on HMDs. The general architecture of such a system includes access to sensor data, an inference engine for the ML model, and the means to visualise the results. The instantiation uses the main camera of the device to access images of the user's surroundings, detects up to twenty different objects in them, and augments the real-world environment with information about location and nature of these objects.
Demonstration and evaluation	The performance is validated by a test run on an object detection benchmark dataset. The utility is evaluated based on feedback from industry experts during an interview, where a live demonstration was given.
Communication	The main result of this research is the implementation of a neural network-based object detection system that runs on a HoloLens. It serves as a basis for the adaptation of other ML models in various problem domains. The results are communicated in this book chapter and as part of a workshop with a technology company.

> A heuristic is a strategy that ignores part of the information, with the goal of making decisions more quickly, frugally, and/or accurately than more complex methods. (Gigerenzer & Gaissmaier, 2011, p. 454)

As Simon (1979) stressed in his Nobel Memorial Lecture more than 40 years ago, the classical model of rationality requires knowledge of all relevant alternatives, their consequences and probabilities, and a predictable world without surprises. Savage (1972) called the availability of such information perfect knowledge. These conditions, however, are rarely met for the problems that individuals and organizations face, for example when developing systems. Rather, and this is what mostly occurs, parts of the relevant information are unknown or have to be estimated from small samples. Thus, the conditions for rational decision theory are not met, making it an impractical norm for optimal reasoning (Binmore, 2008).

According to Gigerenzer and Gaissmaier (2011), there are four classes of heuristics (Table 2.4). The first class exploits recognition memory, the second relies on one good reason only (and ignores all other reasons), the third weights all cues or alternatives equally, and the fourth relies on social information (Gigerenzer & Gaissmaier, 2011).

Traditionally, heuristic decisions were considered the least best decision option. They were only used because of humans' cognitive limitations and therefore, according to common opinion, more information, more computation, and more

Table 2.4 The classes of Heuristics According to Gigerenzer and Gaissmaier (2011)

Class	Heuristic	Description
Recognition-based decision making	Recognition heuristics	If one of two alternatives is recognised and the other is not, then infer that the recognised alternative has the higher value with respect to the criterion
	Fluency heuristics	If both alternatives are recognised but one is recognised faster, then infer that this alternative has the higher value with respect to the criterion.
One-reason based decision making	One-clever-Cue heuristics	Instead of relying on e.g. complex calculations, one is relying simply on one cue to estimate the results efficiently.
	Take-the-best	People infer which of two alterna-tives has a higher value on a criterion, based on binary cue values retrieved from memory
Trade-off heuristics	Tallying	Tallying ignores weights, weight-ing all cues equally. It entails simply counting the number of cues favouring one alternative in comparison to others
	Mapping model	People tally the number of rele-vant cues with an object's positive values. The estimate is the median criterion value of objects with the same number of positive cues.
	1/N rule	Allocate resources equally to each of N alternatives
Social intelligence	Social intelligence	The same heuristics that underlie nonsocial decision making also apply to social decisions (but not vice versa). Exclusive social heuristics include imitation heuristics, tit-for-tat, the social-circle heuristic, and averaging the judgements

time would always be better (Gigerenzer & Brighton, 2009). Nevertheless, it is well-known that rational decision making is often not possible or only feasible with great effort. This results in the *accuracy-effort trade-off* which emphasizes the notion that even though accuracy is preferable, information and calculation cost time and money and could eventually outweigh the gain from increased accuracy. As a result, humans regularly rely on simple heuristics that are less accurate than strategies that use more information and computation (Brink, 1994; Gigerenzer & Brighton, 2009; Tversky, 1972).

But not everyone looked at heuristics this way. For instance, Gigerenzer and Brighton (2009) and Gigerenzer and Gaissmaier (2011) argue that this view of heuristics is a misinterpretation, as *less-is-more effects* may apply: more information or computation can decrease accuracy, therefore humans could also rely on simple heuristics in order to be more accurate than strategies that use more information and time. It is that people satisfice rather than maximize. Maximization means optimization, the process of finding the best solution for a problem, whereas satisficing (a Northumbrian word for 'satisfying') means finding a good-enough solution (Gigerenzer & Brighton, 2009). The latter can be predominantly achieved by using heuristics.

Also, the success of applying heuristics is strongly influenced by the environment in which the decision is made. Environmental structures are, for example, uncertainty, redundancy, sample size, and variability in weights (Todd & Gigerenzer, 2012). For instance, heuristics that rely on only one argument, such as the hiatus heuristic and take-the-best heuristic, tend to succeed (relative to strategies that rely on more reasons) in environments with (a) moderate to high uncertainty and (b) moderate to high redundancy (Gigerenzer & Gaissmaier, 2011).

In conclusion, a heuristic is neither good or bad, nor rational or irrational; its accuracy depends on the structure of the environment (ecological rationality). If the environment is uncertain, applying heuristics may be the only viable alternative since the conditions for rational models do not hold. Indeed, decision making in organizations typically involves heuristics since organizations often operate in an uncertain world (Gigerenzer & Gaissmaier, 2011).

3 Development Process

Next, we analyse the development decisions of an illustrative industrial AR/ML software project. During the development process numerous big and small decisions must be made to create a useful application. We will briefly describe the development of an AR application prototype running exclusively on an HMD that can support quality inspection processes in a manufacturing environment based on ML-based object recognition. Our development included six different stages, namely (1) finding a Use Case, (2) designing a Minimal Viable Product (MVP), (3) achieving a Proof of Concept (PoC), (4) creating a testing environment, (5) conducting quantitative user tests, (6) adapting to real-world environments, and (7) identifying further use cases. We will focus on the parts of the development process where the decision-making process was particularly complex and provide an overall context and outlook for how the development proceeded as a whole.

3.1 Use Case: Finding a Starting Point

Whenever human labor is needed there is the possibility of improvements through the use of support systems. These could, for example, be based on AR and ML. Depending on the industry and the supported process itself, these systems can make processes become more efficient and safer. The manufacturing environment is known for clearly defined, regularly repeated processes with a limited number of influencing and changing variables. Within this steady environment we started searching for a suitable use case.

To begin with, we identified the basic strengths of both technologies, AR and ML. AR is especially useful when the task requires the worker's hands to be free (Angrisani et al., 2020). The technology provides a new way of informing the

Table 2.5 Possible use cases

Context	Workflow	Main ML component
Assembly of industry brakes	Sub-groups are assembled and inspected by hand before assembling the final product. Built-in inspection due to guided assembly.	Object detection/anomaly detection
Welding	Welding is often done manually and thus prone to human errors, strong need for inspection. Automating weld inspection with AR and ML.	Object detection/anomaly detection
3D printing	Printing results vary a lot. Thus, parameters are adjusted on a trial-and-error basis. Detect defects and recommend parameter adjustments (or adjust automatically).	Anomaly detection/recommendation system
Elevators	Regular mandatory maintenance is executed by service personnel. Support in case of errors. Guided maintenance makes sure standards are met.	Object detection/anomaly detection
Picking in logistics	Picking or packaging of goods. Guided picking (route, goods detection) and support for packaging.	Object detection/visual analysis/recommendation system

operator by overlaying digital information onto the real world. This opens ways to reduce required training for specific tasks or to become more agile on the job as information is available in real time. It can also increase the quality and the reproducibility of a given process by reducing human error. ML, on the other hand, opens up new ways of dealing with data and generating information. It is known for its abilities of processing huge amounts of structured and unstructured data. The approach of ML differs from what was previously known as information processing. ML aims to derive knowledge from existing data and learn solving tasks with no or very limited additional support by humans (Çinar et al., 2020).

After assessing the strengths of both technologies, we explored use cases of ML and AR that seemed particularly relevant for industry. Four interviews with industry experts served as a starting point to identify relevant application scenarios. The profession of the interviewed experts varied (e.g. CEO of a brake system manufacturer, regional head of an elevator supplier, and the CEO of a fulfilment service), enabling us to gain an initial market insight.

We derived the following use cases from these interviews (Table 2.5).

It became clear that object detection and anomaly detection are among the most promising and widely applied ML tasks. Moreover, most applications to date use external resources like cloud computing capacities to run the object or anomaly detection (e.g. Del Amo et al., 2018; Freeman, 2020; Miller et al., 2020). This restricts the possible places of the system's installation and usage considerably, as the operator's free moving space is conditional on a good WiFi connection. Especially in a manufacturing environment the provision of such cannot always be guaranteed.

Table 2.6 Example: evaluation of welding

Criteria	Rating	Comment
1	Easy	Different occurrences
2	Reasonable	Some fixed variables, but decide on welded item
3	Medium	e.g. fixed time slot due to production line
4	Average	Sensor technology and object detection models differ
5	Very often	Quite usual task
6	Likely	Found industry partner

To decide on a use case for implementation, we evaluated the identified application scenarios according to the following questions:

1. How accessible is the process to be improved? For example, how can we collect data or test our application?
2. How can we decide on and vary detection complexity?
3. How many influencing factors exist that we cannot control?
4. How easy can the solution be transferred to other use cases?
5. How often is this use case found in industry?
6. Can we find an industry partner with whom we can perform a real-world PoC to validate our solution?

When applying these criteria to all identified use cases (see Table 2.5), "Welding", for example, showed an average performance in most criteria, which was, however, better than most of the other use cases (see Table 2.6).

The interviewed experts indicated that the inspection of welding output represents a promising use case as weld quality is a critical metric (especially in industries with high quality and safety standards like ship building and aerospace) that often cannot be assessed automatically without human intervention. While different ways exist to evaluate welding seams, the most prominent one is the optical surface inspection. Moreover, in depth inspections with either x-ray or ultrasound sensors or by cutting the part can be applied to see the weld's inner quality. Since the quality of welds depends on numerous external factors and their testing is thus complex, we have adopted a sequential approach to technical development. In this regard, we started in a very general way and first implemented an object detection application, which we gradually extended to the specific use case.

3.2 MVP: Getting a First Version

In developing an MVP, we were not yet aiming to design an applicable product, but to become better acquainted with the technologies. Specifically, we were seeking a deeper understanding of AR's and ML's possibilities as well as studying best practices for implementing the technologies. This phase was intense both from an experimental perspective and from a theoretical one. Our goal was to run an object detection algorithm solely on the HoloLens. Since neural networks can be re-trained and thus specified based on more data, we decided that the type of detected objects does not matter initially in order to test the functionality of different neural networks.

The main challenges we faced during the development of the MVP were

- accessing the HoloLens' camera,
- running a neural network on the HoloLens,
- displaying the detection results in the AR and
- reducing computing times.

3.2.1 Camera Feed

In our first iteration we implemented the camera stream access. We had to choose between different built-in and third-party tools. In particular, we considered Unity's WebcamTexture class, the VideoCapture class provided by Windows Mixed Reality's Locatable Camera (Windows, 2021), CameraStream (Vulcan Technologies, 2021), which is a third party plugin using Windows MediaCapture API, and HoloLensForCV (Microsoft, 2021a), Microsoft's answer to rising research efforts in computer vision and robotics (see Table 2.7 for an overview). Even though HoloLensForCV requires the HoloLens' research mode to be activated and therefore increases the power consumption enormously, we decided to use the tool since it enables access to the devices' sensor streams. In our case we only need the camera sensor even though, if necessary, further sensors are also available, giving us a certain flexibility.

3.2.2 Execution Engine and Detection Model

As a second step we decided on our execution engine and the detection model. The HoloLens runs on the Windows 10 Holographic operating system, which uses the Universal Windows Platform (UWP) runtime environment. This confines us to the usage of ML models in the Open Neural Network Exchange (ONNX) (The Linux Foundation, 2021) format and it is not possible to use standard frameworks like TensorFlow (TensorFlow, 2021) or PyTorch (PyTorch, 2021). Yet, models trained in these frameworks can usually be converted to ONNX with tools like

Table 2.7 Accessing the HoloLens Camera in Unity

Camera type	Spatial Info	Parameters	Format	Built-in	Miscellaneous
WebcamTexture	No	Yes	ARGB32	Yes	Easiest to use
VideoCapture	Yes	Yes	N/A	Yes	Camera icon
CameraStream	Yes	Yes	Byte	No	Camera icon
HoloLensForCV	Yes	Yes	Byte	No	Power consumption

TensorFlowToONNX ("TensorFlow to ONNX Converter", 2021) or PyTorch's built-in API.

The integration and execution of these ML models in UWP applications on the device is possible using the Windows ML (WinML) API (Microsoft, 2021b). WinML is an inference engine that allows the use of trained ML models locally on Windows 10 devices. Another way to run these models is to use Barracuda (Puida & Guinier, 2021), Unity Technologies own lightweight inference library. Barracuda works with ONNX models and fully supports CPU and GPU inference on UWP. It is not only easy to use but also part of the Unity ecosystem and thus makes it possible to run it in the editor, which can come in handy for testing purposes. Both options do not currently support all possible ML algorithms. Even though WinML might turn out more flexible in terms of which model to use, the ease of integration and the possibility for in editor testing speak in favour of Barracuda.

Barracuda supports fully convolutional models and the Tiny YOLOv2 object detector (Redmon et al., 2016). Due to the limited computing capacities of the HoloLens we chose the faster Tiny YOLOv2 over its more precise bigger brother YOLOv2.

3.2.3 Image Processing

Before feeding the cameras' data into the neural network, we needed to apply pre-processing. In general pre-processing of images includes a resizing operation and a normalization, as these networks are trained to work best with a special images' size and format. In our case TinyYOLOv2 expects 416x416 pixel images (resizing) and pixel values that are between 0 and 1 (normalization). We thus embedded the pixels in a tensor shape of the shape (1, 416, 416, 3) [(#frames, height, width, #channels)] and handed them over to the neural network for object detection. Tiny YOLOv2 partitions the image into a 13×13 grid and predicts five so-called bounding boxes that represent the rectangular bounds of an object per grid. During the post-processing procedure we removed bounding boxes with a low chance of containing an object and those with a high likelihood to describe the same object, as the object might, for example, extend over multiple grid cells. These remaining bounding boxes describe detected objects and can then be handed over to the visualization module.

3.2.4 Visualization

In this fourth and last step of the MVP development, we aimed at visualizing the calculation results with the help of AR. We were placing two-dimensional bounding boxes to frame the detected object. In later phases of the project, we added three dimensional boxes to surround the objects. The placement of the bounding boxes is done by projecting the bounding boxes on a virtual canvas that is placed between the user and the objects.

3.3 Summary of Steps 3–7: From a Proof of Concept to Future Use Cases

To create an industrially relevant application from this MVP we planned five additional development steps. As a third step, we needed to achieve a real PoC. Since the development of a PoC requires a specific application environment we created a simple and easily adjustable quality inspection scenario our application needs to operate in. In this way we intended to demonstrate that our application can basically be applied in quality inspection and could eventually be transferred to the case of weld inspection. In this case, quality inspection means that in a set of cards a wrong card should be detected. This particular "quality inspection" scenario has the advantage that there is a high variety of possible errors with different degrees of detection difficulty and creating the data set for training purposes of the ML model (a neural network) is relatively easy (see Fig. 2.1).

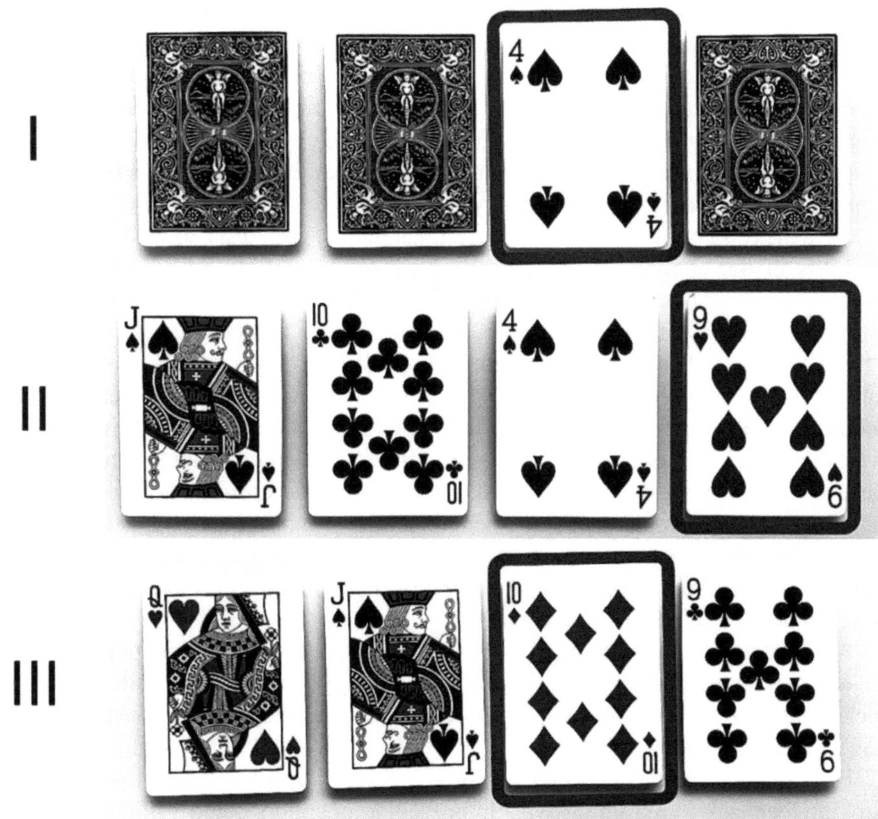

Fig. 2.1 Quality inspection in a set of cards

Firstly, and this is the easiest detectable difference, there is a great contrast between the front- and the backside of a card (see Fig. 2.1, I). It gets a little bit harder if you look at the playing cards suit. This can be done in two stages. The first would be to just differentiate between the red and the black suits (see Fig. 2.1, II). Secondly all four suits (diamonds, clubs, hearts and spades) can be considered. At the highest level of complexity you would be searching for just a single card (combination of suit and rank) in a pile of different cards (see Fig. 2.1, III).

The main research goals for the PoC are:

- Demonstrate that our application works in a quality inspection scenario and
- deepen our understanding on the functionality of ML models, especially neural networks.

As a fourth step, we would need to create a testing environment closer to industry. We deem the inspection of 3D printed models a useful environment, as shapes similar to welds and corresponding defects can be easily printed. Those models can be manufactured directly at the facilities available in Cambridge. Two very reproducible errors in 3D printing are the so-called *stringing* and *warping*. We plan to print different models with and without errors and document the printing process with cameras that are directly attached to the printer. Also, we will randomly change the scale and orientation of the printed model in each print. This way we will be able to create a large amount of raw image data. This data will then be labelled and augmented (through simple and elastic distortions) to create a large data set (Simard et al., 2003). With this data set we can train our neural network to detect errors and apply our application in a first industry-near setting.

As a next step, we aim to conduct quantitative user tests to understand the usability of our application. We already have got a working, reliable application with only minor bugs to fix. By carrying out a user test with a larger group of participants, we aim to achieve a basis for real industry implementation.

Finally, we will be ready to adapt the application to work in a real industry environment. As soon as one is leaving the laboratory environment and entering an industry context the focus needs to shift. Up to this point, the nature of development was very much explorative. We defined requirements and considered constraints, nevertheless meeting them all to satisfaction was neither absolutely necessary, nor possible given the available development time. In the same manner robustness of the application was a plus but not our focus. However, unreliability would devalue any industry use of the application. A high priority will be close consultation and development with industry stakeholders.

As shown above a combination of AR and ML has great potential to detect objects and potential errors. But these are not the only possible applications of these technologies. As a final step, further use cases and extensions of the considered one can be identified.

For example, the technologies could also be used to actively label data sets or provide feedback to retrain the ML. In particular, operators could validate detected errors directly in the AR environment. If the error was correctly identified, the ML has reassurance. If not, the operator can either withdraw the classification as incorrect, redraw the bounding boxes surrounding the error, or rename the label.

On the other hand, an operator could use our application to generate new data points on errors, if there is no sufficient data set available to train the ML. Using our application, the operator could generate new data points directly on spot by drawing bounding boxes to indicate the errors' locations and assigning a label to classify them. Both need only little additional time and are less error-prone than conventional labeling practices because of the integration into existing workflows.

In summary, the development of an AR/ML application must be seen as a process. Even though we followed the Design Science Methodology, which provides guidance on development steps, every development is likely to have moments when developers will not see a clear path forward. In such cases, it is crucial to the development's success to make, ideally effective and efficient, decisions that allow the process to be continued.

4 Decisions and Heuristics During the Development Process

During a development project, such as the one just described, numerous big and small decisions have to be made, which can steer the project into different directions. Some are critical, whereas the importance of others can only be seen when viewed in the project's context. In the following section, we will examine why we have chosen certain paths in developing the ML/AR-based application and the outcome to which they have led. Moreover, we will highlight different types of decision-making situations we have faced and investigate how we actually made the decisions.

4.1 Decision Types

Distinct differentiation between decisions is not always possible as decisions are made in a complex environment containing many links. Yet, we identified three broad categories most decisions fit in, more or less unambiguously. Those categories support us in clustering decisions and understanding their relationship to each other. Namely these are *framework, technological,* and *design decisions*. They are not only distinct in their influence on the project, but they are also characterized by the information underlying, and the methods used to make a decision.

4.1.1 Framework Decisions

Framework decisions cover everything, from organizational to content decisions (e.g choosing the use case). They give direction and guide throughout the development process. In our case, these were, amongst many others, decisions on the use case we aimed to investigate, the project's timeline, or the planning of our next development steps. Some of these decisions, especially the organizational ones, are easy to make. The margin for maneuver is small, as for example in our case the timeline and research goals were set in advance. In addition, best practices for project management tasks already existed. Others are particularly hard to make due to their far-reaching impact on the project. For example, the choice of a suitable use case or a shift of focus due to unforeseen events change the entire development process. Gathering all relevant information to be fully prepared in such decision-making situations is either impossible or would require enormous efforts that are impractical due to resource constraints.

4.1.2 Technological Decisions

Technological decisions inherit a prominent role as they build the backbone of a reliable and efficient software application. They can be as broad as choosing the overall technology of interest and as narrow as deciding on input parameters for our neural network.

We define those far-reaching decisions (e.g. use of plugins, frameworks, devices) as *high level* technological decisions and the narrow ones (e.g. input parameters, image processing) as *low level* decisions accordingly. *High level* decisions need to take into account the defined constraints of the entire project, the overall requirements, and the general technological dependencies. In contrast, *low level* decisions guide concrete ways to actually implement and use the desired technology. For both types of technological decisions, relevant information of good quality is relatively readily available. High level decisions are mostly set by the defined requirements and dependencies, as well as the overall project's orientation. In case of two or more equal alternatives, developers' preferences can be considered. Low level technological decisions can be aligned with industry standards and best practices known from prior projects.

4.1.3 Design Decisions

How do we want the application to feel? What interactions with users do we enable? How can we visualize output of the application? Answering these questions is part of the design decisions along the development process. Design decisions are made in a user-oriented way. They are mainly made relying on assumptions on user's preferences, making a deep understanding of user's needs necessary. Moreover, design decisions need to be intensively tested with potential users in order to evaluate their quality.

4.2 Decision Pyramids

4.2.1 Successive Decisions

As described above, we had to make many decisions during our development process. In fact, one decision often leads to other situations where decisions must be made. We can look at these successive decisions as a pyramid. At the very top level there is one big decision, in our case choosing the use case. With every further step in the development process, we need to take more decisions that depend on the previous ones. For example, after choosing the use case, we investigated what technology (e.g. hardware, frameworks) we wanted to apply. Thereby, the use case provided us with a rough framework within which we had to operate, such as by specifying that the user's hands should remain free when using the application. For this reason, we opted for an HMD as the device on which the application runs. These decisions result in many more decisions that need to be made at a third level (see Fig. 2.2). Every decision made on the levels above and their outcome is influencing the decisions on the current level. Further, they are usually made depending on other open decisions at the same level.

While decisions close to the pyramid's top have a high influence on the overall project's orientation and may shape its outcome significantly, decisions closer to the bottom exert a lower impact that is more of a fine-grained nature (see Fig. 2.2; dashed lines represent the potential expandability of the pyramid). Nevertheless, these decisions should not be underestimated as they could cause the project to fail. For example, selecting wrong input parameters for the neural network could prevent the ML model from detecting any objects.

As projects may be complex, it is not required to create only one decision pyramid. Rather, it is possible to build smaller decision pyramids for each decision strand or even parts of it that refine the larger construct (see Fig. 2.3). In this way, the often extensive set of decision-making situations in a project can be better structured.

Viewing the decisions in a project as a pyramid is certainly an idealized representation. In a real-world application scenario, the pyramid would contain holes, as some decisions are not under the direct influence of the ones above and some

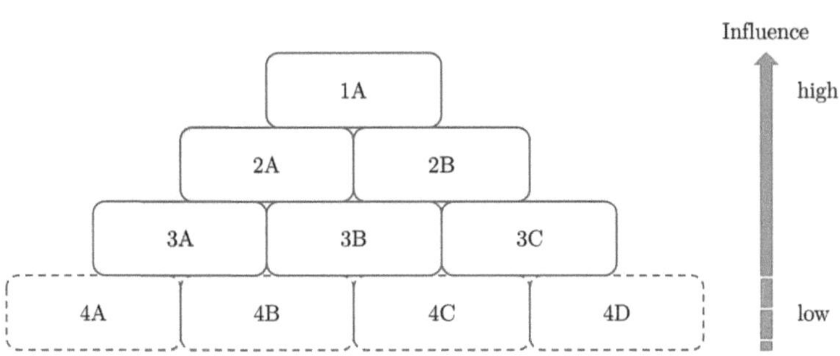

Fig. 2.2 General decision pyramid

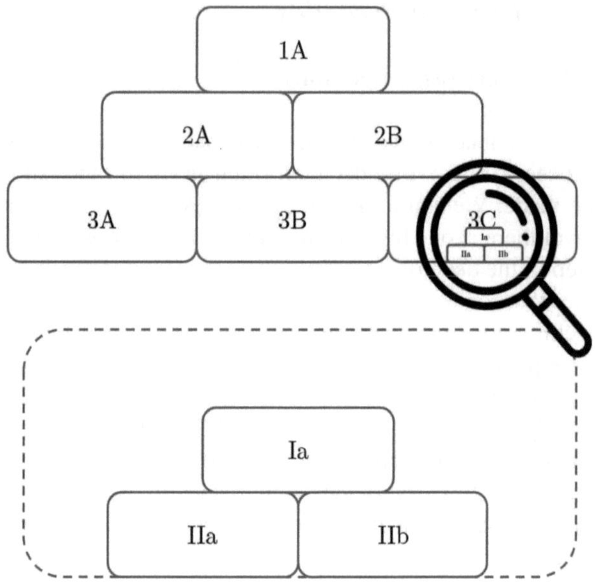

Fig. 2.3 Refinement of the decision pyramid

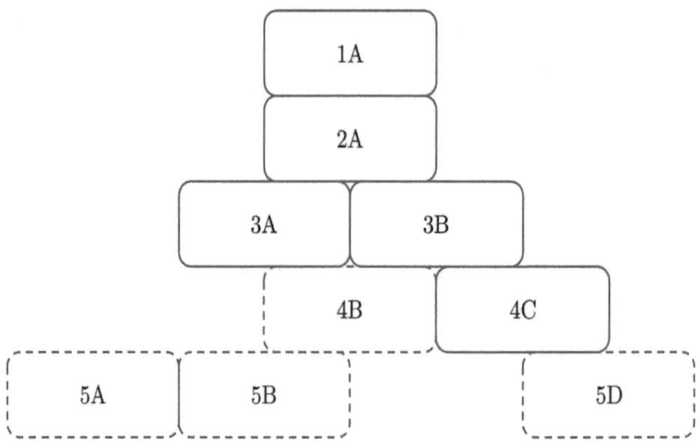

Fig. 2.4 Imperfect decision pyramid

decision strands take more time to accomplish than others. The decision pyramids are also not necessarily regular in their shape as the number of decisions on the current level is not always the number of the levels above plus one (see Fig. 2.4). Moreover, during the development process, decision pyramids need to be adjusted if unforeseen influences occur. Some decisions are later seen as wrong or ineffective, others do not lead to the desired goal. Even though the representation of decisions in the form of a pyramid can prevent us from showing all the edge cases and will have to be corrected over time, it helps us to gain an easily understandable overview of the decisions that need to be made in a project.

4.2.2 Small and Large Worlds

The level in a decision pyramid is also influencing the method used to achieve a decision. As described earlier, decisions that are close to the pyramid's bottom tend to have only a small influence on the overall project course and can often be made fully informed, as variables are limited and information is available. Following Gigerenzer and Gaissmaier (2011) this is what we call a *small world*. However, the closer we get to the pyramid's top, the *larger* our world becomes and the higher the influence of the decision on the project's orientation will be (see Fig. 2.5). Simultaneously, it becomes increasingly difficult to gather all important information and consider every possible eventuality of the decision's outcome. In order to make a rational and informed decision, high investments would have to be made. Therefore, in these decisions of the development project, heuristics are frequently employed.

4.3 *Exemplary Development Decisions*

The above considerations give rise to a framework, which helps to analyze and cluster decisions made during a development process. In the following, we apply the framework onto exemplary decisions that we made during our development process of the ML and AR-based application (see Fig. 2.6).

4.3.1 General Environment

When assessing decisions, a first step is to analyze the general environment in which these decisions were made. In many cases the general conditions of a development project already limit decision possibilities and thus outcomes. In our case, we conducted our research at the Cyber-Human Lab (CHL) at the University of Cambridge, which has a special research focus and technologies of interest such as ML and AR. Specifically, the CHL explores how technology can augment human

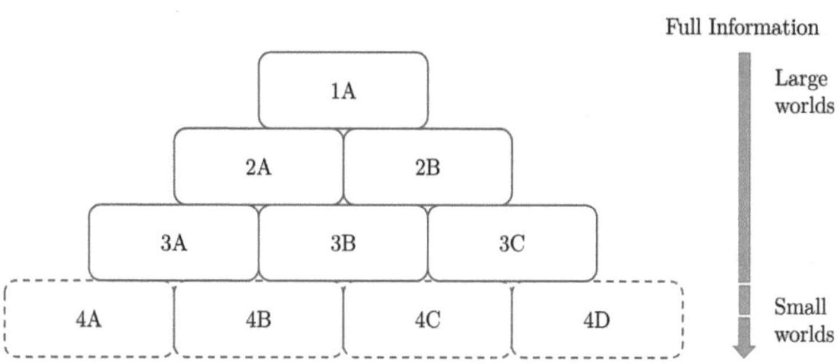

Fig. 2.5 Differentiation between large and small worlds

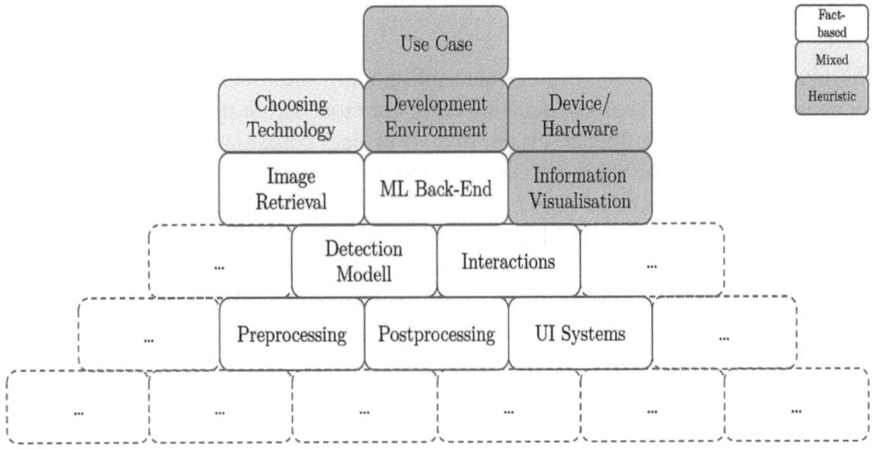

Fig. 2.6 Exemplary decision pyramid

intelligence to improve human performance in industry. The decisions made during the development project were therefore strongly aligned with the objectives of the CHL. This environmental dependency is also present in other organizations that drive development projects. For example, available resources or customer groups can have a major influence on decisions made during the project.

4.3.2 Framework Decisions

In the first step, we decided on the use case that should be investigated throughout the project. We based this decision on expert interviews which provided us with a broad industry overview. Nevertheless, these insights were far away from full information. Due to our limited time capacities, the number of interviews was limited and we had to rely on participants' expertise captured during interviews to identify relevant use cases. We applied a simple heuristic and decided on a use case that most of the interview participants mentioned and deemed to be relevant. We challenged this decision by asking ourselves: can we see a valuable research contribution in this use case? Whether such a decision should be made on the basis of a simple heuristics is highly dependent on the project itself and its timeline. For example, if more time or resources were available more complete information could be gathered, especially for larger decisions at the top of the pyramid.

4.3.3 Technological Decisions

After finding an interesting use case, we needed to decide on the technological setup. When selecting the technologies, we had already been given a rough direction by the general environment, in particular through the research interests of the CHL.

Moreover, the requirements defined by the use case, such as the need for free hands in making and inspecting welds or their high variability, provided a first information basis and influenced our decision to use AR and ML in combination.

Since we had a research budget available, we were somewhat free in our choice of hardware. In order to pick out the most appropriate devices, we relied on a heuristic, a special form of the recognition heuristics (Gigerenzer & Gaissmaier, 2011). Thereby, we chose the HoloLens as it is the current industry standard and most widely recognized. Equivalently, we decided on Unity as our development environment.

From now on, the range of decisions increased drastically as we needed to decide on the actual implementation of the chosen technologies. In particular, we investigated how we want to set up the image retrieval and the ML back-end. To make these technological decisions, we compared different possible technical setups (e.g. HHD vs. HMD, different camera streams) and evaluated them against the given requirements (e.g. no cloud computing or limited battery capacity) (see Chap. 3). As we had reasonably clear ideas about the likely technical requirements of the development project and the possible technical solutions were extensively documented by the providers, we were able to gather all the necessary information relatively easily to make a fact-based decision. Thereby, one combination of technologies emerged as the best overall decision.

4.3.4 Design Decisions

To decide on the design of the application, we needed to collect further information which was not easy to obtain. Specifically, we needed to find a solution to visualize the ML model's output for the potential users. This solution should be as easy to use as possible despite the possibly different requirements imposed by the users. Again, we applied a recognition-based heuristic (Gigerenzer & Gaissmaier, 2011) and looked into best practices of comparable applications that employ, for example, similar technologies. Additionally, as the development project solely aims at a prototype, we were keen to find an implementation that was as simple and resource-efficient as possible.

The decisions described still do not represent the very bottom of the decision pyramid. We have just looked at the first few top layers. For example, directly connected to the decision on the applied ML back end and the respective ML model are decisions on the necessary image pre- and post-processing.

5 Discussion

Human-machine systems to support operators in manufacturing environments and on the shop floor will be of great importance when it comes to the future design of work (Parasuraman et al., 2000). Our ML and AR-based application shows one possible solution to a specified use case, where the augmentation of human intelligence is of value. Nevertheless, this does not paint a full picture. There are countless possible further use cases for the application of ML and AR, in which other key measures can be enhanced. We decided to focus on assuring the welds' quality. Another example could be enabling a cost reduction as for example less skilled operators can perform specific tasks. In addition to the decision for a use case, the course of the development process may also have been deviated by other decisions at other times.

In Sect. 4 we introduced decision pyramids as a framework to analyze our decisions. One may ask why we have chosen to use this representation instead of standard decision trees. In fact, decision trees are very useful to represent linear decision processes, but can be very hard to understand when the decision process becomes complex (Quinlan, 1987). Also, representing decisions that are influenced by more than one previous decision, especially when this decision is associated with another branch, is hardly possible. In contrast decision pyramids are focusing on the overall structure of decisions and how they are influencing each other. They are investigating the decision making rather than considering its outcome. In addition, the complexity of decision pyramids is relatively low as they do not represent possible decision options that have not been chosen. They also allow an accurate guess on how a decision is made (Gigerenzer & Gaissmaier, 2011) as the information availability can be easily estimated. If one needs a more detailed overview of the development process, decision pyramids can be used as a basis to create decision trees. With this in mind they may allow for an interesting perspective on decisions that can complement decision pyramids. Yet, one would preferably use them to visualize less complex decisions or only small sections of the whole decision process due to their higher complexity.

The pyramid framework we created could be applied to our development process. The large world decisions in our project, namely the ones on the use case, development environment, device/hardware and the information visualization were made using heuristics (Savage, 1972). Those were the ones requiring the highest information complexity and gaining full information would have been immensely hard. When choosing the specific software setup, our room for decisions was restricted, since we had already chosen our hardware. Without this constraint, deciding on a specific software solution would have been a classic large world problem. Yet, considering the previously made decision, the space of possible solutions and their corresponding information became smaller.

Recognition-based decision heuristics were the dominant heuristics in our decision process. We applied these heuristics mainly because we were dealing with new technologies and were therefore looking for known approaches to reduce the complexity of our project. Especially during the initial phases of the development process, adhering to industry standards and best practices helped us to decide efficiently. In the further course of the development process, however, other types of heuristics played a more important role. For example, for some technological decisions, their overall impact and resulting constraints could not be foreseen appropriately. That is when trade-off and one-reason based heuristics became handy. Likely one has not dealt with a similar decision before and the problem is too specific to find reliable expert recommendations or standards. In this case, take-the-best and tallying heuristics are especially helpful to make a decision (Gigerenzer & Gaissmaier, 2011). In other cases, using the 1/N rule may be the best solution (Gigerenzer & Gaissmaier, 2011). Understanding and learning to apply these heuristics initially leads to an increase in workload in a development project. However, they may be a useful alternative to further information gathering, especially if collecting the information involves a large amount of resources.

In addition to useful heuristics, it should not be forgotten that people also tend to take subjective influences into account when making decisions. Indeed human decision-making partly relies on emotions and subjective assessments that are not represented by the classic decision methods and theories, which rely on rational reasoning (e.g. Gigerenzer & Brighton, 2009; Gigerenzer & Gaissmaier, 2011; Simon, 1979). For example, it is often difficult to accept sunk costs, i.e. to cancel a development project in which time and money have already been invested.

Retrieving full information is often possible—at least on a theoretical basis with unlimited resources like time, money and access. Yet those resources are often strictly limited in regular development projects and have to be allocated carefully. One of the most important tasks of any project manager is thus to decide on the basis of what and how much information a decision can be made. Even though heuristics may contain inaccuracies and be subjectively shaped, the development of novel systems requires that decisions are made under uncertainty. Therefore, project managers will not be able to avoid applying heuristics, at least at some levels of the decision pyramid.

6 Limitations and Outlook

The description of our development is unavoidably our representation of the development process. Even though we documented the process and described it as accurately as possible, we sorted, filtered and curated the description of what our work actually looked like. The development process was iterative and characterized by many external influences and unforeseeable decision-making situations, which is why strict adherence to an idealized procedure is rarely if ever possible. Especially when working with new technologies, deciding on a detailed problem definition and course of action is not possible upfront (Hevner et al., 2004). As we have not yet implemented all development steps, we cannot rule out—indeed expect—the possibility of further unforeseen events occurring that make a major adjustment of the development process necessary.

We contributed an initial investigation of decision-making during an augmented intelligence development project. We argue that the development of even a fairly objective and seemingly strictly functional technology such as predictive analytics is highly correlated to subjective decisions being made during the process. Of course, this insight is the result of an evaluation of one development process. In order to be able to make more generalizable statements, further development processes would have to be studied from a decision-making perspective.

In conclusion, subjective decision-making situations during an augmented intelligence development project should not be underestimated, as complex requirements and decision-making latitude may exist for implementation. The presented decision pyramid framework and the analysis of heuristics used may provide a starting point for researchers and practitioners to better reflect on and understand the course of development projects and assess their potential impact.

References

Angrisani, L., Arpaia, P., Esposito, A., & Moccaldi, N. (2020). A wearable brain–computer interface instrument for augmented reality-based inspection in industry 4.0. *IEEE Transactions on Instrumentation and Measurement, 69*(4), 1530–1539. https://doi.org/10.1109/TIM.2019.2914712

Bahri, H., Krcmarik, D., & Koci, J. (2019). Accurate object detection system on HoloLens using YOLO algorithm. In *IEEE International Conference on Control, Artificial Intelligence, Robotics & Optimization*, pp. 219–224. https://doi.org/10.1109/ICCAIRO47923.2019.00042.

Binmore, K. (2008). *Rational decisions. The Gorman lectures in economics*. Princeton University Press.

Brink, T. L. (1994). *The adaptive decision maker* (pp. 169–170). Cambridge University Press. https://doi.org/10.1002/bs.3830390207

Çinar, Z. M., Nuhu, A. A., Zeeshan, Q., Korhan, O., Asmael, M., & Safaei, B. (2020). Machine learning in predictive maintenance towards sustainable smart manufacturing in industry 4.0. *Sustainability, 12*(19). https://doi.org/10.3390/su12198211

Dasgupta, A., Manuel, M., Mansur, R. S., Nowak, N., & Gracanin, D. (2020). Towards real time object recognition for context awareness in mixed reality: A machine learning approach. In *IEEE Conference on Virtual Reality and 3D User Interfaces Abstracts and Workshops (VRW)* (pp. 262–268). IEEE. https://doi.org/10.1109/VRW50115.2020.00054.

Del Amo, I. F., Galeotti, E., Palmarini, R., Dini, G., Erkoyuncu, J., & Roy, R. (2018). An innovative user-centred support tool for augmented reality maintenance systems design: A preliminary study. *Procedia CIRP, 70*, 362–367. https://doi.org/10.1016/j.procir.2018.02.020

Eckert, M., Blex, M., & Friedrich, C. M. (2018). Object detection featuring 3D audio localization for Microsoft HoloLens - A deep learning based sensor substitution approach for the blind. In *Proceedings of the 11th International Joint Conference on Biomedical Engineering Systems and Technologies* (pp. 555–561). SCITEPRESS - Science and Technology Publications. https://doi.org/10.5220/0006655605550561.

Engelbart, D. C. (1962). *Augmenting human intellect: A conceptual framework*. Stanford Research Institute.

Farasin, A., Peciarolo, F., Grangetto, M., Gianaria, E., & Garza, P. (2020). Real-time object detection and tracking in mixed reality using Microsoft HoloLens. In *Proceedings of the 15th International Joint Conference on Computer Vision, Imaging and Computer Graphics Theory and Applications*. SCITEPRESS - Science and Technology Publications, pp. 165–172. https://doi.org/10.5220/0008877901650172.

Freeman, J. (2020). Content enhancement with augmented reality and machine learning. *Journal of Southern Hemisphere Earth Systems Science*. https://doi.org/10.1071/ES19046

Gigerenzer, G., & Brighton, H. (2009). Homo Heuristicus: Why biased minds make better inferences. *Topics in Cognitive Science, 1*(1), 107–143. https://doi.org/10.1111/j.1756-8765.2008.01006.x

Gigerenzer, G., & Gaissmaier, W. (2011). Heuristic decision making. *Annual Review of Psychology, 62*, 451–482. https://doi.org/10.1146/annurev-psych-120709-145346

Gregor, S., & Hevner, A. R. (2013). Positioning and presenting design science research for maximum impact. *MIS Quarterly, 37*(2), 337–355. https://doi.org/10.25300/MISQ/2013/37.2.01

Hevner, A. R., March, S. T., Park, J., & Ram, S. (2004). Design science in information systems research. *MIS Quartely, 28*(1), 75–105. https://doi.org/10.2307/25148625

Krenzer, A., Stein, N., Griebel, M., & Flath, C. M. (2019). Augmented intelligence for quality control of manual assembly processes using industrial wearable systems. In *Fortieth International Conference on Information Systems.*

Li, X., Tian, Y., Zhang, F., Quan, S., & Xu, Y. (2020). Object detection in the context of mobile augmented reality. In *IEEE International Symposium on Mixed and Augmented Reality (ISMAR)* (pp. 156–163). IEEE. https://doi.org/10.1109/ISMAR50242.2020.00037.

Liu, L., Li, H., & Gruteser, M. (2019). Edge assisted real-time object detection for Mobile augmented reality. In S. Agarwal, B. Greenstein, A. Balasubramanian, S. Gollakota, & X. Zhang (Eds.), *The 25th Annual International Conference on Mobile Computing and Networking* (pp. 1–16). ACM. https://doi.org/10.1145/3300061.3300116

Mascareñas, D. D. L., Ballor, J. P., McClain, O. L., Mellor, M. A., Shen, C.-Y., Bleck, B., Morales, J., Yeong, L.-M. R., Narushof, B., Shelton, P., Martinez, E., Yang, Y., Cattaneo, A., Harden, T. A., & Moreu, F. (2020). Augmented reality for next generation infrastructure inspections. *Structural Health Monitoring, 2020*, 147592172095384. https://doi.org/10.1177/1475921720953846

Microsoft. (2021a). *HoloLensForCV*. Accessed March 25, 2021, from https://github.com/Microsoft/HoloLensForCV

Microsoft. (2021b). *Windows Machine Learning*. Accessed March 25, 2021, from https://docs.microsoft.com/en-us/windows/ai/windows-ml/

Miller, J., Hoover, M., & Winer, E. (2020). Mitigation of the Microsoft HoloLens' hardware limitations for a controlled product assembly process. *The International Journal of Advanced Manufacturing Technology, 109*(5–6), 1741–1754. https://doi.org/10.1007/s00170-020-05768-y

Palmarini, R., Erkoyuncu, J. A., Roy, R., & Torabmostaedi, H. (2018). A systematic review of augmented reality applications in maintenance. *Robotics and Computer-Integrated Manufacturing, 49*, 215–228. https://doi.org/10.1016/j.rcim.2017.06.002

Parasuraman, R., Sheridan, T. B., & Wickens, C. D. (2000). A model for types and levels of human interaction with automation. *IEEE Transactions on Systems, Man, and Cybernetics. Part A, Systems and Humans: A Publication of the IEEE Systems, Man, and Cybernetics Society, 30*(3), 286–297. https://doi.org/10.1109/3468.844354

Peffers, K., Tuunanen, T., Rothenberger, M. A., & Chatterjee, S. (2008). A design science research methodology for information systems research. *Journal of Management Information Systems, 24*(3), 45–77. https://doi.org/10.2753/MIS0742-1222240302

Puida, M., & Guinier, F. (2021). *Unity-Technologies/Barracuda-Release*. Accessed March 25, 2021, from https://github.com/Unity-Technologies/barracuda-release

Pullan, G., Chuan, T., Wong, D., & Jasik, F. (2019). Enhancing web-based CFD post-processing using machine learning and augmented reality. In *AIAA Scitech 2019 Forum*. American Institute of Aeronautics and Astronautics. https://doi.org/10.2514/6.2019-2223

PyTorch. (2021). *From research to production*. Accessed March 25, 2021, from https://pytorch.org/

Quinlan, J. R. (1987). Simplifying decision trees. *International Journal of Man-Machine Studies, 27*(3), 221–234. https://doi.org/10.1016/S0020-7373(87)80053-6

Redmon, J., Divvala, S., Girshick, R., & Farhadi, A. 2016. You only look once: Unified, real-time object detection. In *IEEE Conference on Computer Vision and Pattern Recognition (CVPR)* (pp. 779–788). IEEE. https://doi.org/10.1109/CVPR.2016.91

Roth, E., Moncks, M., Bohne, T., & Pumplun, L. (2020). Context-aware cyber-physical assistance systems in industrial systems: A human activity recognition approach. In *IEEE International Conference on Human-Machine Systems (ICHMS)* (pp. 1–6). IEEE. https://doi.org/10.1109/ICHMS49158.2020.9209488.

Savage, L. J. (1972). *The foundations of statistics* (2nd ed.). Dover Publications.

Shen, J., Dong, Z., Qin, D., Lin, J., & Li, Y. (2020). IVision: An assistive system for the blind based on augmented reality and machine learning. In M. Antona & C. Stephanidis (Eds.), *Universal access in human-computer interaction. Design approaches and supporting Technologies* (Lecture Notes in Computer Science) (Vol. 12188, pp. 393–403). Springer. https://doi.org/10.1007/978-3-030-49282-3_28

Simard, P. Y., Steinkraum, D., & Platt, J. C. (2003). Best practices for convolutional neural networks applied to visual document analysis. In *Proceedings/Seventh International Conference on Document Analysis and Recognition.*

Simon, H. A. (1979). Rational decision making in business organizations. *The American Economic Review, 69*(4), 493–513.

Su, Y., Rambach, J., Minaskan, N., Lesur, P., Pagani, A., & Stricker, D. (2019). Deep multi-state object pose estimation for augmented reality assembly. In *2019 IEEE International Symposium on Mixed and Augmented Reality Adjunct (ISMAR-Adjunct)* (pp. 222–227). IEEE. https://doi.org/10.1109/ISMAR-Adjunct.2019.00-42

Svensson, J., & Atles, J. (2018). *Object detection in augmented reality.* Lund University.

Szajna, A., Szajna, J., Stryjski, R., & Skasiadek Michałand Woźniak, W. (2019). The application of augmented reality Technology in the Production Processes. In A. Burduk, E. Chlebus, T. Nowakowski, & A. Tubis (Eds.), *Intelligent Systems in Production Engineering and Maintenance* (Advances in intelligent systems and computing) (Vol. 835, pp. 316–324). Springer. https://doi.org/10.1007/978-3-319-97490-3_31

TensorFlow. (2021). *An end-to-end open source machine learning platform.* Accessed March 25, 2021, from https://www.tensorflow.org/

"TensorFlow to ONNX Converter". (2021). Accessed March 25, 2021, from https://github.com/onnx/tensorflow-onnx

The Linux Foundation. (2021). *ONNX.* Accessed March 25, 2021, from https://onnx.ai/

Todd, P. M., & Gigerenzer, G. (2012). *Ecological rationality: Intelligence in the world, evolution and cognition.* Oxford University Press. https://doi.org/10.1093/acprof:oso/9780195315448.001.0001

Trestioreanu, L., Glauner, P., Meira, J. A., Gindt, M., & State, R. (2020). Using augmented reality and machine learning in radiology. In P. Glauner & P. Plugmann (Eds.), *Innovative technologies for market leadership, future of business and finance* (pp. 89–106). Springer. https://doi.org/10.1007/978-3-030-41309-5_8

Tversky, A. (1972). Elimination by aspects: A theory of choice. *Psychological Review, 79*(4), 281–299. https://doi.org/10.1037/h0032955

Vávra, P., Roman, J., Zonča, P., Ihnát, P., Němec, M., Kumar, J., Habib, N., & El-Gendi, A. (2017). Recent development of augmented reality in surgery: A review. *Journal of Healthcare Engineering, 2017*, 4574172. https://doi.org/10.1155/2017/4574172

von Atzigen, M., Liebmann, F., Hoch, A., Bauer, D. E., Snedeker, J. G., Farshad, M., & Fürnstahl, P. (2021). HoloYolo: A proof-of-concept study for marker-less surgical navigation of spinal rod implants with augmented reality and on-device machine learning. *The International Journal of Medical Robotics + Computer Assisted Surgery: MRCAS, 17*(1), 1–10. https://doi.org/10.1002/rcs.2184

Vulcan Technologies. (2021). *CameraStream.* Accessed March 25, 2021, from https://github.com/VulcanTechnologies/HoloLensCameraStream

Windows. (2021). *Mixed reality locatable camera.* Accessed March 25, 2021, from https://docs.microsoft.com/en-us/windows/mixed-reality/develop/platform-capabilities-and-apis/locatable-camera

Wortmann, H. (2020). *Objekterkennung Unter Nutzung von Machine Learning Für Augmented Reality Anwendungen.* Hamburg University of Applied Sciences.

Chapter 3
Judgmental Selection of Forecasting Models (Reprint)

Fotios Petropoulos, Nikolaos Kourentzes, Konstantinos Nikolopoulos, and Enno Siemsen

Keywords Judgmental forecasting · Model selection · Behavioral operations · Decomposition · Combination

1 Introduction

Planning processes in operations—e.g., capacity, production, inventory, and materials requirement plans—rely on a demand forecast. The quality of these plans depends on the accuracy of this forecast. This relationship is well documented (Gardner, 1990; Ritzman & King, 1993; Sanders & Graman, 2009; Oliva & Watson, 2009). Small improvements in forecast accuracy can lead to large reductions in inventory and increases in service levels. There is thus a long history of research in operations management that examines forecasting processes (Seifert et al., 2015; Nenova & May, 2016; van der Laan et al., 2016, are recent examples).

Forecasting model selection has attracted considerable academic and practitioner attention during the last 30 years. There are many models to choose from—different forms of exponential smoothing, autoregressive integrated moving average

Originally published as: Petropoulos, F., Kourentzes, N., Nikolopoulos, K., & Siemsen, E. (2018). Judgmental selection of forecasting models. Journal of Operations Management, 60, 34–46.

F. Petropoulos (✉)
School of Management, University of Bath, Bath, UK
e-mail: fp359@bath.ac.uk

N. Kourentzes
Skövde Artificial Intelligence Lab, School of Informatics, University of Skövde, Skövde, Sweden

K. Nikolopoulos
Durham University Business School, Durham, United Kingdom

E. Siemsen
Wisconsin School of Business, University of Wisconsin, Madison, WI, USA

M. Seifert (ed.), *Judgment in Predictive Analytics*, International Series in Operations Research & Management Science 343,
https://doi.org/10.1007/978-3-031-30085-1_3

(ARIMA) models, neural nets, etc.—and forecasters in practice have to select which one to use. Many academic studies have examined different statistical selection methodologies to identify the best model; the holy grail in forecasting research (Petropoulos et al., 2014). If the most appropriate model for each time series can be determined, forecasting accuracy can be significantly improved (Fildes, 2001), typically by as much as 25–30% (Fildes & Petropoulos, 2015).

In general, forecasting software recommends or selects a model based on a statistical algorithm. The performance of candidate models is evaluated either on in-sample data, usually using appropriate information criteria (Burnham & Anderson, 2002), or by withholding a set of data points to create a validation sample (out-of-sample evaluation, Ord et al., 2017, also known as cross-validated error). However, it is easy to devise examples in which statistical model selection (based either on in-sample or out-of-sample evaluation) fails. Such cases are common in real forecasting applications and thus make forecasting model selection a non-trivial task in practice.

Practitioners can apply judgment to different tasks within the forecasting process, namely:

1. definition of a set of candidate models,
2. selection of a model,
3. parametrization of models,
4. production of forecasts, and
5. forecast revisions/adjustments.

Most of the attention in the judgmental forecasting literature focuses on the latter two tasks. Experts are either asked to directly estimate the point forecasts of future values of an event or a time series (see, for example, Hogarth & Makridakis, 1981; Petropoulos et al., 2017), or they are asked to adjust (or correct) the estimates provided by a statistical method in order to take additional information into account; such information is often called soft data, such as information from the sales team (Fildes et al., 2009).

However, little research has examined the role and importance of human judgment in the other three tasks. In particular, Bunn and Wright (1991) referred to the problem of judgmental model selection (item 2 in the above list), suggesting that the selection of the most appropriate model(s) can be based on human judgment. They also emphasized the dearth of research in this area. Importantly, the majority of the world-leading forecasting support systems allow human judgment as the final arbiter among a set of possible models.[1] Therefore, the lack of research into how well humans perform this task remains a substantive gap in the literature.

In this study, we examined how well human judgment performs in model selection compared with an algorithm using a large-scale behavioral experiment.

---

[1]For example, see the 'Manual Model Selection' feature of SAP Advanced Planning and Optimization (SAP APO), on SAP ERP: https://help.sap.com/viewer/c95f1f0dcd9549628efa8d7d653da63e/7.0.4/en-US/822bc95360267614e10000000a174cb4.html

We analyzed the efficiency of judgmental model selection of individuals as well as groups of participants. The frequency of selecting the best and worst models provides suggestions on the efficacy of each approach. Moreover, we identified the process that most likely will choose models that lead to improved forecasting performance.

The rest of our paper is organized as follows. The next section provides an overview of the literature concerning model selection for forecasting. The design of the experiment to support the data collection is presented in Sect. 3. Section 4 shows the results of our study. Section 5 discusses the implications for theory, practice, and implementation. Finally, Sect. 6 contains our conclusions.

2 Literature

2.1 Commonly Used Forecasting Models

Business forecasting is commonly based on simple, univariate models. One of the most widely used families of models are exponential smoothing models. Thirty different models fall into this family (Hyndman et al., 2008). Exponential smoothing models are usually abbreviated as ETS, which stands for either ExponenTial Smoothing or Error, Trend, Seasonality (the three terms in such models). More specifically, the error term may be either additive (A) or multiplicative (M), whereas trend and seasonality may be none (N), additive (A), or multiplicative (M). Also, the trend can be linear or damped (d). As an example, ETS(M,Ad,A) refers to an exponential smoothing model with a multiplicative error term, a damped additive trend, and additive seasonality. Maximum likelihood estimation is used to find model parameters that produce optimal one-step-ahead in-sample predictions (Hyndman & Khandakar, 2008).

These models are widely used in practice. In a survey of forecasting practices, the exponential smoothing family of models is the most frequently used (Weller & Crone, 2012). In fact, it is used in almost 1/3 of times (32.1%), with averages coming second (28.1%) and naive methods third (15.4%). More advanced forecasting techniques are only used in 10% of cases. In general, simpler methods are used 3/4 times, a result that is consistent with the relative accuracy of such methods in forecasting competitions. Furthermore, an empirical study that evaluated forecasting practices and judgmental adjustments reveals that "the most common approach to forecasting demand in support of supply chain planning involves the use of a statistical software system which incorporates a simple univariate forecasting method, such as exponential smoothing, to produce an initial forecast" (Fildes et al., 2009, p. 4), while it specifies that three out of four companies examined "use systems that are based on variants of exponential smoothing" (Fildes et al., 2009, p. 7).

There are many alternatives to exponential smoothing for producing business forecasts, such as neural networks and other machine learning methods.

Nevertheless, time series extrapolative methods remain very attractive. This is due to their proven track record in practice (Gardner, 2006) as well as their relative performance compared to more complex methods (Makridakis & Hibon, 2000; Armstrong, 2006; Crone et al., 2011). Furthermore, time series methods are fairly intuitive, which makes them easy to specify and use, and enhances their acceptance by the end-users (Dietvorst et al., 2015; Alvarado-Valencia et al., 2017). Complex methods, such as many machine learning algorithms, often appear as black boxes, and provide limited or no insights into how the forecasts are produced and which data elements are important. These attributes of forecasting are often critical for users (Sagaert et al., 2018).

2.2 *Algorithmic Model Selection*

Automatic algorithms for model selection are often built on information criteria (Burnham & Anderson, 2002; Hyndman et al., 2002). Models within a certain family (such as exponential smoothing or ARIMA) are fitted to the data, and the model with the minimum value for a specific information criterion is selected as the best. Various information criteria have been considered, such as Akaike's Information Criterion (AIC) or the Bayesian Information Criterion (BIC). The AIC after correction for small sample sizes (AICc) is often recommended as the default option because it is an appropriate criterion for short time series and it differs only minimally from the conventional AIC for longer time series (Burnham & Anderson, 2002). However, research also suggests that if we focus solely on out-of-sample forecasting accuracy, the various information criteria may choose different models that nonetheless result in almost the same forecast accuracy (Billah et al., 2006).

Information criteria are based on the optimized likelihood function penalized by model complexity. Using a model with optimal likelihood inadvertently assumes that the postulated model is true (Xia & Tong, 2011). In a forecasting context, this assumption manifests itself as follows: The likelihood approach generally optimizes the one-step-ahead errors; for the forecasts to be optimal for multi-step ahead forecasts, the resulting model parameters should be optimal for any longer horizon error distribution as well. This will only occur if the model is true, in which case the model fully describes the structure of the series. Otherwise, the error distributions will vary with the time horizon (Chatfield, 2000). Such time-horizon dependent error distributions are often observed in reality (Barrow & Kourentzes, 2016), providing evidence that any model merely approximates the underlying unknown true process. Not recognizing this can lead to a biased model selection which favors one-step-ahead performance at the expense of longer time horizons that may well be the analyst's real objective.

An alternative to selecting models via information criteria is to measure the performance of different models in a validation set (Fildes & Petropoulos, 2015; Ord et al., 2017). The available data are divided into fitting and validation sets. Models are fitted using the first set, and their performance is evaluated in the second set. The model with the best performance in the validation set is put forward to produce forecasts for the future. The decision maker can choose the appropriate accuracy measure. The preferred measure can directly match the actual cost function that is used to evaluate the final forecasts.

Forecasts for validation purposes may be produced only once (also known as fixed-origin validation) or multiple times (rolling-origin), which is the cross-validation equivalent for time series data. Evaluating forecasts over multiple origins has several advantages, most importantly their robustness against the peculiarities in data that may appear within a single validation window (Tashman, 2000). Model selection on (cross-)validation has two advantages over selection based on information criteria. First, the performance of multiple-step-ahead forecasts can be used to inform selection. Second, the validation approach is able to evaluate forecasts derived from any process (including combinations of forecasts from various models). The disadvantage of this approach is that it requires setting aside a validation set, which may not always be feasible. Given that product life cycles are shortening, having a validation sample available can be an out of reach luxury for forecasters.

A final category for automatic model selection involves measurement of various time series characteristics (such as trend, seasonality, randomness, skewness, intermittence, variability, number of available observations) as well as consideration of decision variables (such as the forecast horizon). Appropriate models are selected based on expert rules (Collopy & Armstrong, 1992; Adya et al., 2001) or meta-learning procedures (Wang et al., 2009; Petropoulos et al., 2014). However, such approaches are very sensitive to the selected rules or meta-learning features. No widely accepted set of such rules exists.

Regardless of the approach used for the *automatic* selection of the best model, all processes outlined above (information criteria, validation, and selecting based on rules) are based on statistics or can be implemented through an algorithmic process. A commonality among all algorithmic model selection approaches is that selection is based on historical data. None of these algorithms can evaluate forecasts when the corresponding actual values (for example, actually realized demand) are not yet available. These statistical selection approaches have been adopted from non-time series modeling problems in which the predictive aspect of a model may not be present. Therefore, forecasting is only implicitly accounted for in these algorithmic approaches. Good forecasts, rather than good descriptions of the series, are done with a "leap of faith".

2.3 Model Selection and Judgment

Despite the fact that the automatic selection of forecasting models has been part of many statistical packages and commercial software, what is often observed in practice is that managers select a model (and in some cases parameters) in a *judgmental* way. Automatic selection procedures are often hard to understand and communicate within companies. In that sense, managers lack trust in automatic statistical forecasting (Alvarado-Valencia et al., 2017). A standard issue is that automatic selection methods tend to change models between successive planning periods, substantially altering the shape of the series of forecasts. This issue reduces the users' trust in the system, especially after the statistical selection makes some poor choices (Dietvorst et al., 2015). Users then eventually resort to either fully overriding the statistical selection or implementing custom ad-hoc judgmental "correction" rules.

Moreover, managers often believe firmly that they better understand the data and the business context that created the data. For example, even if the result of an algorithm suggests that the data lacks any apparent seasonality (and as such a seasonal model would not be appropriate), managers may still manually select a seasonal model because they believe that this better represents the reality of their business. Lastly, the sense of ownership of forecasts can drive experts to override statistical results because more often than not evaluation of their work performance is associated with taking actions about the forecasts (Önkal & Gönül, 2005; Ord et al., 2017), or is influenced by organizational politics (Kolassa & Siemsen, 2016; Ord et al., 2017).

To the best of our knowledge, the efficacy of judgmental model selection has not been studied. We expect that when forecast models are presented in a graphical environment (actual data versus fitted values plus forecasts, as is the case in the majority of forecasting software), forecasters may pay little attention to the actual fit of the model in the in-sample data (or the respective value of the AIC if provided). However, critical to the judgmental selection will be the matching of the out-of-sample forecasts with the expected reality. Harvey (1995) observed that participants in a laboratory experiment made predictions so that the noise and patterns in the forecasts were representative of the past data. This finding leads us to believe that forecasters perform a mental extrapolation of the available in-sample data, rejecting the models that result in seemingly unreasonable forecasts and accepting the ones that represent a possible reality for them. Thus, in contrast to algorithmic model selection, forecasters will attempt to evaluate the out-of-sample forecasts, even if the future realized values of the forecasted variable are not yet available.

As the amount of information increases, decision makers are unable to process it efficiently and simultaneously (Payne, 1976). Accordingly, research in decision analysis and management judgment has established that decomposition methods, which divide a task into smaller and simpler ones, lead to better judgment. Such methods have also been found to be useful in judgmental forecasting tasks, especially for forecasts that involve trends, seasonality and/or the effect of special events such as promotions. (Edmundson, 1990) examined the performance of judgmental forecasting under decomposition. Similar to the way exponential smoothing works (Gardner, 2006), forecasters were asked to estimate the structural components of the time series (level, trend, and seasonality) separately. The three estimates were subsequently combined. Edmundson (1990) found that estimating the components independently resulted in superior performance compared with producing judgmental forecasts directly. In another study, Webby et al. (2005) observed similar results when the effects of special events were estimated separately. Also, Lee and Siemsen (2017) demonstrated the value of task decomposition on order decisions, especially when coupled with decision support.

We expect that these insights may be applied to judgmental model selection. When judgmentally selecting between forecasting models (through a graphical interface), we expect a model-build approach to outperform the simple choice between different models. In a model-build approach, forecasters are asked to verify the existence (or not) of structural components (trend and seasonality). This changes the task from identifying the best extrapolation line to determining whether the historical information exhibits specific features that the expert believes will extend into the future.

2.4 Combination and Aggregation

Forecast combinations can result in significant improvement in forecast accuracy (Armstrong, 2001). There is also ample evidence that combining the output of algorithms with the output of human judgment can confer benefits. Blattberg and Hoch (1990) used a simple (50–50%) combination and found significant gains in combination methods compared with the separate use of algorithms and judgment. Their results have been repeatedly confirmed in the forecasting literature. Franses and Legerstee (2011) found that a simple combination of forecasts outperformed both statistical and judgmentally adjusted forecasts. Petropoulos et al. (2016) demonstrated that a 50–50 combination of forecasts in the period after a manager's adjustments have resulted in significant losses can indeed increase accuracy by 14%. Wang and Petropoulos (2016) found that a combination is as good, if not better, than selecting between a statistical or an expert forecast. Trapero et al. (2013) demonstrated further gains with more complex combination schemes. We anticipate that a combination will also be beneficial in the context of forecast model selection.

The concept of the wisdom of crowds refers to the aggregation of the judgments of a group of decision makers/stakeholders. Surowiecki (2005) provided several cases in which this concept has been found to increase performance compared with individual judgments. Ferrell (1985) also argued for the importance of combining individual judgments and discussed the significantly improved performance of judgmental aggregation. He suggested that the process of the combination itself is of little significance. However, a later study added that aggregation done mechanically is better than if it is done by one of the forecasters to avoid the possibility of biased weights (Harvey & Harries, 2004). In any case, we expect that the aggregation of judgmental model selections will lead to improved performance compared with selecting a single model, either judgmentally or statistically.

Table 3.1 The four forecasting models considered in this study

Model description	ETS model	Trend	Seasonality
Simple exponential smoothing (SES)	A,N,N	✗	✗
SES with additive seasonality	A,N,A	✗	✓
Damped exponential smoothing (DES)	A,Ad,N	✓	✗
DES with additive seasonality	A,Ad,A	✓	✓

3 Design of the Behavioral Experiment

3.1 Selecting Models Judgmentally

Specialized forecasting software lists forecast methods and models. The users of such systems must choose one from the list to extrapolate the data at hand. In some cases, this list of choices is complemented by an option that, based on an algorithm, automatically identifies and applies the best of the available methods. However, in the context of the current paper, we assumed that forecasters do not have such a recommendation available and instead rely solely on their own judgment. We also assumed that the choice set is constrained to four models able to capture various data patterns (level, trend, and seasonality). This is not an unreasonable setup, with some established systems offering such specific options (such as the well established SAP APO-DP system). We resorted to the exponential smoothing family of models (Hyndman et al., 2008) and focused on the four models presented in Table 3.1 (mathematical expressions are provided in 7).

To emulate the simple scenario implied by standard forecasting support systems (choose one of the available forecasting models), we used radio buttons to present the different model-choices as a list, as depicted in the left part of Fig. 3.1. A user can navigate across the different choices and examine the forecasts produced by each method. Once the forecasts produced by a method are considered satisfactory, then a manager can submit the choice and move to the next time series. We call this approach "judgmental model selection".

We also considered a second approach where the user builds a model instead of selecting between models. In the "model-build" condition, we ask a user to identify the existence of a trend and/or seasonality in the data; the response can be used to select the respective model from the Table 3.1. For example, identification of a trend implies damped exponential smoothing; identification of seasonality without a trend implies SES with seasonality. This can be implemented in the software design by including two check-boxes (right panel of Fig. 3.1). Once a change in one of these two check-boxes has been made, the forecasts of the respective model are drawn. To facilitate identification, we provide trend and seasonal plots (with usage instructions) in an attempt to aid the users with the pattern identification task.

In both cases, once a participant submits his or her decisions, we use the selected forecasting method to produce one-year-ahead (12 months) forecasts for that time series. The forecasts are compared with the actual future values, which were withheld, to find the forecast accuracy of the submitted choice.

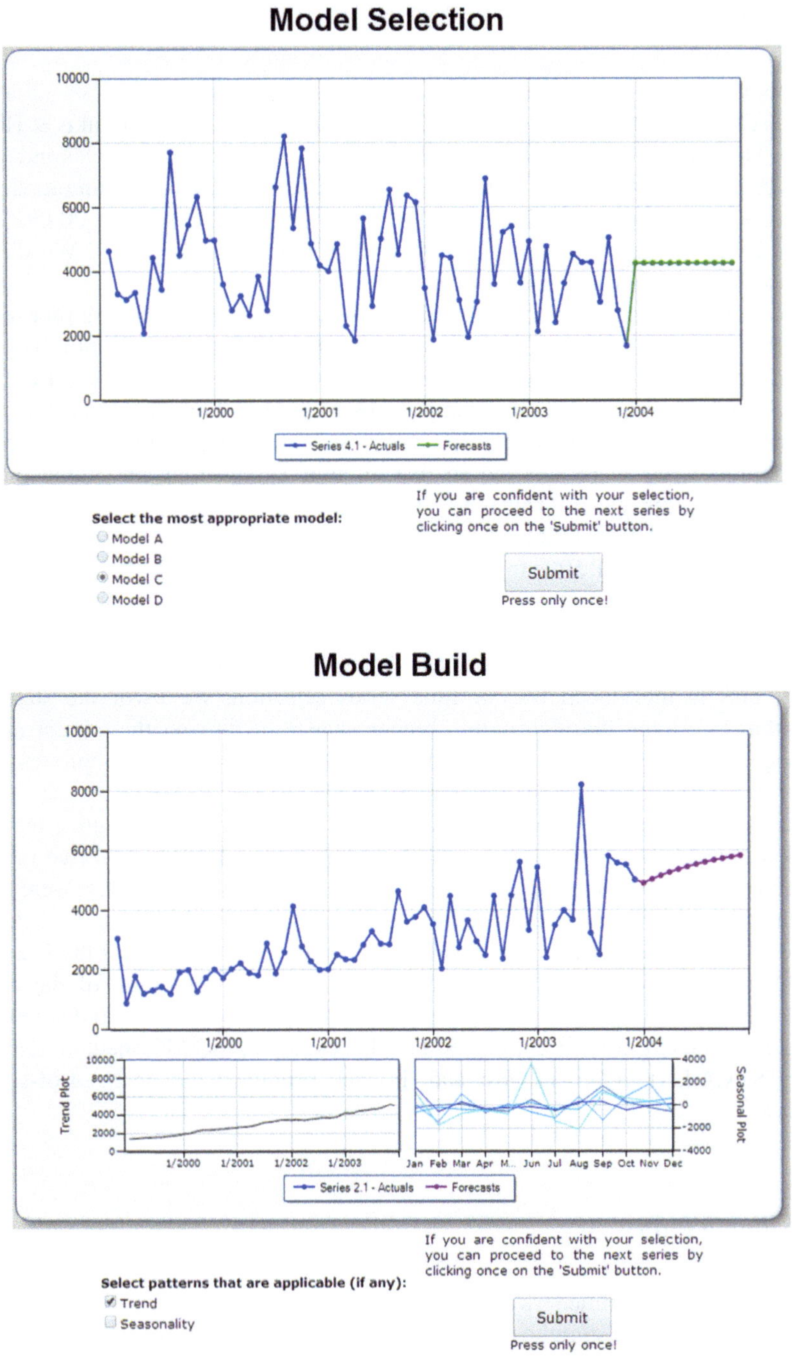

Fig. 3.1 Screens of the Web-based environment of the behavioral experiment. The left panel shows the implementation of model selection; the right panel presents the model-build

3.2 Data

To compare the performance of statistical versus judgmental model selection, we used a subset of time series from the M3-Competition dataset (Makridakis & Hibon, 2000). This dataset consists of 3003 real time series of various frequencies and types. It has been used many times in empirical evaluations of new forecasting models or processes (Hyndman et al., 2002; Taylor, 2003; Hibon & Evgeniou, 2005; Crone et al., 2011; Athanasopoulos et al., 2017; Petropoulos et al., 2018). We did not disclose the data source to the participants.

We focused on series with a monthly frequency and handpicked 32 time series. We selected series so that in half of them, the statistical model selection based on minimizing the value of the AIC succeeds in identifying the best model as evaluated in the hold-out sample (out-of-sample observations). For the other half, this minimum-AIC model fails to produce the best out-of-sample forecast. Moreover, the 32 time series were selected so that all four exponential smoothing models considered in this paper (Table 3.1) are identified as best in some time series according to the AIC criterion.

This success rate of 50% for the statistical algorithm to pick the correct model probably overestimated (but not by much) the true success rate. When the four models presented in Sect. 3.1 were applied on the 1428 monthly series of the M3-competition, selection based on AIC is accurate in 36% of the cases. As such, if any bias is introduced by our time series selection, we favor the statistical algorithm by giving the algorithm a higher chance of picking the correct model. Consequently, the true effect size with which human judgment improves upon performance may be underestimated in our analysis.

Because time series from the M3-Competition are of various lengths, we truncated all selected series to a history of 72 months (six years). The first five years of data (60 months) were treated as the in-sample data, on which the models were fitted. The last year (12 months) was used for out-of-sample evaluation. Figure 3.2 depicts a typical time series used in this behavioral experiment. Along with the historical data that cover five years, we also draw the (unobserved) future of the series. Moreover, we show the statistical point forecasts of the four exponential smoothing models considered in different colors. For this example, the AIC method identifies the ETS(A,Ad,A) model as best and, in fact, produced the best out-of-sample forecasts.

3.3 Participants

The behavioral experiment was introduced as an elective exercise to groups of undergraduate and postgraduate students studying at various universities (Institutions with at least 20 participants include Bangor University, Cardiff University, Lancaster University, the National Technical University of Athens, and Universidad

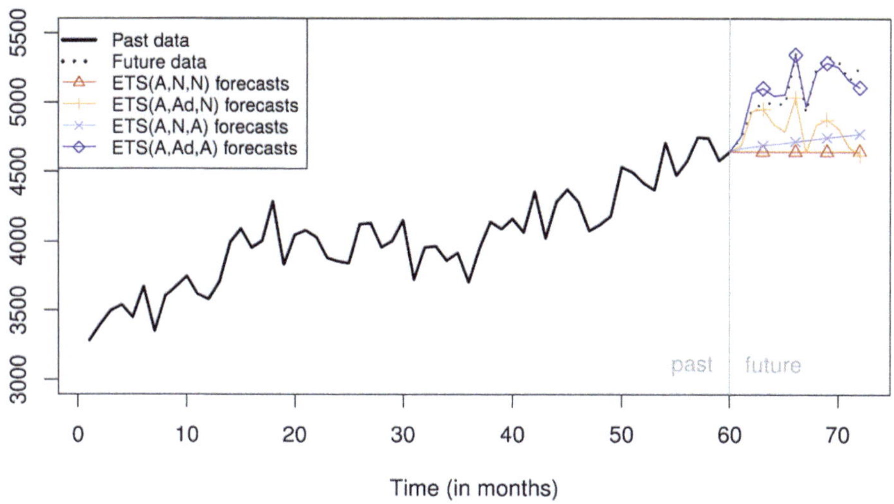

Fig. 3.2 A typical time series used in this research, along with the forecasts from the four models

de Castilla-La Mancha). Details regarding the modules where the experiment was introduced as an elective exercise are provided in Table 3.6 of 8 in Appendix. We ran the exercise as a seminar (workshop) session during the respective modules. The experiment was also posted to several relevant groups on LinkedIn and three major forecasting blogs. As an incentive, the participants were told they would receive £ 50 if their performance ranked within the top-20 across all the participants (see Lacetera et al., 2014, for some positive effects in rewarding volunteers).

We recruited more than 900 participants; 693 of them completed the task. Upon commencing the task, the participants were asked to self-describe themselves as undergraduate/postgraduate students, researchers, practitioners, or other. At the same time, each participant was randomly assigned to either the "model selection" or "model-build" condition. Table 3.2 presents the distribution of participants across roles (rows) and experimental conditions (columns).

Most previous behavioral studies of judgmental forecasting were limited to students' participation (Lee et al., 2007; Thomson et al., 2013), which is common in behavioral experiments (Deck & Smith, 2013). In this study, our sample of student participants was complemented by a sample of practitioners (90 forecasting experts). Practitioner participants come from a variety of industries, as depicted in Table 3.7 of 8 in Appendix. Our analysis also considered this sub-sample separately to check for differences and similarities between the practitioners and the students.

The completion rate was high for student participants (at 83%). This was expected, as the experiment was conducted in a lab for several cohorts of this population. The rate was lower for the other groups (practitioners 67%, researchers

Table 3.2 Participants per role and experiment

Role	Model selection	Model-build	Total
UG students	139	137	276
PG students	103	108	211
Researchers	13	31	44
Practitioners	46	44	90
Other	40	32	72
Total	341	352	693

57%, and other participants 56%). These observed completion rates are slightly lower than other professional web-based surveys (78.6%[2]); this difference may be explained by the duration of this behavioral experiment (around 30 minutes, as opposed to the recommended 15 minutes[3]).

3.4 *The Process of the Experiment*

After being randomly assigned to experimental conditions, participants were given a short description of the experimental task. The participants assigned to the model-build condition were given brief descriptions of the trend and seasonal plots.

The 32 time series we selected from the M3 Competition were divided into four groups of 8 time series each (the same for all participants), and the actual experiment consisted of 4 rounds. In each round, the participants were provided with different information regarding the forecasts and the fit derived from each model. The purpose was to investigate how different designs and information affect judgmental model selection or model-build. The information provided in each round is as follows:

- Only the out-of-sample forecasts (point forecasts for the next 12 months) were provided.
- The out-of-sample forecasts and the in-sample forecasts (model fit) were provided.
- The out-of-sample forecasts and the value of the AIC, which refers to the fit of the model penalized by the number of parameters, were provided.
- The out-of-sample forecasts, the in-sample forecasts, and the value of the AIC were provided.

The order of the rounds, as well as the order of the time series within each round, was randomized for each participant. Attention checks (Abbey & Meloy, 2017) were not performed. To maximize participant attention, round-specific instructions were given at the beginning of each round so that the participants were able to identify and potentially use the information provided.

[2]http://fluidsurveys.com/university/response-rate-statistics-online-surveys-aiming/

[3]http://fluidsurveys.com/university/finding-the-correct-survey-length/

Our experiment has both a between-subjects factor (model selection vs. model-build) as well as a within-subjects factor (information provided). Since the latter produced little meaningful variation, our analysis focuses on the former. We chose this design since we believed that the difference between model selection vs. model-build could introduce significant sequence effects (making this more suited for between-subjects analysis), but we did not believe that differences in information provided would lead to sequence effects (making this more suited for within-subjects analysis).

3.5 *Measuring Forecasting Performance*

The performance of both algorithmic (based on AIC) and judgmental model selection is measured on the out-of-sample data (12 monthly observations) that were kept hidden during the process of fitting the models and calculating the AIC values. Four metrics were used to this end. The first metric was a *percentage score* based on the ranking of the selections. This was calculated as follows: A participant receives 3 points for the best choice (the model that leads to the best forecasts) for a time series, 2 points for the second best choice, and 1 point for the third best choice. Zero points were awarded for the worst (out of four) choices. The same point scheme can be applied to both judgmental forecasting approaches (model selection and model-build) once the identified patterns are translated to the respective model. The mean absolute error (*MAE*) was used as the cost function for evaluation. The range of points that anyone could collect is (given the number of time series) 0–96, which was then standardized to the more intuitive scale of 0–100. The *percentage score* of each participant, along with a pie chart presenting the distribution of best, second best, third best, and worst selections, was presented at the very last page of the experiment.

Apart from the percentage score based on the selections, we also use three formal measures of forecasting performance: (1) Mean Percentage Error (*MPE*) is a measure suitable for measuring any systematic bias in the forecasts, (2) Mean Absolute Percentage Error (*MAPE*) and (3) Mean Absolute Scaled Error (*MASE*; Hyndman & Koehler, 2006) are suitable for measuring the accuracy of the forecasts. Although the *MAPE* suffers from several drawbacks (Goodwin & Lawton, 1999), it is intuitive, easy to interpret, and widely used in practice. *MASE* is the Mean Absolute Error scaled by the in-sample Mean Absolute Error of the naive method that uses the last observed value as a forecast. The intuition behind this scaling factor is that it can always be defined and only requires the assumption that the time series has no more than one unit root, which is almost generally true for real time series. Other scaling factors, such as the historical mean, impose additional assumptions, such as stationarity. *MASE* has desirable statistical properties and is popular in the literature. In particular, *MASE* is scale independent without having the computational issues of *MAPE*. It is always defined and finite, with the only exception being the extreme case where all historical data would be equal. Note that *MAE* and *MASE* would both give the same rankings of the models within a series and, as a result, the same *percentage scores*. However, *MAE* is a scale-dependent error measure and not suitable for summarizing across series. For all three measures, *MPE*, *MAPE*, and *MASE*, values closer to zero are better. Moreover, whereas *MPE* can take both positive and negative values, the values for *MAPE* and *MASE* are always non-negative.

The values of *MPE*, *MAPE*, and *MASE* for a single time series across forecast horizons are calculated as

$$MPE = \frac{100}{H}\sum_{i=1}^{H}\frac{y_{n+i} - f_{n+i}}{y_{n+i}},$$

$$MAPE = \frac{100}{H}\sum_{i=1}^{H}\frac{|y_{n+i} - f_{n+i}|}{|y_{n+i}|},$$

$$MASE = \frac{(n-1)\sum_{i=1}^{H}|y_{n+i} - f_{n+i}|}{H\sum_{j=1}^{n}|y_j - f_{j-1}|},$$

where y_t and f_t refer to the actual and the forecasted value at period t, n is the size of the training sample, and H is the forecast horizon.

4 Analysis

4.1 Individuals' Performance

We next examined the performance of the judgmental model selection and judgmental model-build approaches. We contrasted their performance with the algorithmic model selection by AIC (Hyndman & Khandakar, 2008).

How do judgmental model selection and judgmental model-build perform based on the percentage score? The left panel of Fig. 3.3 presents the percentage scores of the participants under the two approaches. The performance of each participant is depicted with a dot marker (blue for the practitioner participants, gray for all other participants), and the respective box-plots are also drawn. The square symbol represents the arithmetic mean of the percentage score for each approach. The horizontal (red) dashed line refers to the statistical benchmark (performance of automatic model selection based on AIC). Generally, participants performed better under the model-build approach than the model selection approach. In essence, the average participant in model-build performs as well as the participant in the 75tth percentile of the model selection approach. More importantly, participants under the model-build approach perform on average as well as the statistical selection. However, the differences in scores between individuals are large, with the range spanning between 32% and 83%.

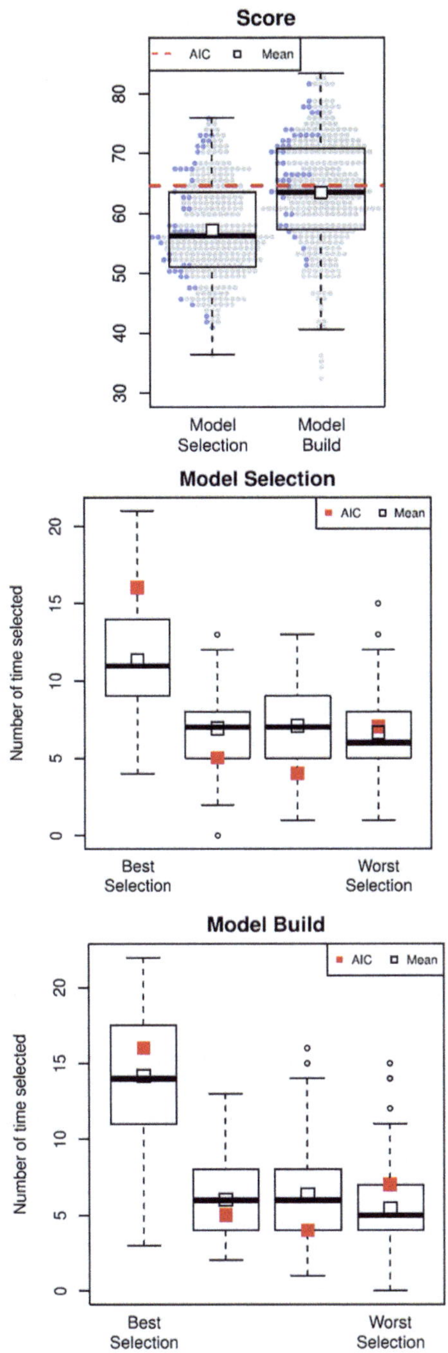

Fig. 3.3 Performance of model selection and model-build in terms of scores and distributions of best/worst selections

Table 3.3 Frequencies of selected models

Selection method	SES	Seasonal SES	DES	Seasonal DES
Selection based on AIC	46.88%	31.25%	6.25%	15.62%
Judgmental model selection	17.27%	33.68%	17.64%	31.41%
Judgmental model-build	18.44%	27.85%	24.48%	29.23%
Best out-of-sample performance	34.38%	21.88%	21.88%	21.88%

Do humans select similarly to the algorithm? The middle and right panels of Fig. 3.3 present, respectively, how many times the participants selected the best, second best, third best, and worst models under the model selection and model-build approaches. The differences in performance (in terms of percentage scores) between model selection and model-build derives from the fact that model-build participants were able to identify the best model more frequently. By comparing how frequently algorithms (red squares) and humans make the best and worst model selection, we can observe that humans are superior to algorithms in avoiding the worst model, especially in the model-build case. The differences are statistically significant according to *t*-tests for both best and worst selections and both strategies, model selection and model-build ($p < 0.01$). The frequencies that algorithms and humans selected each model are presented in Table 3.3, along with how many times each model performs best in the out-of-sample data. We observe that algorithms generally select the level models (SES and SES with seasonality) more often compared to their trended counterparts. AIC, as with other information criteria, attempts to balance the goodness-of-fit of the model and its complexity as captured by the number of parameters. More uniform distributions are observed for human selection and out-of-sample performance.

How do the judgmental approaches perform based on error measures? Figure 3.4 presents the performance of all 693 participants for *MPE*, *MAPE*, and *MASE*, the three error measures considered in this study. In both the model selection and the model-build treatments, human judgment is significantly better (less biased and more accurate) than statistical selection in terms of *MPE* and *MAPE*. At the same time, although the judgmental model-build performs on a par with statistical selection for *MASE*, the judgmental model selection performs worse than the statistical selection. Differences in the insights provided by the different error measures can be attributed to their statistical properties.

It is noteworthy that statistical selection is, on average, positively biased (negative values for *MPE*). However, this is not the case for all participants. In fact, slightly more than 30% of all participants (42% of those that were assigned in the model selection experiment) are, on average, negatively biased. The positive bias of statistical methods is consistent with the results of Kourentzes et al. (2014) who investigated the performance of statistical methods on all M3-competition data.

Do the results differ if only the practitioners' subgroup is analyzed? Independent sample *t*-tests were performed to compare the performance of practitioners and students. The results showed that differences are not statistically significant, apart from the case of model-build and *MASE* where practitioner participants perform

Fig. 3.4 Performance of model selection and model-build in terms of error measures when all participants are considered

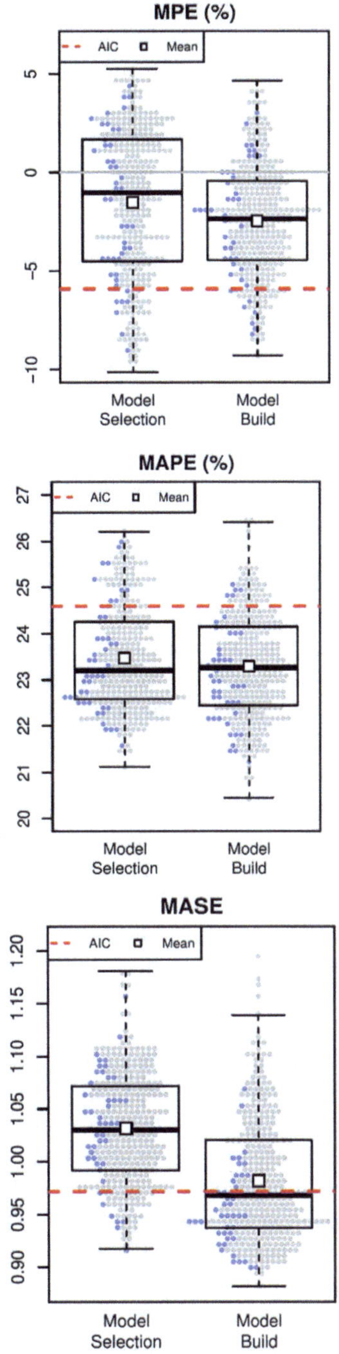

significantly better. The similarities in the results between student and practitioner participants are relevant for the discussion of the external validity of behavioral experiments using students (Deck & Smith, 2013). Similar to Kremer et al. (2015), our results suggest that student samples may be used for (at least) forecasting behavioral experiments.

4.2 Effects of Individuals' Skill and Time Series Properties

Going beyond the descriptives presented thus far, we constructed a linear mixed effects model to account for the variability in the skills of individual participants, as well as for the properties of each time series that was used. We model values of *MASE* (containing the performance of the participants on each individual response: 693 participants × 32 time series) considering the following fixed effects: (i) experimental condition (model selection or model-build); (ii) interface information (out-of-sample only, in- and out-of-sample and these options supplemented with fit statistics; see Sect. 3.4); and (iii) the role of the participants (see Table 3.2 in which both under- and postgraduate students were grouped together). We accounted for the variation between participants as a random effect that reflects any variability in skill. Similarly, we consider the variability between time series as a second random effect, given that they have varying properties.

To conduct the analysis we use the lme4 package (Bates et al., 2015) for R statistical computing language (R Core Team, 2016). We evaluated the contribution of each variable in model by using two information criteria (AIC and BIC). To facilitate the comparison of the alternative model specification using information criteria, we estimated the model using maximum likelihood. We found only marginal differences between the models recommended by the two criteria, a result suggesting that the most parsimonious option was the better choice.

We concluded that only the effect of the experimental condition was important, and that the interface information and role of participant did not explain enough variability to justify the increased model complexity. Consequently, these two effects were removed from the model. Both random effects were deemed useful. We investigated for random slopes as well, but found that such slopes did not explain any additional variability in *MASE*. Therefore, they were not considered further.

The resulting fully crossed random intercept model is reported in Table 3.4. The estimated model indicates that model-build improves *MASE* by 0.0496 over model select, which is consistent with the analysis so far. The standard deviations of the participant and series effects, respectively, are 0.0314 and 0.5551 (the intraclass correlations are 0.0026 and 0.8178). This shows that the skills of the participants account for only a small degree of the variability of *MASE*. This helps explain the insignificant role of the participant background. To put these values into perspective, the standard deviation of *MASE* on individual time series responses is 0.6144.

Table 3.4 Linear mixed effects model output

Mixed effect model statistics			
Fixed effects			
	Estimate	Standard Error	
Intercept	1.0313	0.0982	
Experiment setup	−0.0496	0.0042	
Random effects			
	Standard Deviation		
Participant (intercept)	0.0314		
Time series (intercept)	0.5551		
Residual	0.2601		
Model statistics			
	AIC	BIC	Log Likelihood
	3742.3	3782.3	−1866.1

4.3 50% Statistics + 50% Judgment

The seminal work by Blattberg and Hoch (1990) suggested that combining the outputs of statistical models with managerial judgment will provide more accurate outputs than each single-source approach (model *or* manager), while a 50–50% combination is "a nonoptimal but pragmatic solution" (Blattberg & Hoch, 1990, p. 898). Their result has been confirmed in many subsequent studies. In this study, we considered the simple average of the two predictions, i.e., the equal-weight combination of the forecasts produced by the model selected by the statistics (AIC) and the model chosen by each participant.

How does a 50–50 combination of statistical and judgmental selection perform? Figure 3.5 presents the performance of the simple combination of statistical and judgmental selection for the three error measures considered in this study and categorized by the two judgmental approaches. We observed that performance for both judgmental model selection and model-build was improved significantly compared with using statistical selection alone (horizontal dashed red line). In fact, the combination of the statistical + judgmental selection is less biased than statistical selection in 86% of the cases and produces lower values for *MAPE* and *MASE* for 99% and 90% of the cases, respectively. Moreover, the differences in the performance of the two approaches are now minimized.

Does a 50–50 combination bring robustness? On top of the improvements in performance, an equal-weight combination also reduces the between-subject variation in performance. Focusing, for instance, on *MASE* and the judgmental model-build approach, Fig. 3.4 suggests a range of 0.314 (between 0.882 and 1.196) between best and worst performers. The comparable range according to Fig. 3.5 is 0.136 (between 0.864 and 1). Therefore, a 50–50 combination renders the judgmental selection approaches more robust.

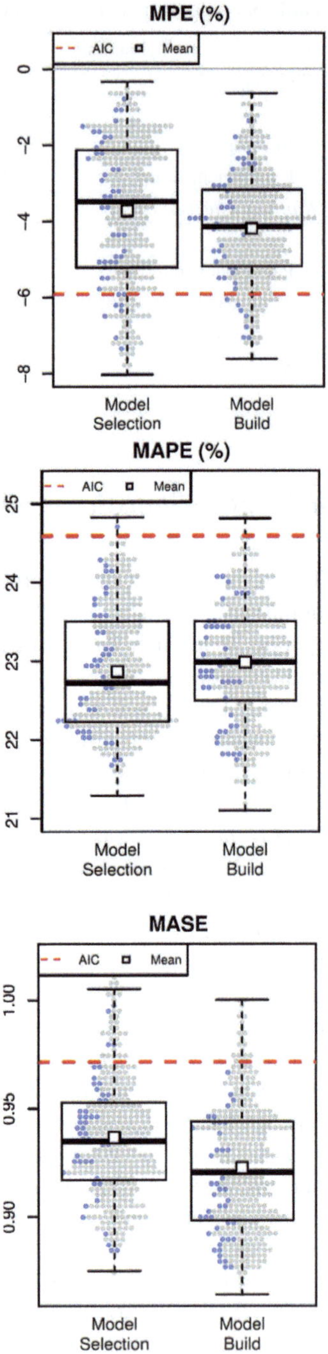

Fig. 3.5 Performance of 50–50% combination of statistical and judgmental selection

4.4 Wisdom of Crowds

An alternative to combining the statistical and judgmental selections is judgmental aggregation—a combination of the judgmental selections of multiple participants. The concept of the "wisdom of crowds" is not new (Surowiecki, 2005) and has repeatedly been shown to improve forecast accuracy as well as the quality of judgments in general.

For example, consider that a group of 10 experts is randomly selected. Given their selections regarding the best model, we can derive the frequencies that show how many times each model is identified as best. In other words, experts preferences are equally considered (each expert has exactly one vote, and all votes carry the same weight). This procedure leads to a weighted combination of the four models for which the performance can be measured. We consider groups of 1 to 25 experts randomly re-sampled 1000 times.

Does judgmental aggregation improve forecasting performance? Figure 3.6 presents the results of judgmental aggregation. The light blue area describes the range of the performance of judgmental aggregation for various group sizes when the model-build approach is considered. The middle-shaded blue area refers to the 50% range of performances, and the dark blue line refers to the median performance. In other words, if one considers a vertical line (i.e., for particular group size), the points at which this line intersects provide the minimum, first quartile, median, third quartile, and maximum descriptive summary of the performance. For the judgmental model selection approach, only the median is drawn in the black dotted line, and the performance of statistical selection is represented by a red dashed horizontal line. We observed significant gains in performance as the group size increased, coupled with lower variance in the performance of different equally sized groups. We also observed the convergence of performance, meaning that no further gains were noticed in the average performance for group sizes higher than 20. Judgmental aggregation outperforms both statistical and individual selection.

How many experts are enough? A careful examination of Fig. 3.6 reveals, on top of the improvements if aggregation is to be used, the critical thresholds for deciding on the optimal number of experts in groups. We observed that if groups of five participants are considered, then their forecasting performance was almost always better than that of the statistical selection on all three measures, regardless of their role (undergraduate/postgraduate students, practitioners, researchers, or other). Of even more interest, the third quartile of groups of size two always outperforms the statistical benchmark. It is not the first time that the thresholds of two and five appear in the literature. These results confirm previous findings: "only two to five individuals' forecasts must be included to achieve much of the total improvement" (Ashton & Ashton, 1985, p. 1499). This result holds even if we only consider specific sub-populations of our sample, e.g. practitioners only. Once judgmental aggregation is used, the results from practitioners are all-but identical to results from students and other participants.

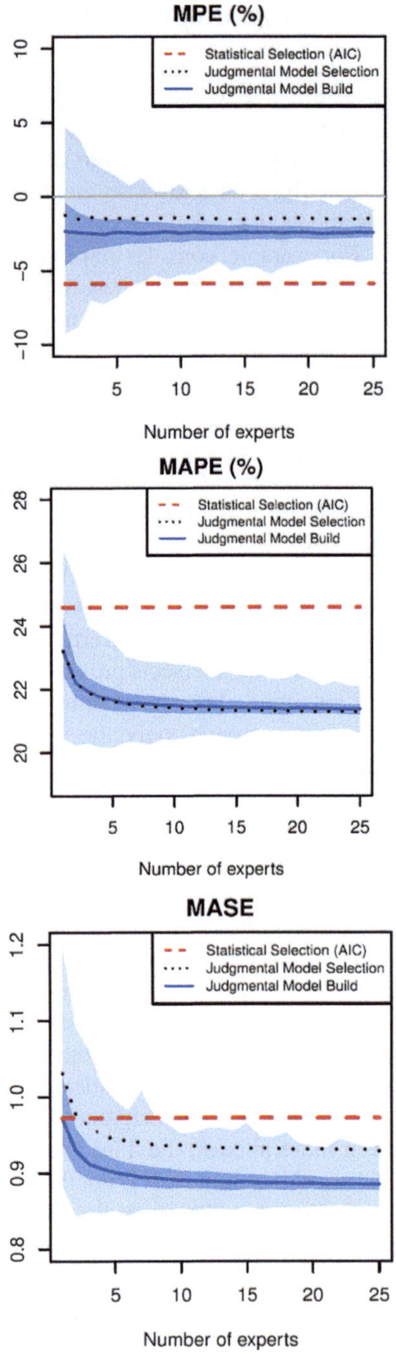

Fig. 3.6 Wisdom of crowds' performance for different numbers of experts

Note that in this study we consider only unweighted combinations of the judgmental model selections. If the same exercise were carried dynamically (over time), then one could also consider, based on past performance, weighted combinations that have been proven to enhance performance in other forecasting tasks (Tetlock & Gardner, 2015).

How does the model with the most votes perform? Instead of considering a weighted-combination forecast based on votes across the four models, we have also examined a case in which the aggregate selection is the model with the most votes. In case two (or more) models are tied for first place in votes, then an equal-weight combination among them was calculated. The performance of this strategy is worse, in terms of the accuracy measures considered, than the wisdom of crowds with weighted combinations; moreover, the quality of the performance depends heavily on the sample selected (the variance is high even for groups with a large number of experts). However, on average, merely choosing the most-popular model still outperforms statistical selection.

4.5 Evaluation Summary and Discussion

Table 3.5 summarizes the results presented in the above sections. As a sanity check, we also provide the performance of four additional statistical benchmarks: (noitemsep, nolistsep)

- *Random selection* refers to a performance obtained by randomly selecting one of the four available choices. We have reported the arithmetic mean performance when such a procedure is repeated 1000 times for each series. This benchmark is included to validate the choice of the time series and to demonstrate that non-random selection is indeed meaningful (in either a statistical or a judgmental manner).

Table 3.5 Summary of the results; top-method is underlined; top-three methods are in boldface

Method	MPE (%)	MAPE (%)	MASE
Individual selection			
Random selection	−2.91	24.52	1.104
Selection based on AIC	−5.93	24.59	0.971
Judgmental model selection	−1.52	23.48	1.031
Judgmental model-build	−2.45	23.30	0.982
Combination			
Equal-weight combination	−2.90	21.96	0.985
Weighted combination based on AIC	−4.84	23.39	0.931
Combination of best two based on AIC	−4.65	23.12	0.921
50–50% combination of AIC and judgment	−3.96	22.93	0.930
Wisdom of crowds: 5 humans (model-build)	−2.43	21.68	0.903

- *Equal-weight combination* refers to the simple average of the forecasts across all four models. In other words, each model was assigned a weight of one-quarter. It is included as a benchmark of the performance of 50–50 combinations and the wisdom of crowds.
- *Weighted combinations based on AIC* were proposed by Burnham and Anderson (2002) and evaluated by Kolassa (2011). This approach showed improved performance over selecting the model with the lowest AIC. It is used in this study as a more sophisticated benchmark for the wisdom of crowds.
- *Combination of best two based on AIC* refers to the equal-weight combination of the best and second-best models based on the AIC values.

Focusing on the first four rows of Table 3.5 that refer to selecting a single model with different approaches, random selection performs poorly compared with all other approaches. This is especially true in terms of *MASE*. Moreover, although it seems to be less biased than statistical selection, random selection's absolute value of *MPE* is larger than any of the two judgmental approaches.

The last five rows of Table 3.5 present the performance of the various combination approaches. First, we observe that the equal-weight combination performs very well according to all metrics, apart from *MASE*. A weighted combination based on AIC improves the performance of the statistical benchmark, confirming the results by Kolassa (2011); however, it is always outperformed by both 50–50 combinations of statistical and judgmental selection and by the wisdom of crowds. Combination of the best two models based on AIC performs slightly better than weighted combination based on AIC.

We also present in boldface the top three performers for each metric. The top performer is underlined. We observe that the wisdom of crowds (which is based on model-build) is always within the top three and ranked first for two of the metrics. The wisdom of crowds based on model selection also performs on par. We believe that this is an exciting result because it demonstrates that using experts to select the appropriate method performs best against state-of-the-art benchmarks.

5 Implications for Theory, Practice, and Implementation

This work provides a framework for judgmental forecasting model selection, and highlights the conditions for achieving maximal gains. We now discuss the implications of our work for theory and practice as well as issues of implementation.

Statistical model selection has been dominated by goodness-of-fit derived approaches, such as information criteria, and by others based on cross-validated errors (Fildes & Petropoulos, 2015). The findings of our research challenge these approaches and suggest that an alternative approach based on human judgment is feasible and performs well. Eliciting how experts perform the selection of forecasts may yield still more novel approaches to statistical model selection. Our research provides evidence that a model-build approach works better for humans. We postulate that model-build implies a suitable structure (and a set of restrictions) that aid selection. Statistical procedures could potentially benefit from a similar framing.

Our findings are aligned with the literature on judgmental forecasting as well as research in behavioral operations. The good performance of the 50–50 combination (of judgment and algorithm) and judgmental aggregation resonates with findings in forecasting and cognitive science (Blattberg & Hoch, 1990; Surowiecki, 2005).

Our research looks at an everyday problem that experts face in practice. Planners and managers are regularly tasked with the responsibility of choosing the best method to produce the various forecasts needed in their organizations. We not only benchmark human judgment against a state-of-the-art statistical selection but also provide insights into how to aid experts. Another exciting aspect of this research is that it demonstrates that expert systems that rely on algorithms to select the right model, such as Forecast Pro, Autobox, SAP Advanced Planning and Optimization—Demand Planning (APO-DP), IBM SPSS Forecasting, SAS, etc., may be outperformed by human experts, if these experts are supported appropriately in their decision making. This has substantial implications for the design of algorithms for both expert systems algorithms and the user interfaces of forecasting support systems.

Judgmental model selection is used in practice because it has some endearing properties. It is intuitive: A problem that necessitates human intervention is always more meaningful and intellectually and intuitively appealing for users. It is interpretable: Practitioners understand how this process works. The version of model-build that is based on judgmental decomposition is easy to explain and adapt to real-life setups. This simplicity is a welcome property (Zellner et al., 2002). In fact, the configuration used in our experiment is already offered in a similar format in popular software packages. For example, SAP APO-DP provides a manual (judgmental) forecasting model selection process, providing clear guidance of a judgmental selection to be driven from prevailing components (most notably trend and seasonality) as perceived by the user (manager). Specialized off-the-shelf forecasting support systems like Forecast Pro also allow their optimal algorithmic selection to be overridden by the user.

At its most basic form, implementing judgmental model selection requires no investment. Nonetheless, to obtain maximum gains, existing interfaces will need some redesign to allow incorporation of the model-build approach. However, a crucial limitation is the cost of using human experts. Having an expert going through all items that need to be forecasted may not be feasible for many organizations, such as large retailers that often require millions of forecasts. Of course, using the judgmental aggregation approach requires even more experts.

In a standard forecasting and inventory setting, ABC analysis is often used to classify the different stock keeping units (SKUs) into importance classes. In this approach, the 20% most important and the 50% least important items are classified as A and C items, respectively. Additionally, XYZ analysis is used in conjunction with ABC analysis to further classify the products into "easy-to-forecast" (X items) to "hard-to-forecast" (Z items). As such, nine classes are considered, as depicted in Fig. 3.7 (Ord et al., 2017). We propose the application of the wisdom of crowds for

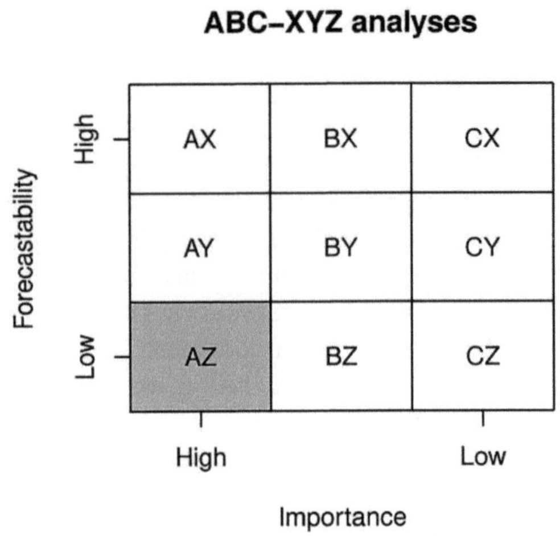

Fig. 3.7 A visual representation of ABC-XYZ analyses; resources for judgmental model selection should be allocated towards the important and low-forecastable items

judgmental model selection/build on the AZ items (those of high importance but difficult to forecast). This class is shaded with gray color in Fig. 3.7. In many cases, these items represent only a small fraction of the total number of SKUs. Thus, judgmentally weighted selections across the available models of a forecasting support system should be deduced by the individual choices, either in terms of models or patterns, of a small group of managers; our analysis showed that selections from five managers would suffice.

A potential limitation of our current study, especially in the big data era, is the number of series and respective contexts examined. As a future direction, we suggest the extension of our study by increasing the amount of time series represented in each context and the number of contexts to allow the evaluation of the robustness of the superior performance of the judgmental model selection in each context. This could include more and higher frequencies and exogenous information. Such extensions would also benefit the ongoing limitation of controlled experimental studies towards more generalizable results.

6 Conclusions

Model selection of appropriate forecasting models is an open problem. Optimal ex-ante identification of the best ex-post model can bring significant benefits regarding forecasting performance. The literature has so far focused on automatic/statistical approaches for model selection. However, demand managers and forecasting practitioners often tend to ignore system recommendations and apply judgment when selecting a forecasting model. This study is the first, to the best of our knowledge, to investigate the performance of judgmental model selection.

We devised a behavioral experiment and tested the efficacy of two judgmental approaches to select models, namely simple model selection and model-build. The latter one was based on the judgmental identification of time series features (trend and seasonality). We compared the performance of these methods against that of a statistical benchmark based on information criteria. Judgmental model-build outperformed both judgmental and statistical model selection. Significant performance improvements over statistical selection were recorded for the equal-weight combination of statistical and judgmental selection. Judgmental aggregation (weighted combinations of models based on the selections of multiple experts) resulted in the best performance of any approaches we considered. Finally, an exciting result is that humans are better, compared to statistics, in avoiding the worst model.

The results of this study suggest that companies should consider judgmental forecasting selection as a complementary tool to statistical model selection. Moreover, we believe that applying the judgmental aggregation of a handful of experts to the most important items is a trade-off between resources and performance improvement that companies should be willing to consider. However, forecasting support systems that incorporate simple graphical interfaces and judgmental identification of time series features are a prerequisite to the successful implementation of do-it-yourself (DIY) forecasting. This does not seem too much to ask for software in the big data era.

Given the good performance of judgment in model selection forecasting tasks, the emulation of human selection processes through artificial intelligence approaches seems a natural way forward toward eventually deriving an alternative statistical approach. We leave this for future research. Furthermore, we expect to further investigate the reasons behind the difference in the performance of judgmental model selection and judgmental model-build. To this end, we plan to run a simplified version of the experiment of this study that will be coupled with the use of an electroencephalogram (EEG) to record electrical brain activity. Future research could also focus on the conditions (in terms of time series characteristics, data availability, and forecasting horizon) under which judgmental model selection brings more benefits. Finally, field experiments would provide further external validity for our findings.

Acknowledgments FP and NK would like to acknowledge the support for conducting this research provided by the Lancaster University Management School Early Career Research Grant MTA7690.

Appendix

Forecasting Models

We denote:

α: smoothing parameter for the level ($0 \leq \alpha \leq 1$).

β: smoothing parameter for the trend ($0 \le \beta \le 1$).
γ: smoothing parameter for the seasonal indices ($0 \le \gamma \le 1$).
ϕ: damping parameter (usually $0.8 \le \phi \le 1$).
y_t: actual (observed) value at period t.
l_t: smoothed level at the end of period t.
b_t: smoothed trend at the end of period t.
s_t: smoothed seasonal index at the end of period t.
m: Number of periods within a seasonal cycle (e.g., 4 for quarterly, 12 for monthly).
h: forecast horizon.
$\widehat{y}_{t+h}$: forecast for h periods ahead from origin t.
SES, or ETS(A,N,N), is expressed as:

$$l_t = \alpha y_t + (1-\alpha) l_{t-1}, \tag{3.1}$$

$$\widehat{y}_{t+h} = l_t. \tag{3.2}$$

SES with additive seasonality, or ETS(A,N,A), is expressed as:

$$l_t = \alpha(y_t - s_{t-m}) + (1-\alpha) l_{t-1}, \tag{3.3}$$

$$s_t = \gamma(y_t - l_t) + (1-\gamma) s_{t-m}, \tag{3.4}$$

$$\widehat{y}_{t+h} = l_t + s_{t+h-m}. \tag{3.5}$$

DES, or ETS(A,Ad,N), is expressed as:

$$l_t = \alpha y_t + (1-\alpha)(l_{t-1} + \phi b_{t-1}), \tag{3.6}$$

$$b_t = \beta(l_t - l_{t-1}) + (1-\beta)\phi b_{t-1}, \tag{3.7}$$

$$\widehat{y}_{t+h} = l_t + \sum\nolimits_{i=1}^{h} \phi^i b_t. \tag{3.8}$$

DES with additive seasonality, or ETS(A,Ad,A), is expressed as:

$$l_t = \alpha(y_t - s_{t-m}) + (1-\alpha)(l_{t-1} + \phi b_{t-1}), \tag{3.9}$$

$$b_t = \beta(l_t - l_{t-1}) + (1-\beta)\phi b_{t-1}, \tag{3.10}$$

$$s_t = \gamma(y_t - l_t) + (1-\gamma) s_{t-m}, \tag{3.11}$$

$$\widehat{y}_{t+h} = l_t + \sum\nolimits_{i=1}^{h} \phi^i b_t + s_{t+h-m}. \tag{3.12}$$

Participants Details

Table 3.6 University modules with at least 20 participants where the behavioral experiment was introduced as an elective exercise

University	Module (and keywords)	Level
Bangor University	Applied business projects: Operations management (operations, strategy, competitiveness, supply chain, capacity, planning, inventory, forecasting)	PG
Cardiff University	Logistics modelling (business statistics, forecasting, stock control, system dynamics, bull-whip effect, queuing analysis, simulation)	PG
Lancaster University	Business forecasting (time series, forecasting, regression, evaluation, model selection, judgment)	UG
Lancaster University	Forecasting (time series, forecasting, univariate and causal models, evaluation, model selection, judgment)	PG
National Technical University of Athens	Forecasting techniques (time series, forecasting, decomposition, univariate and causal models, evaluation, support systems, judgment)	UG
Universidad de Castilla-La Mancha	Manufacturing planning and control (planning, forecasting, manufacturing, just-in-time, stock control, inventory models)	UG

Table 3.7 Industries associated with the practitioner participants

Industry	Participants
Consulting (including analytics)	14
Banking & Finance	11
Software (including forecasting software)	9
Advertising & Marketing	9
Retail	8
Health	8
Government	6
Manufacturing	5
Food & Beverage	4
Energy	3
Logistics	3
Telecommunications	3
Automotive	2
Other	5

References

Abbey, J. D., & Meloy, M. G. (2017). Attention by design: Using attention checks to detect inattentive respondents and improve data quality. *Journal of Operations Management, 53-56*, 63–70.

Adya, M., Collopy, F., Armstrong, J. S., & Kennedy, M. (2001). Automatic identification of time series features for rule-based forecasting. *International Journal of Forecasting, 17*(2), 143–157.

Alvarado-Valencia, J., Barrero, L. H., Önkal, D., & Dennerlein, J. T. (2017). Expertise, credibility of system forecasts and integration methods in judgmental demand forecasting. *International Journal of Forecasting, 33*(1), 298–313.

Armstrong, S. J. (2001). Combining forecasts. In *Principles of forecasting* (International Series in Operations Research & Management Science) (pp. 417–439). Springer.

Armstrong, J. S. (2006). Findings from evidence-based forecasting: Methods for reducing forecast error. *International Journal of Forecasting, 22*(3), 583–598.

Ashton, A. H., & Ashton, R. H. (1985). Aggregating subjective forecasts: Some empirical results. *Management Science, 31*(12), 1499–1508.

Athanasopoulos, G., Hyndman, R. J., Kourentzes, N., & Petropoulos, F. (2017). Forecasting with temporal hierarchies. *European Journal of Operational Research, 262*(1), 60–74.

Barrow, D. K., & Kourentzes, N. (2016). Distributions of forecasting errors of forecast combinations: Implications for inventory management. *International Journal of Production Economics, 177*, 24–33.

Bates, D., Mächler, M., Bolker, B., & Walker, S. (2015). Fitting linear mixed-effects models using lme4. *Journal of Statistical Software, 67*(1), 1–48.

Billah, B., King, M. L., Snyder, R., & Koehler, A. B. (2006). Exponential smoothing model selection for forecasting. *International Journal of Forecasting, 22*(2), 239–247.

Blattberg, R. C., & Hoch, S. J. (1990). Database models and managerial intuition: 50% model + 50% manager. *Management Science, 36*(8), 887–899.

Bunn, D., & Wright, G. (1991). Interaction of judgemental and statistical forecasting methods: Issues & analysis. *Management Science, 37*(5), 501–518.

Burnham, K. P., & Anderson, D. R. (2002). *Model selection and multi-model inference: A practical information-theoretic approach*. Springer.

Chatfield, C. (2000). *Time-series forecasting*. CRC Press.

Collopy, F., & Armstrong, J. S. (1992). Rule-based forecasting: Development and validation of an expert systems approach to combining time series extrapolations. *Management Science, 38*(10), 1394–1414.

Crone, S. F., Hibon, M., & Nikolopoulos, K. (2011). Advances in forecasting with neural networks? Empirical evidence from the NN3 competition on time series prediction. *International Journal of Forecasting, 27*(3), 635–660.

Deck, C., & Smith, V. (2013). Using laboratory experiments in logistics and supply chain research. *Journal of Business Logistics, 34*(1), 6–14.

Dietvorst, B. J., Simmons, J. P., & Massey, C. (2015). Algorithm aversion: People erroneously avoid algorithms after seeing them err. *Journal of Experimental Psychology: General, 144*(1), 114.

Edmundson, R. H. (1990). Decomposition: a strategy for judgemental forecasting. *Journal of Forecasting, 9*(4), 305–314.

Ferrell, W. R. (1985). Combining individual judgments. In G. Wright (Ed.), *Behavioral decision making* (pp. 111–145). Springer.

Fildes, R. (2001). Beyond forecasting competitions. *International Journal of Forecasting, 17*, 556–560.

Fildes, R., & Petropoulos, F. (2015). Simple versus complex selection rules for forecasting many time series. *Journal of Business Research, 68*(8), 1692–1701.

Fildes, R., Goodwin, P., Lawrence, M., & Nikolopoulos, K. (2009). Effective forecasting and judgmental adjustments: An empirical evaluation and strategies for improvement in supply-chain planning. *International Journal of Forecasting, 25*(1), 3–23.

Franses, P. H., & Legerstee, R. (2011). Combining SKU-level sales forecasts from models and experts. *Expert Systems with Applications, 38*(3), 2365–2370.
Gardner, E. S. (1990). Evaluating forecast performance in an inventory control system. *Management Science, 36*(4), 490–499.
Gardner, E. S. (2006). Exponential smoothing: The state of the art - Part II. *International Journal of Forecasting, 22*(4), 637–666.
Goodwin, P., & Lawton, R. (1999). On the asymmetry of the symmetric MAPE. *International Journal of Forecasting, 15*(4), 405–408.
Harvey, N. (1995). Why are judgments less consistent in less predictable task situations? *Organizational Behavior and Human Decision Processes, 63*(3), 247–263.
Harvey, N., & Harries, C. (2004). Effects of judges' forecasting on their later combination of forecasts for the same outcomes. *International Journal of Forecasting, 20*(3), 391–409.
Hibon, M., & Evgeniou, T. (2005). To combine or not to combine: Selecting among forecasts and their combinations. *International Journal of Forecasting, 21*, 15–24.
Hogarth, R. M., & Makridakis, S. (1981). Forecasting and planning: An evaluation. *Management Science, 27*(2), 115–138.
Hyndman, R. J., & Khandakar, Y. (2008). Automatic time series forecasting: The forecast package for R. *Journal of Statistical Software, 27*(3), 1–22.
Hyndman, R. J., & Koehler, A. B. (2006). Another look at measures of forecast accuracy. *International Journal of Forecasting, 22*(4), 679–688.
Hyndman, R. J., Koehler, A. B., Snyder, R. D., & Grose, S. (2002). A state space framework for automatic forecasting using exponential smoothing methods. *International Journal of Forecasting, 18*(3), 439–454.
Hyndman, R., Koehler, A. B., Ord, J. K., & Snyder, R. D. (2008). *Forecasting with exponential smoothing: The state space approach*. Springer.
Kolassa, S. (2011). Combining exponential smoothing forecasts using akaike weights. *International Journal of Forecasting, 27*(2), 238–251.
Kolassa, S., & Siemsen, E. (2016). *Demand forecasting for managers*. Business Expert Press.
Kourentzes, N., Petropoulos, F., & Trapero, J. R. (2014). Improving forecasting by estimating time series structural components across multiple frequencies. *International Journal of Forecasting, 30*(2), 291–302.
Kremer, M., Siemsen, E., & Thomas, D. J. (2015). The sum and its parts: Judgmental hierarchical forecasting. *Management Science, 62*(9), 2745–2764.
Lacetera, N., Macis, M., & Slonim, R. (2014). Rewarding volunteers: A field experiment. *Management Science, 60*(5), 1107–1129.
Lee, Y. S., & Siemsen, E. (2017). Task decomposition and newsvendor decision making. *Management Science, 63*(10), 3226–3245.
Lee, W. Y., Goodwin, P., Fildes, R., Nikolopoulos, K., & Lawrence, M. (2007). Providing support for the use of analogies in demand forecasting tasks. *International Journal of Forecasting, 23*(3), 377–390.
Makridakis, S., & Hibon, M. (2000). The M3-competition results, conclusions and implications. *International Journal of Forecasting, 16*(4), 451–476.
Nenova, Z., & May, J. H. (2016). Determining an optimal hierarchical forecasting model based on the characteristics of the dataset: Technical note. *Journal of Operations Management, 44*(5), 62–88.
Oliva, R., & Watson, N. (2009). Managing functional biases in organizational forecasts: A case study of consensus forecasting in supply chain planning. *Production and Operations Management, 18*(2), 138–151.
Önkal, D., & Gönül, M. S. (2005). Judgmental adjustment: A challenge for providers and users of forecasts. *Foresight: The International Journal of Applied Forecasting, 1*, 13–17.
Ord, J. K., Fildes, R., & Kourentzes, N. (2017). *Principles of business forecasting* (2nd ed.). Wessex Press Publishing.

Payne, J. W. (1976). Task complexity and contingent processing in decision making: An information search and protocol analysis. *Organizational Behavior and Human Performance, 16*(2), 366–387.

Petropoulos, F., Makridakis, S., Assimakopoulos, V., & Nikolopoulos, K. (2014). 'Horses for courses' in demand forecasting. *European Journal of Operational Research, 237*, 152–163.

Petropoulos, F., Fildes, R., & Goodwin, P. (2016). Do 'big losses' in judgmental adjustments to statistical forecasts affect experts' behaviour? *European Journal of Operational Research, 249*(3), 842–852.

Petropoulos, F., Goodwin, P., & Fildes, R. (2017). Using a rolling training approach to improve judgmental extrapolations elicited from forecasters with technical knowledge. *International Journal of Forecasting, 33*(1), 314–324.

Petropoulos, F., Hyndman, R. J., & Bergmeir, C. (2018). Exploring the sources of uncertainty: Why does bagging for time series forecasting work? *European Journal of Operational Research.*

R Core Team. (2016). *R: A language and environment for statistical computing*. R Foundation for Statistical Computing. Retrieved from https://www.R-project.org/

Ritzman, L. P., & King, B. E. (1993). The relative significance of forecast errors in multistage manufacturing. *Journal of Operations Management, 11*(1), 51–65.

Sagaert, Y. R., Aghezzaf, E.-H., Kourentzes, N., & Desmet, B. (2018). Tactical sales forecasting using a very large set of macroeconomic indicators. *European Journal of Operational Research, 264*(2), 558–569.

Sanders, N. R., & Graman, G. A. (2009). Quantifying costs of forecast errors: A case study of the warehouse environment. *Omega, 37*(1), 116–125.

Seifert, M., Siemsen, E., Hadida, A., & Eisingerich, A. (2015). Effective judgmental forecasting in the context of fashion products. *Journal of Operations Management, 36*(1), 33–45.

Surowiecki, J. (2005). The wisdom of crowds: Why the many are smarter than the few. *Abacus.*

Tashman, L. J. (2000). Out-of-sample tests of forecasting accuracy: An analysis and review. *International Journal of Forecasting, 16*(4), 437–450.

Taylor, J. W. (2003). Exponential smoothing with a damped multiplicative trend. *International Journal of Forecasting, 19*(4), 715–725.

Tetlock, P., & Gardner, D. (2015). *Superforecasting: The art and science of prediction*. Crown.

Thomson, M. E., Pollock, A. C., Gönül, M. S., & Önkal, D. (2013). Effects of trend strength and direction on performance and consistency in judgmental exchange rate forecasting. *International Journal of Forecasting, 29*(2), 337–353.

Trapero, J. R., Pedregal, D. J., Fildes, R., & Kourentzes, N. (2013). Analysis of judgmental adjustments in the presence of promotions. *International Journal of Forecasting, 29*(2), 234–243.

van der Laan, E., van Dalen, J., Rohrmoser, M., & Simpson, R. (2016). Demand forecasting and order planning for humanitarian logistics: An empirical assessment. *Journal of Operations Management, 45*(7), 114–122.

Wang, X., & Petropoulos, F. (2016). To select or to combine? The inventory performance of model and expert forecasts. *International Journal of Production Research, 54*(17), 5271–5282.

Wang, X., Smith-Miles, K., & Hyndman, R. (2009). Rule induction for forecasting method selection: Meta-learning the characteristics of univariate time series. *Neurocomputing, 72*(10–12), 2581–2594.

Webby, R., O'Connor, M., & Edmundson, B. (2005). Forecasting support systems for the incorporation of event information: An empirical investigation. *International Journal of Forecasting, 21*(3), 411–423.

Weller, M., & Crone, S. (2012). *Supply chain forecasting: Best practices & benchmarking study.* Lancaster Centre For Forecasting, Technical Report.

Xia, Y., & Tong, H. (2011). Feature matching in time series modeling. *Statistical Science, 2011*, 21–46.

Zellner, A., Keuzenkamp, H. A., & McAleer, M. (Eds.). (2002). *Simplicity, inference and modelling: Keeping it sophisticatedly simple.* Cambridge University Press.

Chapter 4
Effective Judgmental Forecasting in the Context of Fashion Products (Reprint)

Matthias Seifert, Enno Siemsen, Allègre L. Hadida, and Andreas E. Eisingerich

Keywords Judgmental forecasting · Fashion products · Lens model design · Demand uncertainty · Music industry · New product forecasting

1 Introduction

The accurate prediction of the commercial success of newly launched products or services represents a crucial managerial problem (Steenkamp et al., 1999; Stremersch & Tellis, 2004; Van den Bulte & Stremersch, 2004). Generating such forecasts can be extremely difficult, particularly in environments involving fashion-oriented consumer products (hereafter referred to as "fashion products"), where the nature of the products may contain a substantial creative, artistic component, and consumer taste constantly changes (Christopher et al., 2004; Hines & Bruce, 2007; Hirsch, 1972). In more conventional forecasting domains, for example when predicting the demand of machine spare parts in manufacturing (e.g. Sani & Kingsman, 1997) or when estimating electricity demand (e.g. Taylor & Buizza,

Seifert, M., Siemsen, E., Hadida, A. L., & Eisingerich, A. B. (2015). Effective judgmental forecasting in the context of fashion products. *Journal of Operations Management*, *36*, 33–45. Copyright Elsevier.

M. Seifert (✉)
IE Business School – IE University, Madrid, Spain
e-mail: matthias.seifert@ie.edu

E. Siemsen
Wisconsin School of Business, University of Wisconsin, Madison, WI, USA

A. L. Hadida
Judge Business School, Cambridge University, Cambridge, UK

A. E. Eisingerich
Imperial College Business School, London, UK

M. Seifert (ed.), *Judgment in Predictive Analytics*, International Series in Operations Research & Management Science 343,
https://doi.org/10.1007/978-3-031-30085-1_4

2003), large amounts of historical data are often available to calibrate decision support models and achieve high levels of model accuracy. Forecasts about the demand of fashion products, on the other hand, often lack such integral information as the demand pattern tends to be highly uncertain (Choi et al., 2014; Green & Harrison, 1973; Sichel, 2008; Sun et al., 2008).

Companies producing fashion products for which consumer tastes and preferences cannot be tracked continuously often perceive themselves more as trend-setters than trend-followers (Eliashberg et al., 2008). For instance, in the apparel industry, firms face the challenge of quickly commercializing new designs that are introduced during the New York or Paris Fashion Week in order to create and satisfy new consumer demand. Yet, forecasts about future sales of new clothing designs are highly volatile. They depend both on managers' ability to accurately anticipate uncertain consumer preferences and on their firm's time-to-market capability relative to its competitors. In such supply-driven environments (Moreau & Peltier, 2004), conventional time series methods typically cannot be employed to predict demand with reasonable accuracy (Eliashberg & Sawhney, 1994; Moe & Fader, 2001; Sawhney & Eliashberg, 1996). Instead, researchers have proposed several approaches to overcome the problem of model calibration when only limited and/or unreliable data are available (for an extensive review, please see Nenni et al., 2013). In particular, these approaches have relied on Bayesian estimation when new sales data becomes available (e.g. Green & Harrison, 1973), Fourier analysis (Fumi et al. 2013), binomial distribution models (Cachon & Fisher, 2000), the Croston method (Snyder, 2002), two-stage dynamic sales forecasting models (Ni & Fan, 2011), artificial neural networks (Au et al., 2008; Gutierrez et al., 2008; Yu et al., 2011), fuzzy logic (Thomassey et al., 2005), extreme learning machines (Sun et al., 2008; Xia et al., 2012), and hybrid intelligent models (Choi et al., 2014; Wong & Guo, 2010). Furthermore, quick response manufacturing strategies have frequently been proposed as efficient means to shorten production lead times and gather "early sales" signaling data to reduce demand uncertainty (e.g. Cachon & Swinney, 2011; Fisher & Raman, 1996; Iyer & Bergen, 1997). However, despite the growing variety of managerial approaches and quantitative models at our disposal, the practice of forecasting product and service success continues to crucially depend on human judgment (Sanders & Manrodt, 2003; Boulaksil & Franses, 2009).

Past research has therefore emphasized the importance of understanding how and when managerial judgment contributes to improving forecasting accuracy. For example, in the context of newsvendor decision making, accurately estimating product demand before the selling period is extremely important in order to minimize inventory costs and avoid lost sales (Bolton & Katok, 2008; Bostian et al., 2008; Bendoly et al., 2006; Schweitzer & Cachon, 2000). Lee and Siemsen (2015) showed that decomposing the forecasting task from the ordering task in these contexts can improve judgmental accuracy under certain conditions. Task decomposition can therefore be seen as a possible means to *debias* the so-called pull-to-center effect that has often been found in behavioral studies of the newsvendor problem (e.g. Ho et al., 2010; Ren & Croson, 2013; Su, 2008). Moreover, Blattberg and Hoch (1990) demonstrated in the context of catalogue fashion sales how buyers' demand predictions could be improved by relying on an equally weighted combination of model and expert forecasts. Gaur et al. (2007) emphasized the effectiveness of using dispersion among expert judgments to predict the actual uncertainty of demand for fashion products. Eliashberg et al. (2008) also concluded that managerial judgment is essential for the scheduling of motion pictures and the prediction of box office results. Furthermore, Lawrence et al. (2006) argued that managerial judgment is likely to prove valuable whenever the "ecological validity" of formal models is low, which is indicated by a low fit between model and forecasting environment.

The purpose of this paper is to study the conditions that determine the effectiveness of judgmental forecasting in environments involving fashion products. The sense-making mechanism underlying expert judgment is often viewed as a pattern-matching process during which forecasters perceive informational stimuli of a forecasting event and make their inference by comparing them to similar situations experienced in the past. As forecasting performance is limited by judges' ability to retrieve comparable situations from their memory, providing more *historical cases* to increase the possibility of finding a good match may therefore support their judgmental processes (Hoch & Schkade, 1996). A second approach to aid decision making is to support the strengths of expert judgment by providing forecasters with better *contextual data*. Numerous studies have demonstrated that the strengths of expert judges lie in particular in their ability to diagnose new variables, recognize abnormal "broken leg cues", and evaluate information that is otherwise difficult to quantify (Blattberg & Hoch, 1990; Einhorn, 1974). Providing judges with more contextual data may therefore enable them to make better sense of a forecasting event. Such contextual data might comprise product-specific information relative to promotional activities, manufacturing data, and more general domain knowledge including competitor data or macroeconomic forecasts (Lawrence et al., 2006).

We extend Hoch and Schkade's (1996) work by offering empirical insights into the usefulness of decision support approaches relying on historical cases and on contextual data in a setting characterized by high uncertainty. Whereas research on the general accuracy and appropriateness of judgmental forecasts has a long-standing tradition, only few studies have attempted to decompose the forecasting context into different types of knowledge components in order to examine their effect on judgmental performance. Among these, Blattberg and Hoch (1990), Stewart et al. (1997), and Seifert and Hadida (2013) have studied whether judgmental forecasts can add value beyond the predictions of linear models. However, these studies rest on the assumption that human judgment is capable of approximating the linear regression model of the environment fairly well. This assumption appears questionable, at least in some forecasting contexts, when considering that the information processing capacity of forecasters is limited. In addition, while Lee and Siemsen (2015) as well as Seifert and Hadida (2013) acknowledge that differences in forecasting effectiveness may depend on task structure, a more systematic decomposition of the forecasting environment is required to fully understand how task characteristics fundamentally influence judgmental performance and how decision support systems should be designed to improve predictive accuracy.

To study how historical demand anchors and contextual anchors interactively influence the performance of *human judgment*, we employ a judgment analysis approach (Cooksey, 1996; Hammond, 1996). Judgment analysis allows us to analyze managerial predictions beyond forecasting accuracy by decomposing judgments into two different components. First, we focus on the degree to which a manager's interpretation of a forecasting event matches the efficiency of a linear model. Second, we measure the extent to which a managerial judgment can reduce the residual variance of the linear model by interpreting contextual knowledge surrounding a forecasting event and, hence, add predictive value over and beyond the linear model. Our empirical context is the music industry, which, due to its creative and artistic nature, can be understood as a typical market for fashion products (Santagata, 2004). Specifically, we study forecasts about the Top 100 chart entry positions of upcoming pop music singles. While past research has primarily focused on sales predictions in the domain of fashion retailing, the music sector appears to be a particularly interesting forecasting domain where product success is highly contingent on the subjective evaluation of a multitude of independent industry gatekeepers (Vogel, 2007).[1] Our study reveals that when the primary concern is to maximize predictive accuracy and forecasts are based on human judgment *only*, providing both types of decision support data is likely to minimize forecast errors. However, more interestingly, our results also showcase the ambivalent nature of decision support anchors as the presence of both types of data appears to improve forecasters' ability to exploit nonlinearities, while impairing their effectiveness of interpreting linearities in the task environment. In fact, our field data indicate that the exploitation of nonlinearities is easiest for human judgment if contextual data are present but historical data are absent. Thus, if the objective is to use human judgment to maximize the explanation of nonlinearities surrounding the forecasting task, we suggest that decision support should be restricted to contextual data anchors and subsequently combined with the prediction of a statistical linear model to optimize combined forecasting performance. The next section introduces the theoretical background of our research. Section 3 provides an overview of the data collection context and methods. In Sects. 4 and 5, we report the model results, and discuss them in light of existing literature. Section 6 concludes the study, discusses its limitations and offers directions for future research.

[1] In the music industry, rank predictions are highly correlated with actual sales forecasts. Specifically, our secondary data analysis of Top 40 *album* chart ranks in the UK between 1996 and 2003 and their associated average weekly sales levels reveal a strong negative correlation of $r = -0.74$, indicating that smaller ranks are associated with higher sales levels. A corresponding analysis of the Top 40 *singles* charts resulted in a similarly strong negative correlation of $r = -0.68$.

2 Theoretical Background

2.1 Judgment Analysis

Judgment analysis is rooted in the assumption that forecasting environments can be described in terms of the probabilistic relationships between relevant informational cues (i.e., the predictor variables) and an ecological criterion (i.e., the forecasting event) (e.g., Dougherty & Thomas, 2012; Einhorn, 1970, 1971; Hammond & Summers, 1972). Past research has frequently tested a forecaster's ability to exploit linear relations between predictor variables and forecasting event (hereafter called "linearities") by analyzing how well a linear regression model of the environment fits subjects' judgments (e.g., Dhami & Harries, 2001; Karelaia & Hogarth, 2008). Because a linear model describes the task fairly well in *stable* environments, judgment analysis studies have often assumed any type of nonlinear relation between predictor variables and forecasting event (hereafter referred to as "non-linearities") to be negligible random noise in their research designs (e.g., Dunwoody et al., 2000; Hammond & Summers, 1972). However, in many organizational settings, reducing the forecasting context to linear cue-criterion relations may produce an overly simplistic representation of the task environment (Seifert & Hadida, 2013). When forecasting the demand for an upcoming novel in the publishing industry, a linear model would fail to exploit important nonlinear relationships, for instance between demand and the book's story line, genre and/or author information. Yet, such informational cues may prove extremely useful for reducing the residual variance of a linear model as they could help the forecaster to understand, for instance, whether the book satisfies the demand arising from a recent hype surrounding Scandinavian crime stories.

In this article, we regard the residual variance of linear models as an important source of information that judges may exploit by relying on their familiarity with the task and their domain-specific expertise (Lawrence et al., 2006). Similarly, in a study of temperature and precipitation forecasts, judgment analysis was employed to demonstrate how systematically exploiting nonlinear relations between predictor variables and forecasting event could lead to improved forecasting accuracy (Stewart et al., 1997). The same study also revealed a relationship between the predictive value of nonlinearities and the type of information available to the forecaster. In particular, when the informational cues available to forecasters were highly complex, the study not only demonstrated the limited usefulness of linear model forecasts, but also the increasing importance of nonlinearities for improving forecasting accuracy. We next discuss the extant literature on forecasting fashion products and then turn to the relationship between task properties and forecasting accuracy.

2.2 Forecasting the Demand of Fashion Products

Fashion products are sometimes referred to as *semaphoric* goods due to their creative and artistic nature and to the subjective, symbolic value associated with them (Santagata, 2004). Markets for fashion goods can generally be described in terms of four distinctive features (Christopher et al., 2004; Nenni et al., 2013): (1) short (seasonal) product life-cycles, which can usually be measured in months or weeks; (2) high volatility due to instable, nonlinear product demand; (3) low week-by-week and/or item-by-item predictability; and (4) highly impulsive purchasing decisions of consumers, which crucially require products to be available in the store. Given these characteristics, firms operating in fashion markets need to maintain a high level of flexibility during the manufacturing process and very short lead times (Choi et al., 2014). Consequently, conventional organizational structures and decision support approaches often prove inadequate (Christopher et al., 2004).

The music industry matches the characteristics of a market for fashion products. Similar to firms in the apparel, publishing and movie industries, music companies continuously face the challenge of creating and satisfying consumer demand for new *fads and fashions* in order to secure organizational survival (Hines & Bruce, 2007; Hirsch, 1972; Peterson & Berger, 1971). Typically, music firms do not only forecast ordinary sales targets as the basis for their production planning and marketing strategies. They also generate predictions about the entry positions of new music releases in the Top 100 charts. Chart entry predictions are extremely complex, because record companies can only promote the release of a single to a limited extent. In particular, promotion depends on several independent gatekeepers such as radio and tv stations, press, retailers, and online communities. These gatekeepers rely on their own selection criteria, processes and time lines to decide to feature, review and/or include a product in playlists prior to its official release date (Hadida & Paris, 2014; Vogel, 2007). In addition, chart predictions are further complicated by the fact that record companies are required to publicly announce all upcoming single launches several months in advance. Release schedules may consequently partly reflect a company's strategy to avoid or capitalize on competing product launches that were known at the time of scheduling. The chart entry position itself represents a direct measure of sales performance, since the overall 100 best-selling singles in any given week are included in the charts of the following week.[2] Because sales typically peak during the release week, the chart performance of a single tends to be highest at the time of chart entry and then declines during the subsequent weeks until the single

[2] Since the chart entry of a new single is always relative to the sales performance of other, previously launched singles, it does not represent an independent event. However, our interviewees noted that the influence that other singles included in the charts exert is not well understood because their actual sales level is not observable at the time of the forecast. For modeling purposes, we therefore treat the interaction between singles in the charts as random noise.

drops out of the top 100-selling products.[3] In general, separate music charts exist for both single song releases in a given week and for full album releases. The chart position of a music single may hereby serve as an early sales signal for an upcoming album release, as it allows companies to test the market response and gather information for estimating the demand of a full album. Singles therefore play an important role in the quick response inventory management strategies of music firms (Choi et al., 2006; Serel, 2009). Taking this into account, although the advent of digital distribution has led to a drastic decrease in the cost of forecast errors, the music industry context has the advantage of providing very frequent introductions of fashion products with publicly observable market outcomes. This advantage is important for our study.

2.3 Hypotheses

In this paper, we study forecasts in the domain of fashion products by focusing on how the presence of two types of data, *historical data* and *contextual data*, supports a judge's ability to process linearities and nonlinearities in the forecasting environment. Although each of the two types of data individually has received considerable attention in the judgmental forecasting literature (e.g., Gaur et al., 2007; Hoch & Schkade, 1996; Lawrence et al., 2006), to the best of our knowledge, none of the existing studies has systematically examined their joint effect on a manager's judgmental forecasting performance.

Historical Data Prior research has established that historical demand data can reduce judgmental forecasting uncertainty (Gaur et al., 2007; Lawrence et al., 2006). Consequently, historical information has been argued to increase the transparency of means-end relationships (Fellner, 1961; Frisch & Baron, 1988) by shedding more light on (a) the probability distributions that link predictive variables and forecasting event (Camerer & Weber, 1992; Wood, 1986); (b) the transparency of path sequences; and/or (c) the appropriateness of organizing principles, which could be utilized to assess the relative importance of informational cues and, hence, systematically exploit them (Hamm, 1987; Hammond, 1996; Steinmann, 1976; Wood, 1986). In time series forecasting, however, where historical information mostly relates to the length of a time series, research findings have been somewhat contradictory: Although there is some evidence indicating forecasters' proficiency in using historical data for detecting trends, seasonality, randomness and discontinuities (Lawrence et al., 2006; Lawrence & Makridakis, 1989), others have argued that quite minor changes in a series and in the presentation of a task can substantially impair judges' forecasting ability (Andreassen & Kraus, 1990; Goodwin & Wright, 1993). Behavioral experiments based on the newsvendor problem have shown that

[3] For further details about the industry context please see Seifert and Hadida (2013), in which part of the same dataset has been used to analyze the performance of forecast aggregations.

managerial judgments can be severely influenced by the recent history of a time series (Bostian et al., 2008; Schweitzer & Cachon, 2000). Furthermore, Kremer et al. (2011) observed that in contexts of stable time series, forecasters have a tendency to engage in "demand chasing": They predict the next step in a time series according to simple error-response learning, despite having enough data available to determine that errors are mostly driven by noise in the time series.

Taking this into account, research has directly related judgmental forecasting performance to the level of noise in the underlying data series (Harvey, 1995; Harvey et al., 1997). In particular, in uncertain environments forecasters appear to reproduce noise levels in their judgments. This can be understood as a reflection of their inability to disentangle useful informational signals from random variability. The observation that human judges do not perform well as "intuitive statisticians" in noisy environments is not new (Brehmer, 1978; Harvey et al., 1997; Kahneman & Tversky, 1973). In fact, when regressing forecasters' judgments onto informational cues, some residual variance typically remains unexplained (Harvey et al., 1997). In his study of multiple-cue probability learning tasks, Brehmer (1978) examined whether the residual variance component of forecasters' judgments contained systematic deviations from linearity given different levels of data noise. His study showed that judgmental inconsistencies were unsystematic and therefore could not be explained by forecasters' use of nonlinear functional rules in any of the tested settings.

In light of this discussion, when forecasting the commercial performance of fashion products, the level of noise underlying historical data is likely to have implications for forecasters' ability to interpret linearities as well as nonlinearities in the environment. Specifically, Brehmer's (1978) study indicates that noise appears to deteriorate judgmental performance in general and, hence, is likely to affect the effectiveness of both information processing mechanisms. Thus, data noise appears to create uncertainty about how to systematically weight informational cues in the task environment. Furthermore, data noise decreases the reliability of human judgments, because it creates ambiguity regarding the *type* of linear or nonlinear function that could best describe the relationship between predictor variables and forecasting event (Einhorn, 1971; Harvey, 1995; Hogarth, 1987). Research in another market for fashion products, cinema, also found no significant impact of past historical demand anchors such as the prior box office performance of leading actors (Zuckerman & Kim, 2003) or directors and producers (Elberse & Eliashberg, 2003; Zuckerman & Kim, 2003) on a film's box office receipts. We therefore expect that the provision of historical demand anchors neither improves judges' ability to better utilize linear models, nor does it enhance their ability to process nonlinearities among predictor variables in a more effective manner.

Hypothesis H_1: When forecasting the commercial performance of fashion products, the presence of historical demand data neither improves forecasters' ability to process linearities (i.e. by approximating the accuracy of a best fitting linear model) nor does it improve forecasters' ability to process nonlinearities in the task environment.

Contextual Data Contextual data surrounding a prediction task can serve as an important anchoring point, which is then up- or down-adjusted by the forecaster to arrive at a judgment (Lawrence et al., 2006). These data refer to all non-historical information on a forecasting event, including predictions derived from market research, past and future promotional plans, and competitor, manufacturing and macroeconomic data. Such non-historical information can often have a more subjective, qualitative nature and enable judges to obtain a more comprehensive understanding of the *nonlinear* relationships that describe a forecasting event (Blattberg & Hoch, 1990; Seifert & Hadida, 2013). Contextual data therefore allows forecasters to develop a more holistic mental representation of the prediction task; and may ultimately result in higher predictive accuracy. In addition, prior research has shown that when contextual data are available, cognitive shortcuts, such as pattern matching or the use of an equal weighting procedure for assessing informational cues, are frequently adopted to efficiently integrate task information (Clemen, 1989; Einhorn & Hogarth, 1975; Hoch & Schkade, 1996). Therefore, the presence of contextual data is likely to improve judgmental accuracy by allowing forecasters to better exploit nonlinearities.

However, because informational cues in highly uncertain environments can represent imperfect signals of a real-world event (Olshavsky, 1979; Payne, 1976; Slovic & Lichtenstein, 1971), contextual anchors may also lead to systematic errors if they entice forecasters to see patterns where none exist. In particular, previous studies have shown that people frequently fail to judge randomly generated patterns in a sequence of items as originating from true randomness (Harvey, 1995). Instead, they adjust their forecasting judgments in an attempt to make them become more representative of the underlying data structure. In fact, the representativeness heuristic has often served as an explanation for why people fail to make regressive predictions (Harvey et al., 1997; Kahneman & Tversky, 1973). Contextual anchors thus have an ambivalent nature: although they enrich forecasters' understanding of the environment, they are also likely to increase the complexity of a forecasting task. This happens particularly through (a) an increase in the number of informational cues available for making a judgment, (b) a decrease in cue inter-correlations, (c) the necessity of applying nonlinear functions and unequal cue weightings, and (d) the absence of a simple organizing principle for integrating information (Hamm, 1987; Steinmann, 1976). An increase in the amount of contextual data is therefore likely to come primarily at the expense of forecasters' ability to process linearities in the environment. Specifically, previous research has demonstrated that the consistency of exploiting linear relationships between predictor variables and forecasting event decreases as the complexity of a forecasting task increases (Einhorn, 1971; Lee & Yates, 1992). In sum, we expect contextual anchors to have a positive effect on forecasters' ability to process nonlinearities and a negative effect on their ability to process linearities in the task environment.

Hypothesis H_2: When forecasting the commercial performance of fashion products, the presence of contextual anchors is negatively associated with forecasters' ability to process linearities, and positively associated with their ability to process nonlinearities in the task environment.

Joint Effects of Historical Demand and Contextual Anchors We now turn to how historical demand anchors and contextual anchors may jointly affect a judge's ability to process linearities and nonlinearities in the task environment. Hypothesis H_2 states that when forecasting the commercial performance of fashion products, contextual data emit signals that are primarily useful for processing nonlinearities. As per Hypothesis H_1, historical demand data, on the other hand, are unlikely to improve forecasters' ability to exploit linear or nonlinear relationships between predictor variables and forecasting events. If no historical demand data are available, it seems likely that the impact of contextual anchors on processing linearities will be smaller than if both historical and contextual anchors are provided. In the joint presence of the two types of data, forecasters may be able to search for (misleading) confirmatory evidence between signals. Thus, in the case of linearities, the negative effect of contextual data anchors may be amplified in the presence of historical data. Similarly, the two anchors also interactively influence forecasters' efficiency in processing nonlinearities in the task environment. Efficiently exploiting nonlinearities is relatively more difficult than exploiting linearities, because there may be many different types of nonlinear functions that could potentially describe the forecasting data. Stewart et al. (1997) and Seifert and Hadida (2013) have shown that judges are likely to be most proficient in exploiting nonlinearities when the predictability of the forecasting environment is low. The presence of both types of anchors, however, is likely to lower forecasters' ability to extract nonlinearities, because the historical demand anchor may provide (false) unrepresentative clues about forecasters' accurate perception of how nonlinearities should be evaluated. In the absence of historical demand data, we thus expect the positive impact of contextual anchors on processing nonlinearities to be greater.

Hypothesis H_{3a}: In the presence of historical demand data, contextual data will have a greater negative effect on forecasters' ability to exploit linearities than in the absence of historical demand data.

Hypothesis H_{3b}: In the presence of historical demand data, contextual data will have a smaller positive effect on forecasters' ability to exploit nonlinearities than in the absence of historical demand data.

3 Methods

Our approach is based on a Brunswikian lens model (Brunswik, 1956) that is frequently used in judgment analysis. We develop two distinct explanatory variables to assess a judge's ability to integrate linear and nonlinear forecasting information. Specifically, our dependent variables represent effectiveness measures that capture the degree to which judgmental forecasts deviate from the predictions of the best fitting linear model between the available data (cues) and the product success variable the data should predict (criterion). Our first dependent variable indicates a judge's effectiveness in analyzing linear cue-criterion relations. The variable focuses

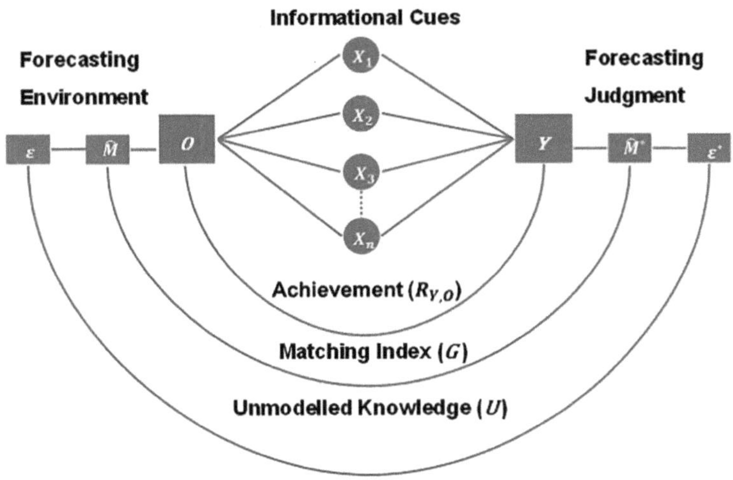

Fig. 4.1 Overview of the Judgment Analysis approach (Adapted from Cooksey, 1996)

on the extent to which the linear model of the expert judge approximates the best-fitting linear model of the forecasting environment. Our second dependent variable measures a judge's ability to process nonlinearities. The variable refers to the amount of variance that is unaccounted for by the linear model of the environment, and which instead can be explained by respondents' individual judgments. We further derive two independent sets of predictor variables from the empirical context to operationalize historical data and contextual data. A multivariate analysis of variance is used to conduct our hypotheses tests.

An overview of the full judgment analysis method is provided in Fig. 4.1. In particular, the models are constructed around the ecological criterion (*O*), which represents a probabilistic function of a number of observable environmental cues (*Xi*) (Cooksey, 1996). These informational cues may be redundant, and a full model of the ecological criterion may be specified as follows:

$$O = \widehat{M}(X_1, X_2 \ldots X_n) + \varepsilon, \tag{a}$$

where $\widehat{M}$ denotes the linear, best-fitting model between cues and the criterion.[4] The model can be interpreted as a benchmark to which linearities in the task environment

[4]While ordered logistic regression is typically used when the task involves the prediction of rank orders, it would prove inefficient in the context of our forecasting task. Specifically, since our dependent variable contains 100 levels, the resulting model would require an estimation of 100-1 equations, which would make a meaningful analysis become extremely challenging. Moreover, such a model would impose major restrictions on the characteristics of the data sample needed for making meaningful inferences. Instead, scholars have argued that linear regression models can approximate the predictions of logit models very well if few outliers exist (e.g. Iman & Conover, 1979; Lattin et al., 2003). For this reason, we only included new singles that actually entered the

can be integrated.[5] The error ε represents the residual, unmodeled variance; it comprises random error and nonlinear relations between the informational cues and the criterion (Blattberg & Hoch, 1990). Our participants observe the predictor variables before they make a forecasting judgment (Y) regarding O. When regressing the environmental cues onto Y, we can derive a second linear model $\widehat{M}^*$, which captures the extent to which forecasting judgments can be explained as a function of the informational cues given to our participants. Therefore, a participant's judgment regarding the ecological criterion can be modeled as follows:

$$Y = \widehat{M}^*(X_1, X_2 \ldots X_n) + \varepsilon^* \tag{b}$$

where the error term ε^* represents the residual variance unexplained by the judge's model $\widehat{M}^*$. Given that judges rely on the same informational cues to generate individual predictions (Y) as the best-fitting linear model, the following lens model equation ($R_{Y,O}$) can be used to establish the relationship between judge and task environment (Stewart et al., 1997; Stewart, 2001; Tucker, 1964):

$$R_{Y,O} = R_{O,X} G R_{Y,X} + U\sqrt{1 - R_{O,X}^2}\sqrt{1 - R_{Y,X}^2}.\mathrm{Y} = \gamma\widehat{\mathrm{M}} + \widehat{\mathrm{U}} \tag{c}$$

In Eq. (c), $R_{O,\ X}$ denotes the correlation between target event O and the best-fitting linear model of the environment $\widehat{M}$, whereas $\mathrm{R}_{\mathrm{Y},X}$ measures the correlation between the judge's model $\widehat{M}^*$ and judgments Y. Our two focal points of analysis, however, are G and U. Coefficient G is often called a matching index (Cooksey, 1996; Tucker, 1964) because it describes judges' effectiveness in processing linear cue-criterion relations. More specifically, we use G^2 as a measure for the amount of task variance that is jointly explained by $\widehat{M}$ and $\widehat{M}^*$. Moreover, U represents a semipartial correlation coefficient that indicates the strength of association between the unmodeled components of a target event and judgmental forecasts ($r_{\varepsilon,\varepsilon*}$). We interpret U as an important proxy of judges' skill in analyzing nonlinearities in the forecasting environment. We note, however, that U may be somewhat distorted by random noise as well as omitted variables that are neither captured by the linear model of the judge nor by the best-fitting linear model. When squared, U^2 includes the unique contribution of judges' forecasts Y in explaining the unmodeled variance in the task environment (Cohen & Cohen, 1975). Although our interpretation of the

Top 100 singles charts, which led to an exclusion of 9 observations from the sample. The relatively small number of outliers can be partially explained by the fact that music singles generally enter the charts at their peak position (followed by a decline in the subsequent weeks) and by companies' scheduling strategies for upcoming single releases, which aim at avoiding cannibalization effects with other singles released during the same week.

[5] In the context of our study, we only included linear task relations in our ecological models in order to clearly distinguish between judges' ability to process linear and nonlinear cue-criterion relations. Future studies could also extend our models to include nonlinearities as a more conservative threshold for judges' ability to pick up unmodeled residual variance.

residual component U^2 cannot be used to precisely isolate forecasters' skill from the deficiencies of the linear model, several studies have shown that it does represent a useful indicator for the validity of judges' ability to analyze nonlinearities (e.g., Blattberg & Hoch, 1990; Seifert & Hadida, 2013). Together, G^2 and U^2 measure the effectiveness of judgments by decomposing them into forecasters' linear and nonlinear achievements.

4 Empirical Setting

Our field data consist of judgments about new product releases in the music industry. More specifically, we analyze judgmental predictions about the entry positions of new, previously unreleased pop music singles on the national Top 100 singles sales charts. We began our data collection by conducting 23 semi-structured interviews with senior managers at the "big 4" major record companies, which together account for more than 80% of the global music market share (IFPI, 2014). We collected our interview data in the two largest domestic markets within the European Union: Germany and the United Kingdom. Our interview participants can be regarded as experts in the industry, given the "up-or-out" hiring structure in major record companies, which ensures that only managers with the most successful track records of placing "hits" are employed (Vogel, 2007). The average industry tenure of our interviewees was 10 years, and all held divisional or regional responsibilities. The main purpose of the interviews was to elicit potential predictor variables surrounding our forecasting event, which would enable us to specify the best-fitting linear model of the environment. We used an iterative process, during which interviewees were given the opportunity to revise their initial list of predictor variables to reduce the possibility of omitting relevant variables in the subsequent model building process. This revision strengthened the validity of our measure for forecasters' ability to analyze nonlinearities in the environment (U^2).

The second phase of our data collection involved quantitative predictions made by 92 Artist & Repertoire (A&R) managers from major and medium-sized record companies: approximately one-half in Germany, and the remaining in the United Kingdom. All managers were contacted through the leading international A&R managers' association "HitQuarters". Participants included in the sample had placed 21.74 hits on average on previous Top 100 charts, and had a mean industry experience of 7 years.

Participants generated forecasting judgments by completing four online questionnaires over a period of 12 weeks. The questionnaires were based on the predictor variables identified during the initial interviews and contained the profiles of a set of

Table 4.1 Overview of stimuli included in the study

Predictor type	Level of analysis	Decision support type C−/H−	C−/H+	C+/H−	C+/H+
Artist					
Type	Band versus single artist?	✓	✓	✓	✓
Online fanbase	Log of *Myspace* members	✓			
Prior recognition	No of sales awards received		✓		✓
Chart history: Album	Mean chart position		✓		✓
Chart history: Single	Mean chart position		✓		✓
Product					
Music genre	Mainstream pop/rock? (Y/N)	✓	✓	✓	✓
Album	Has an album been released? (Y/N)	✓	✓	✓	✓
Producer	No of previous top 100 hits	✓	✓	✓	✓
International success	Included US charts? (Y/N)	✓	✓	✓	✓
Radio	Highest airplay chart position			✓	✓
Printed media	Mean rating in key press reviews			✓	✓
Television	Highest music video chart position			✓	✓
Marketing	Log of expenditure in US dollars	✓		✓	✓

"✓" indicates that predictor variable was included in best fitting linear model
C^{-}: Contextual anchor absent C^{+}: Contextual anchor present
H^{-}: Historical data anchor absent H^{+}: Historical data anchor present

yet unreleased pop music singles.[6] Participants were asked to estimate the imminent chart entry position of the singles, and to indicate their confidence in the judgment they provided.[7] The cue profiles were generated by drawing on information obtained from marketing research firms, chart compilation companies, retailers, key media firms, record companies, and the Internet (Table 4.1).

Our data collection period took place in four batches and allowed for a period of two to three weeks between prediction and entry in the music charts, so that every pop-music single was exposed to the media for an equal duration. Altogether, we generated a sample of 210 prediction cases, and each participant provided 40 forecasts about the pop music singles. Because participants made judgments about real (rather than artificially generated) prediction cases, it was not possible to assign singles in a truly randomized manner. Instead, singles included in each questionnaire were dictated by the record labels' release schedules in a particular week. We created

[6]The appendix provides two examples of the cue profiles used.

[7]We excluded the analysis of participants' judgmental confidence from the scope of this paper. However, the data are available from the authors upon request.

four batches of judges (two for each country sample); and alternated their participation in two-week intervals. With each round of participation, subjects provided up to 10 forecasts about the upcoming releases in a particular week. They had one week to return the questionnaires, and response dates were recorded to control for any time-based effects on the accuracy of participants' predictions.

Historical Demand and Contextual Data It became clear from our initial interviews that the task of forecasting music singles' chart entry positions may depend on two important factors: (1) the intensity and success of the promotional efforts conducted one month ahead of the product release and (2) the distinction between established artists and new, unknown artists.

We use factor (1) as a proxy for contextual anchors. They include the following informational cues: the peak position of the single in the radio airplay charts; the production of a music video and its performance in the music video charts; critical reviews of the upcoming single in the most influential industry magazines; and the amount of money spent by the record label to finance general retail campaigns. We also considered whether the single formed part of a previously released album, which may provide additional information about the single's potential chart success. Even so, each media outlet employs its own criteria and institutional procedure for selecting and ranking music singles, and signals regarding a single's success potential are typically imperfect and conflicting (e.g., achieving a high ranking in the radio airplay charts and, at the same time, a low rating in the music press).

We use factor (2) as a proxy for the presence of historical demand anchors. New artists lack an existing track record of previous successes or an established fan base, both of which could be valuable indicators of future success. As a consequence, they decrease the predictability of the environment. We included the following historical chart success variables in our study: *the upcoming single's inclusion in the US billboard charts as a key reference market*; and *the mean chart-entry positions of previously released singles and albums in domestic and foreign markets*. In addition to the two identified categories of environmental cues, we also collected general information about the single itself, such as: a *picture of the artist*; *information about the producer and record label, the size of the artists' online fan base* and a 30-second *audio sample*.

5 Results

Consistent with previous research decomposing forecasting task environments (Dunwoody et al., 2000), Table 4.2 provides an overview of the descriptive statistics associated with our four decision support conditions. The table includes information about (1) the number of cues available for making a judgment, (2) the degree of redundancy among the cues (measured as the mean inter-correlation (r_c) among all relevant cues), and (3) the degree to which the environmental cues are equally weighted (measured by the standard deviation of β weights derived from the

Table 4.2 Characteristics of decision support conditions

Measurement	Decision support type			
	C^-/H^-	C^-/H^+	C^+/H^-	C^+/H^+
Number of predictors included in model	7	8	9	12
Intercorrelation of predictors (*r*)	−0.03	0.04	−0.03	0.11
Standard deviation of β weights	3.71	3.25	3.88	2.45
Sample size (*N*)	43	47	59	61

C^-: Contextual anchor absent C^+: Contextual anchor present
H^-: Historical data anchor absent H^+: Historical data anchor present
Intercorrelations and standard deviations are averaged across all pairs of predictor variables

environmental model). Inter-correlation between predictors is highest when both contextual and historical data anchors are present, whereas all other decision support conditions exhibit a negligible strength of association between predictor variables. In addition, the two conditions in which no historical data anchor is available are associated with the greatest standard deviations of regression weights.

To build linear best-fitting models of our task environment, we first compiled a broad list of informational cues that "domain experts" perceived as relevant for generating predictions of pop music singles' chart success. Next, we tested a series of candidate models, using stepwise selection to investigate the regression fit of all possible combinations of variables (Mentzer & Moon, 2005). The resulting models contain 7 predictor variables in the absence of both contextual and historical anchors, 8 variables when only historical information is available, 9 variables when only contextual data are present and 12 variables when both types of decision support are available. Because the participants did not all generate forecasts regarding the identical subsamples of cue profiles, we also adjusted the resulting models for model shrinkage. Table 4.3 provides an overview of the cross-validated results. The results indicate a relatively good model fit, despite average model shrinkage of 13%.[8] No significant difference was observed between the two country samples. In addition, our models indicate an unequal, compensatory distribution of cue weights, in that a low score on some of the higher weighted cues (e.g., airplay chart position or previous number of hits) could be compensated for by a high score on some of the less important cues (e.g., whether an album had been released at the same time as the single) (Karelaia & Hogarth, 2008).[9]

[8]We performed cross-validation analyses for all linear models and judgments obtained from the participants to control for model shrinkage (Blattberg & Hoch, 1990). In the case of the statistical model, 10 random samples were drawn in which one-half of the dataset was used to simulate predictions about the remaining data. The participants were tested pair-wise using a double cross-validation method, and averaged to represent their associated participant category (Cooksey, 1996).

[9]To ensure that our final models did not omit potentially important predictor variables, we also compared the variables included in these final models to the relative importance ratings of the cues that we elicited while interviewing senior executives in the industry. When variables were highly redundant and when their individual inclusion in the models did not significantly change the total

We constructed separate linear models for the environment ($\widehat{M}$) and each of the judges ($\widehat{M}^*$), both of which show a higher fit when historical data are available ($\Delta R^2_{O,X} = 0.26$; $\Delta (GR_{O,X})^2 = 0.23$). We then tested whether the lower model fit in absence of historical data could simply be explained by the smaller number of predictor variables in the model or merely by the fact that the artist is "new". For this purpose, we re-modeled forecasting judgments about established artists, but this time without including the historical demand information. Although the absence of such data does indeed substantially reduce model fit ($C^+/H^{(+)}$: $1-R^2 = 0.55$ and $C^-/H^{(+)}$: $1-R^2 = 0.63$), judgments about brand new artists still remain less predictable in comparison with those about established artists. Hence, although historical demand data explain a large part of the difference in predictability between established and new artists, they don't fully account for it.

As a next step, we computed the mean values of the judges' ability to process linear (G^2) and nonlinear (U^2) cue-criterion relations in each of the four decision support conditions and for all forecasters in our sample. The linear model of both environment and judges reveals the highest forecasting accuracy when both historical demand and contextual anchors are present (C^+/H^+: $R_{O,X} = 0.82$; $R_{Y,O} = 0.78$), followed by the condition in which only historical demand anchors are available (C^-/H^+: $R_{O,X} = 0.71$; $R_{Y,O} = 0.72$), the case in which only contextual anchors are present (C^+/H^-: $R_{O,X} = 0.62$; $R_{Y,O} = 0.68$), and the condition in which neither anchor is provided (C^-/H^-: $R_{O,X} = 0.53$; $R_{Y,O} = 0.57$). Because our hypotheses rest on the premise that the forecasting event can be at least partially characterized in terms of nonlinear relationships between environmental cues and chart positions, we conducted a series of Ramsey reset tests (Ramsey, 1969) to investigate whether such nonlinearities exist in the forecasting residuals. In particular, we used power functions to create new predictor variables from the linear model predictions and added them stepwise to our models. The results show significant *F*-tests for both forecasting conditions in which historical data anchors are absent (C^-/H^-: $F = 5.834$, $p < 0.05$; C^+/H^-: $F = 10.981, p < 0.01$). Moreover, when historical demand anchors are present, we only find a significant test result when no contextual anchor is available (C^-/H^+: $F = 6.619$, $p < 0.05$); but not when both types of anchors are provided to judges (C^+/H^+: $F = 0.309$, $p =$ n.s.). Hence, our analysis indicates that the unexplained model variance contains both random error *and* nonlinearities in three out of the four conditions under investigation. In addition, we also created a plot of the correlation coefficients between *G* and *U* in each of the four decision support conditions (Fig. 4.2). The plot shows that the correlation between linear and nonlinear judgmental achievement is strongest when only contextual decision support, but no historical data are available (C^+/H^-; $R_{GU} = 0.39$). This is followed by the decision support condition, in which neither of the two types of data anchors are present (C^-/H^-; $R_{GU} = 0.32$). Finally, linear and nonlinear forecasting components show the lowest level of redundancy when either both types of decision support are

---

variance explained, we selected those variables that received the highest subjective importance rating.

Table 4.3 Summary of regression results

Decision support type	Best linear fit model[a]			Judgment model[b]			Judgment decomposition	
	Model accuracy (R_{OX})	Linear model Fit ($\widehat{M}$)	Unexplained variance	Judgment accuracy ($R_{Y,O}$)	Linear model Fit ($\widehat{M}^*$)	Unexplained variance	Exploiting linearities (G^2)	Exploiting nonlinearities (U^2)
C^-/H^-	.53	.28	.72	.57 (.09)[c]	.21	.68	.92	.11
C^-/H^+	.71	.50	.50	.72 (.11)	.46	.48	.96	.06
C^+/H^-	.62	.38	.62	.68 (.12)	.31	.53	.93	.16
C^+/H^+	.83	.68	.32	.78 (.06)	.52	.40	.84	.08

Contextual data anchor: C^+ = present, C^- = absent; Historical data anchor: H^+ = present, H^- = absent.
[a]Cross-validated results
[b]Mean values across participants
[c]Standard deviation in parentheses

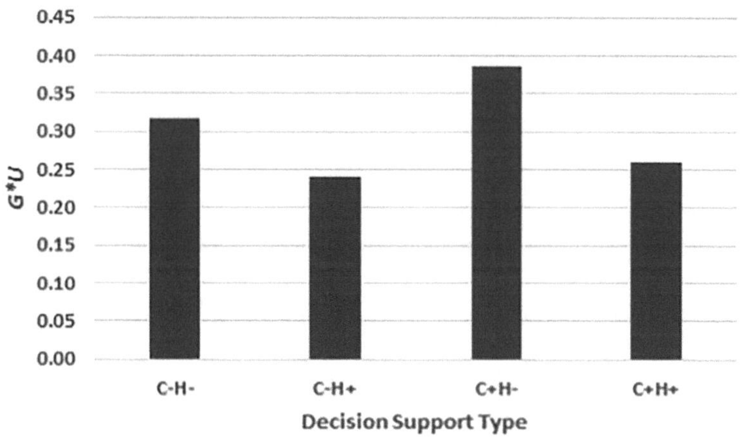

Fig. 4.2 Correlations of linear and nonlinear forecasting efficiency (G*U)

available (C^+/H^+; $R_{GU} = 0.26$) or when only historical data anchors are provided to the judge (C^+/H^-; $R_{GU} = 0.24$).

We tested our hypotheses by conducting a multivariate analysis of variance, in which contextual anchors and historical demand anchors serve as predictors of forecasters' ability to process linear and nonlinear cue-criterion relations. MANOVA was used to enable us to detect any kind of systematic relationship between the two dependent variables under investigation, which would otherwise be unobservable when relying on separate univariate analyses of variance. The MANOVA results reveal significant Pillai's Trace coefficients for both independent variables (for contextual anchors: $T = 0.56$, $p < 0.01$, and for historical demand anchors: $T = 0.68$, $p < 0.01$). Moreover, the multivariate analysis also shows a significant interaction term with a Pillai's Trace value of $T = 0.56$ and $p < 0.01$. Based on these findings, we then tested our hypotheses by conducting six significance tests at a *p*-level of $p < 0.008$ to control for potentially inflated Type I errors.

Hypothesis H_1 postulates that the provision of historical demand anchors neither improves forecasters' ability to process linearities nor their ability to interpret non-linearities in the task environment. Our MANOVA results provide support for this hypothesis in the following way: First, the *absence* of historical demand data leads to a slight improvement in exploiting linearities ($\Delta G^2 = 0.03$, $F = 22.379$, $p < 0.001$, $\eta^2 = 0.15$). We further explored the statistical power of this effect by utilizing the G*Power program (Faul et al., 2007). In particular, based on the F-tests underlying our MANOVA analysis, we used *Cohen's* f^2 as our baseline effect measure as well as an alpha level of $\alpha = 0.01$ and conducted a series of post hoc tests in order to determine the odds (1-β) that the identified effect was indeed present. While a coefficient value of $f^2 = 0.01$ indicates a small effect size, coefficient values of $f^2 = 0.06$ and $f^2 = 0.16$ are generally considered to represent moderate and large effect sizes, respectively (Cohen, 2013). In the case of our music data, the results of the conducted post hoc tests show a large *Cohen's* f^2 coefficient of $f^2 = 0.18$ with a probability of $(1 - \beta) = 0.99$ that the effect was not resulting from a Type II error.

Second, the data reveal that in the absence of historical demand data forecasters also appear to be more effective in exploiting nonlinearities ($\Delta U^2 = 0.06$, $F = 216.726$, $p < 0.001$, $\eta^2 = 0.64$). This finding is strongly supported by our post hoc power analyses ($f^2 = 1.78$; $(1 - \beta) = 1.0$). In particular, when historical data are present, judges are only capable of picking up 7% of the unexplained variance in the best-fitting linear model, compared to 14% if no such anchors are available.

We then tested the impact of contextual anchors (Hypothesis H_2), which we predict to be negative when processing linearities and positive when interpreting nonlinearities in the forecasting environment. Our MANOVA results indicate a highly significant, negative main effect for the relationship between the presence of contextual anchors and processing linearities ($\Delta G^2 = -0.06$, $F = 130.138$, $p < 0.001$, $\eta^2 = 0.51$), which is also associated with a large post hoc effect size ($f^2 = 1.04$; $(1 - \beta) = 1.0$). This finding implies that judges are less efficient in analyzing task linearities when contextual information is available. Thus, in the absence of such anchors, forecasters' performance is more likely to approximate the accuracy of the best-fitting linear model of the environment. When considering the effectiveness of processing nonlinearities, our data show a significant, positive main effect ($\Delta U^2 = 0.03$, $F = 39.973$, $p < 0.001$, $\eta^2 = 0.24$), which indicates that the presence of contextual anchors facilitates forecasters' interpretation of nonlinearities. Our power analyses indicate a moderate effect size associated with this relationship ($f^2 = 0.32$; $(1 - \beta) = 1.0$), which lends support to Hypothesis H_2.

Hypotheses H_{3a} and H_{3b} test the interactive effects of both contextual and historical demand data on the effectiveness of processing linear and nonlinear cue-criterion relations. Hypothesis H_{3a} posits that the presence of contextual data influences forecasters' ability to analyze linearities to a greater (negative) extent when historical demand data are available. Conversely, Hypothesis H_{3b} predicts that in the presence of historical demand data, contextual data have a smaller (positive) effect on forecasters' ability to process nonlinearities. To test both hypotheses, we analyzed the significance levels of each interaction term obtained in the MANOVA model and plotted two interaction graphs, with historical demand data on the x-axis and contextual data illustrated by separate lines (Fig. 4.3).

We observe significant interaction effects for both dependent variables (G^2: $F = 141.593$, $p < 0.001$, $\eta^2 = 0.53$; U^2: $F = 9.141$, $p < 0.003$, $\eta^2 = 0.07$). Specifically, in the absence of historical data, the first interaction graph shows that contextual data have a negligible impact on a judge's effectiveness in exploiting linear cue-criterion relations, whereas in the presence of historical data, providing contextual data leads to a difference in G^2 of almost 12%. The statistical power of this finding is strong as our post hoc tests show considerably large effect size measures ($f^2 = 1.13$; $(1 - \beta) = 1.0$). Therefore, the first interaction graph validates the relationship proposed in Hypothesis H_{3a}. Consistent with H_{3b}, our analysis indicates that the unavailability of historical demand data significantly amplifies the effect of contextual data on forecasters' ability to process nonlinearities

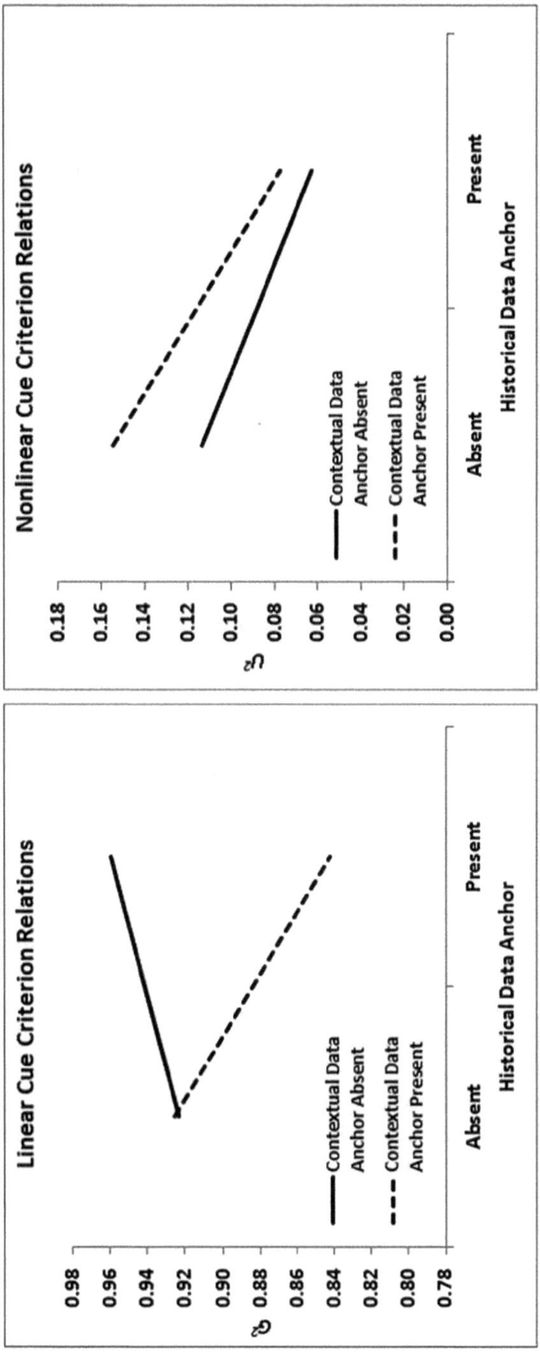

Fig. 4.3 Effect of contextual anchors and historical demand anchors on judges' effectiveness in exploiting linear and nonlinear cue-criterion relations

($\Delta U^2 = 0.04$ and $\Delta\ U^2 = 0.01$, respectively), which is supported by observing moderate effect sizes in our post hoc examination ($f^2 = 0.08$; $(1 - \beta) = 0.96$). In other words, historical data moderates the effect of contextual data on forecasting performance. Although historical data lowers the variability of outcomes resulting from contextual data in processing linear cue-criterion relations, they heighten the variability of outcomes in interpreting nonlinear cue-criterion relations.

When relying on hierarchical regression[10] instead of MANOVA to re-analyze our data, the moderation effect is also reflected in the beta weights associated with the regression model. If we focus on U^2 as the dependent variable and convert beta values into relative weights, historical data account for 77% of the variance explained by the model. This result shows that nonlinear information processing primarily represents a function of the amount of historical data involved. Similarly, when computing such relative weights for G^2, historical data account for 78% of the model weights, whereas contextual data are associated with a relative beta weight of 32%. These results stress the importance of accounting for the detrimental effect of ambiguous historical data when generating judgmental forecasts about fashion products. Our empirical findings therefore provide support for Hypothesis H_{3b}.

Absolute Forecast Deviations To further assess the meaningfulness of our findings for music forecasting practice, we focused on the extent to which providing or removing the two decision support types would impact forecasting effectiveness. More specifically, we started out by calculating absolute deviations of predictions from the actual chart positions for both best-fitting linear models and judges (Fig. 4.4).

One-sided ANOVA tests indicated that all mean differences across decision support types were significant at the $p < 0.01$ level. In general, absolute forecast deviations were slightly larger for judges, ranging between 12.78 and 23.55 chart positions compared to 10.96 and 23.66 chart positions for the best-fitting linear model. For both the best-fitting linear model and judges the joint presence of historical and contextual data (C^+/H^+) led to the smallest absolute deviations, whereas removing the two decision support anchors resulted in the largest deviation. However, in the latter case, judgmental forecasts resulted in slightly narrower prediction intervals than model predictions. We then constructed a series of hypothetical scenarios, in which we used our regression estimates to adjust absolute forecast deviations to simulate the removal or addition of a specific decision support anchor. Adjustments were made by accounting for the proportion of residual variance that was not explained by judges in each of the four conditions. Figure 4.5 provides an overview of our analysis. In particular, it indicates that the forecast deviations can be reduced on average by 4.28 chart positions when one additional decision support type is offered and 9.72 positions when both contextual and historical data are provided.

[10] This analysis is made possible by the fact that our predictor variables are only associated with two categorical levels.

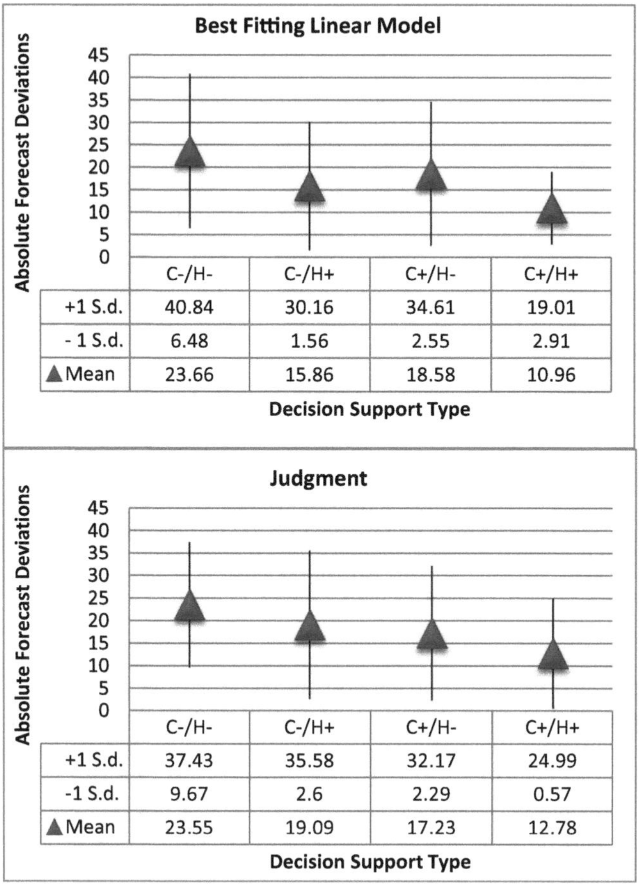

Fig. 4.4 Absolute deviations from forecasting event

6 Discussion

We conducted this study to obtain a better understanding of the conditions that drive judgmental effectiveness when generating forecasts in the domain of fashion products. The dynamic nature of such environments makes it difficult to use conventional forecasting tools. We specifically focused on historical and contextual data, because they frequently form integral components of decision support systems. Our findings will help forecasters design better decision support systems, by outlining the conditions under which such data are likely to improve or reduce their ability to detect and interpret critical information.

Specifically, we show that when forecasters are concerned with predictive accuracy and only managerial judgments are employed, providing both historical and contextual data is beneficial. Moreover, our analyses indicate that if judgmental forecasts are combined with other methods, decision support provided to forecasters

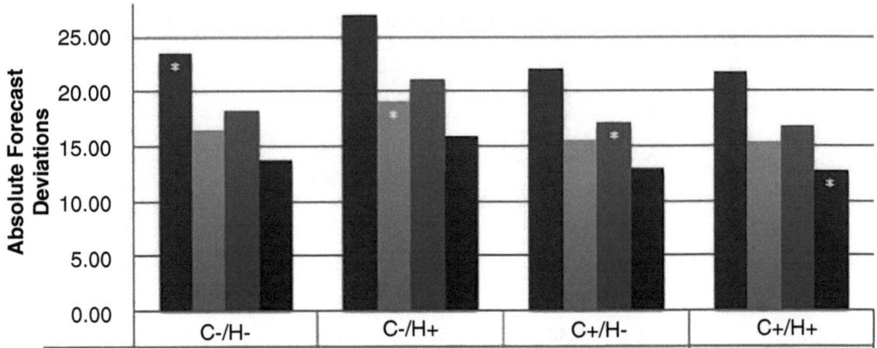

	C-/H-	C-/H+	C+/H-	C+/H+
■ C-/H-	23.55*	27.04	22.11	21.73
■ C-/H+	16.62	19.09*	15.60	15.34
■ C+/H-	18.36	21.08	17.23*	16.93
■ C+/H+	13.85	15.91	13.00	12.78*

Decision Support Type

Fig. 4.5 Forecasting effectiveness with and without contextual and historical data anchors ("*" indicates actual deviation)

should be restricted to contextual anchors. In particular, we find that the exploitation of nonlinearities is easiest for human judgment if contextual, but no historical data are present. Thus, if the role of managerial judgment is to detect these nonlinearities (and the linearities are taken care of by some statistical model with which managerial judgments are combined), then a restriction of the decision support data provided to decision makers makes sense.

The music data this research builds on reinforce the conclusions of prior research demonstrating that contextual data make it harder for forecasters to perform on a par with the best-fitting linear model of the ecology (Karelaia & Hogarth, 2008). We extend these findings by suggesting that, at the same time, the presence of contextual data may actually facilitate forecasters' ability to process nonlinearities. In addition, historical data significantly affect forecasters' effectiveness in analyzing nonlinear cue-criterion relations. In the absence of historical data, forecasters explain far more of the linear model's residual variance than when such information is available. Thus, our findings provide empirical support for Dane and Pratt's (2007) theoretical proposition that efficiently exploiting nonlinearities becomes more important as tasks become less predictable. We further investigated the interactive relationship between historical demand and contextual anchors and showed supporting evidence for the proposition that historical data moderate the effect of contextual anchors when analyzing linearities and nonlinearities.

Our empirical data indicate that forecasters' ability to exploit nonlinearities only fully unfolds when they can rely on contextual data and when judgmental biases due to misleading historical data are kept at a minimum. Prior research suggests that in these types of task environments, decision makers frequently rely on heuristic shortcuts such as pattern matching to help them process task information in a more efficient way (Gigerenzer et al., 1999; Hoch & Schkade, 1996). Because judgment analysis studies frequently model forecasts *as if* judges utilize a linear model (Hogarth & Karelaia, 2007), one interesting path for future research could be to adopt Klein's (2003) Recognition-Primed Decision Models (RPDM) to analyze forecasters' underlying cognitive processes. RPDM assume that decision makers are capable of identifying cue patterns in highly uncertain environments because of a variety of *action scripts* developed over years of training. Typically, RPDM have been investigated in naturalistic decision-making settings in which actors (such as firefighters and emergency service units) have little time to process incomplete, highly uncertain task information (Weick, 1988). The findings of our research warrant further studies of RPDM, which could systematically decompose the properties of a forecasting task and then employ more qualitative research designs to investigate how expert forecasters process relevant information to form mental representations of the underlying forecasting environment.

From the perspective of general systems theory (Churchman, 1971; Flood & Carson, 1988), our empirical findings may also provide a better understanding of the way judges tend to engage in individual learning. Cooksey (2000) argues that managerial decision behavior can be understood as a direct consequence of negative and positive feedback loops. Negative system feedback refers to the gap between a system's current position and the goal that it tries to attain. It triggers managerial actions aimed at reaching a level of convergence between current position and targeted goal. In contrast, positive system feedback may lead to a fundamental revision of managers' existing system goals by encouraging them to try out new paths that could result in the attainment of even higher-level objectives (Cooksey, 2000). The underlying logic of negative and positive system feedback is therefore consistent with Argyris' (1990) notion of single (emphasizing negative feedback) and double (emphasizing positive feedback) loop learning. When considering forecasters' ability to process linear cue-criterion relations, we speculate that our observed interaction primarily triggers learning from negative system feedback. More specifically, the presence of both historical demand and contextual anchors will increase the need to take corrective measures to approximate the performance of the best-fitting linear model. Conversely, positive system feedback seems more likely to improve judges' effectiveness in processing nonlinear cue-criterion relations as forecasters need to identify and employ new organizing principles for improving forecasting accuracy. Our empirical results illustrate that learning from positive system feedback is most likely to occur in the absence of historical demand data, when the consideration of contextual knowledge leads to the greatest variability in judges' performance. We believe that our study opens the path to a number of potentially interesting research projects which could specifically test the relationship between learning from system feedback and forecasting effectiveness. Lastly, our findings have clear implications for the effective design of quick response policies in music companies (Cachon & Swinney, 2011; Fisher & Raman, 1996; Iyer & Bergen, 1997). Specifically, as pop music singles are commonly used as pre-seasonal sales indicators of an upcoming full album release, efficiently predicting chart positions of music singles is likely to further reduce lead time by making performance signals become available even before the official publication of the music singles charts.

Additional research on the impact of historical demand and contextual anchors is crucially needed. Indeed, being able to employ the mode of information processing most likely to be effective in a given situation has recently been argued to represent a key skill in managerial decision-making (Seifert & Hadida, 2013). Our findings shed further light on the circumstances that favor reliance on model prediction or expert judgment, particularly since many organizations utilize linear model forecasts as a fundamental tool for coping with uncertainty. Prior research in this domain suggested that forecast combinations relying on a simple 50:50 split between model and manager judgment efficiently exploited nonlinear cue-criterion relations in the task environment (Blattberg & Hoch, 1990). By specifically demonstrating how the presence of different types of decision support anchors can harm forecasters' ability to interpret informational cues, our research provides a more refined view of the drivers behind judgmental performance.

To conclude, we hope that future research will provide additional tests of the robustness of our results by adopting alternative research designs, prediction tasks, and operationalizations of key constructs. Specifically, the extent to which the reported relationship between task-level characteristics and judgmental performance is robust and extends to other types of forecasting contexts involving fashion products (e.g., the textile or movie industries) remains unclear. Despite our efforts to reduce systematic error in our chosen methodology, our operationalization of forecasters' nonlinear information processing skill could not fully rule out the possibility that omitted variables contaminated our findings. We therefore encourage replication of our study in alternative contexts. Moreover, further insight into the effectiveness of judges' sense-making processes may also be gained by developing more sophisticated models that incorporate key nonlinearities from the task environment (such as the visual appearance or sound sample in our study). Finally, future research could examine the extent to which the "wisdom of the crowd" associated with collective forecasting judgments may compensate for the performance limitations of group members when facing different types of task characteristics. In sum, we believe that the results of our study raise a number of intriguing questions for follow-up research, and offer promising opportunities to further clarify the role of decision support systems in judgmental forecasting.

References

Andreassen, P. B., & Kraus, S. J. (1990). Judgmental extrapolation and the salience of change. *Journal of Forecasting, 9*(4), 347–372.

Argyris, C. (1990). *Overcoming organizational Defenses: Facilitating organizational learning.* Allyn & Bacon.

Au, K. F., Choi, T. M., & Yu, Y. (2008). Fashion retail forecasting by evolutionary neural networks. *International Journal of Production Economics, 114*(2), 615–630.

Bendoly, E., Donohue, K., & Schultz, K. L. (2006). Behavior in operations management: Assessing recent findings and revisiting old assumptions. *Journal of Operations Management, 24*(6), 737–752.
Blattberg, R. C., & Hoch, S. J. (1990). Database models and managerial intuition: 50% model + 50% manager. *Management Science, 36*, 887–899.
Bolton, G. E., & Katok, E. (2008). Learning by doing in the newsvendor problem: A laboratory investigation of the role of experience and feedback. *Manufacturing & Service Operations Management, 10*(3), 519–538.
Bostian, A. A., Holt, C. A., & Smith, A. M. (2008). Newsvendor "pull-to-center" effect: Adaptive learning in a laboratory experiment. *Manufacturing & Service Operations Management, 10*(4), 590–608.
Boulaksil, Y., & Franses, P. H. (2009). Experts' stated behavior. *Interfaces, 39*(2), 168–171.
Brehmer, B. (1978). Response consistency in probabilistic inference tasks. *Organizational Behavior and Human Performance, 22*(1), 103–115.
Brunswik, E. (1956). *Perception and the representative design of psychological experiments* (2nd ed.). The University of California Press.
Cachon, G. P., & Fisher, M. (2000). Supply chain inventory management and the value of shared information. *Management Science, 46*(8), 1032–1048.
Cachon, G. P., & Swinney, R. (2011). The value of fast fashion: Quick response, enhanced design, and strategic consumer behavior. *Management Science, 57*(4), 778–795.
Camerer, C., & Weber, M. (1992). Recent developments in modeling preferences: Uncertainty and ambiguity. *Journal of Risk and Uncertainty, 5*, 325–370.
Choi, T. M. J., Li, D., & Yan, H. (2006). Quick response policy with Bayesian information updates. *European Journal of Operational Research, 170*(3), 788–808.
Choi, T. M., Hui, C. L., Liu, N., Ng, S. F., & Yu, Y. (2014). Fast fashion sales forecasting with limited data and time. *Decision Support Systems, 59*, 84–92.
Christopher, M., Lowson, R., & Peck, H. (2004). Creating agile supply chains in the fashion industry. *International Journal of Retail and Distribution Management, 32*(8), 367–376.
Churchman, C. W. (1971). *The Design of Inquiring Systems.* Basic Books, Perseus Books Group.
Clemen, R. T. (1989). Combining forecasts: A review and annotated bibliography. *International Journal of Forecasting, 5*, 559–583.
Cohen, J. (2013). *Statistical power analysis for the behavioral sciences.* Routledge Academic.
Cohen, J., & Cohen, I. (1975). *Applied multiple regression/correlation analysis of the behavioral sciences.* Erlbaum.
Cooksey, R. (1996). *Judgment analysis: Theory, methods and applications.* Academic Press.
Cooksey, R. W. (2000). Mapping the texture of managerial decision making: A complex dynamic decision perspective. *Emergence: A Journal of Complexity Issues in Organizations and Management, 2*, 102–122.
Dane, E., & Pratt, M. G. (2007). Exploring intuition and its role in managerial decision making. *Academy of Management Review, 32*, 33–54.
Dhami, M. K., & Harries, C. (2001). Fast and frugal versus regression models of human judgment. *Thinking and Reasoning, 7*, 5–27.
Dougherty, M. R., & Thomas, R. P. (2012). Robust decision making in a nonlinear world. *Psychological Review, 119*(2), 321–344.
Dunwoody, P. T., Haarbauer, R., Marino, C., & Tang, C. (2000). Cognitive adaptation and its consequences: A test of cognitive continuum theory. *Journal of Behavioral Decision Making, 13*, 35–54.
Einhorn, H. J. (1970). The use of nonlinear noncompensatory models in decision making. *Psychological Bulletin, 73*, 221–230.
Einhorn, H. J. (1971). Use of nonlinear, noncompensatory models as a function of task and amount of information. *Organizational Behavior and Human Performance, 6*, 1–27.
Einhorn, H. J. (1974). Cue definition and residual judgment. *Organizational Behavior and Human Decision Processes, 12*, 30–49.

Einhorn, H. J., & Hogarth, R. M. (1975). Unit weighting schemes for decision-making. *Organizational Behavior and Human Decision Processes, 13*, 171–192.
Elberse, A., & Eliashberg, J. (2003). Demand and supply dynamics of sequentially released products in international markets: The case of motion pictures. *Marketing Science, 22*(3), 329–354.
Eliashberg, J., & Sawhney, M. S. (1994). Modeling goes to Hollywood: Predicting individual differences in movie enjoyment. *Management Science, 40*(9), 1151–1173.
Eliashberg, J., Weinberg, C. B., & Hui, S. K. (2008). Decision models for the movie industry. In *Handbook of marketing decision models* (pp. 437–468). Springer.
Faul, F., Erdfelder, E., Lang, A. G., & Buchner, A. (2007). G* Power 3: A flexible statistical power analysis program for the social, behavioral, and biomedical sciences. *Behavior Research Methods, 39*(2), 175–191.
Fellner, W. (1961). Distortion of subjective probabilities as a reaction to uncertainty. *The Quarterly Journal of Economics, 75*, 670–694.
Fisher, M., & Raman, A. (1996). Reducing the cost of demand uncertainty through accurate response to early sales. *Operations Research, 44*(1), 87–99.
Flood, R. L., & Carson, E. R. (1988). *Dealing with complexity: An introduction to the theory and application of systems science*. Plenum.
Frisch, D., & Baron, J. (1988). Ambiguity and rationality. *Journal of Behavioral Decision Making, 1*, 149–157.
Fumi, A., Pepe, A., Scarabotti, L., & Schiraldi, M. M. (2013). Fourier analysis for demand forecasting in a fashion company. *International Journal of Engineering Business Management, 5*(Godište 2013), 5–30.
Gaur, V., Kesavan, S., Raman, A., & Fisher, M. L. (2007). Estimating demand uncertainty using judgmental forecasts. *Manufacturing and Service Operations Management, 9*(4), 480–491.
Gigerenzer, G., Todd, P. M., & ABC Research Group. (1999). *Simple heuristics that make us smart*. Oxford University Press.
Goodwin, P., & Wright, G. (1993). Improving judgmental time series forecasting: A review of the guidance provided by research. *International Journal of Forecasting, 9*(2), 147–161.
Green, M., & Harrison, P. J. (1973). Fashion forecasting for a mail order company using a Bayesian approach. *Journal of the Operational Research Society, 24*(2), 193–205.
Gutierrez, R. S., Solis, A. O., & Mukhopadhyay, S. (2008). Lumpy demand forecasting using neural networks. *International Journal of Production Economics, 111*(2), 409–420.
Hadida, A. L., & Paris, T. (2014). Managerial cognition and the value chain in the digital music industry. *Technological Forecasting and Social Change, 83*, 84–97.
Hamm, R. M. (1987). Clinical intuition and clinical analysis: Expertise and the cognitive continuum. In J. Dowie & A. Einstein (Eds.), *Professional judgment*. Cambridge University Press.
Hammond, K. R. (1996). *Human judgment and social policy*. Oxford University Press.
Hammond, K. R., & Summers, D. A. (1972). Cognitive control. *Psychological Review, 79*, 58–67.
Harvey, N. (1995). Why are judgments less consistent in less predictable task situations? *Organizational Behavior and Human Decision Processes, 63*(3), 247–263.
Harvey, N., Ewart, T., & West, R. (1997). Effects of data noise on statistical judgement. *Thinking & Reasoning, 3*(2), 111–132.
Hines, T., & Bruce, M. (Eds.). (2007). *Fashion marketing*. Routledge.
Hirsch, P. M. (1972). Processing fads and fashions: An organization-set analysis of cultural industry systems. *American Journal of Sociology, 77*(4), 639–659.
Ho, T.-H., Lim, N., & Cui, T. H. (2010). Reference dependence in multilocation newsvendor models: A structural analysis. *Management Science, 56*, 1891–1910.
Hoch, S. J., & Schkade, D. A. (1996). A psychological approach to decision support systems. *Management Science, 42*(1), 51–64.
Hogarth, R. M. (1987). *Judgment and choice* (2nd ed.). Wiley.
Hogarth, R. M., & Karelaia, N. (2007). Heuristic and linear models of judgment: Matching rules and environments. *Psychological Review, 114*(3), 733–758.

IFPI. (2014). Accessed from http://www.ifpi.org

Iman, R. L., & Conover, W. J. (1979). The use of the rank transform in regression. *Technometrics, 21*(4), 499–509.

Iyer, A. V., & Bergen, M. E. (1997). Quick response in manufacturer-retailer channels. *Management Science, 43*(4), 559–570.

Kahneman, D., & Tversky, A. (1973). On the psychology of prediction. *Psychological Review, 80*(4), 237.

Karelaia, N., & Hogarth, R. M. (2008). Determinants of linear judgment: A meta-analysis of lens model studies. *Psychological Bulletin, 134*, 404–426.

Klein, G. (2003). *Intuition at work*. Doubleday.

Kremer, M., Moritz, B., & Siemsen, E. (2011). Demand forecasting behavior: System neglect and change detection. *Management Science, 57*(10), 1827–1843.

Lattin, J. M., Carroll, J. D., & Green, P. E. (2003). *Analyzing multivariate data*. Thomson Brooks/Cole.

Lawrence, M., & Makridakis, S. (1989). Factors affecting judgmental forecasts and confidence intervals. *Organizational Behavior and Human Decision Processes, 43*(2), 172–187.

Lawrence, M., Goodwin, P., O'Connor, M., & Önkal, D. (2006). Judgmental forecasting: A review of progress over the last 25 years. *International Journal of Forecasting, 22*, 493–518.

Lee, Y. S., & Siemsen, E. (2015). *Task decomposition and newsvendor decision making*. Working Paper. Accessed from http://ssrn.com/abstract=2232073

Lee, J. W., & Yates, J. F. (1992). How quantity judgment changes as the number of cues increases: An analytical framework and review. *Psychological Bulletin, 112*, 363–377.

Mentzer, J. T., & Moon, M. A. (2005). *Sales forecasting management* (2nd ed.). Sage.

Moe, W. W., & Fader, P. S. (2001). Modeling hedonic portfolio products: A joint segmentation analysis of music compact disc sales. *Journal of Marketing Research, 2001*, 376–385.

Moreau, F., & Peltier, S. (2004). Cultural diversity in the movie industry: A cross-national study. *Journal of Media Economics, 17*(2), 123–143.

Nenni, M. E., Giustiniano, L., & Pirolo, L. (2013). Demand forecasting in the fashion industry: A review. *International Journal of Engineering Business Management, 5*(37), 1–6.

Ni, Y., & Fan, F. (2011). A two-stage dynamic sales forecasting model for the fashion retail. *Expert Systems with Applications, 38*(3), 1529–1536.

Olshavsky, R. (1979). Task complexity and contingent processing in decision-making: A replication and extension. *Organizational Behavior and Human Performance, 24*, 300–316.

Payne, J. (1976). Task complexity and contingent processing in decision-making: An information search and protocol analysis. *Organizational Behavior and Human Performance, 16*, 366–387.

Peterson, R. A., & Berger, D. G. (1971). Entrepreneurship in organizations: Evidence from the popular music industry. *Administrative Science Quarterly, 16*(1), 97–106.

Ramsey, J. B. (1969). Tests for specification errors in classical linear least squares regression analysis. *Journal of the Royal Statistical Society, Series B., 31*(2), 350–371.

Ren, Y., & Croson, R. (2013). Overconfidence in newsvendor orders: An experimental study. *Management Science, 59*, 2502–2517.

Sanders, N. R., & Manrodt, K. B. (2003). Forecasting software in practice: Use, satisfaction, and performance. *Interfaces, 33*(5), 90–93.

Sani, B., & Kingsman, B. G. (1997). Selecting the best periodic inventory control and demand forecasting methods for low demand items. *Journal of the Operational Research Society, 48*(7), 700–713.

Santagata, W. (2004). Creativity, fashion and market behavior. In D. Power & A. J. Scott (Eds.), *Cultural industries and the production of culture* (pp. 75–90). Routledge.

Sawhney, M. S., & Eliashberg, J. (1996). A parsimonious model for forecasting gross box-office revenues of motion pictures. *Marketing Science, 15*(2), 113–131.

Schweitzer, M. E., & Cachon, G. P. (2000). Decision bias in the newsvendor problem with a known demand distribution: Experimental evidence. *Management Science, 46*(3), 404–420.

Seifert, M., & Hadida, A. L. (2013). On the relative importance of linear model and human judge (s) in combined forecasting. *Organizational Behavior and Human Decision Processes, 120*, 24–36.

Serel, D. A. (2009). Optimal ordering and pricing in a quick response system. *International Journal of Production Economics, 121*(2), 700–714.

Sichel, B. (2008). Forecasting demand with point of sales data: A case study of fashion products. *Journal of Business Forecasting Methods and Systems, 27*(4), 15–16.

Slovic, P., & Lichtenstein, S. (1971). Comparison of Bayesian and regression approaches to the study of information processing in judgment. *Organizational Behavior and Human Performance, 6*, 649–744.

Snyder, R. (2002). Forecasting sales of slow and fast moving inventories. *European Journal of Operational Research, 140*(3), 684–699.

Steenkamp, J.-B. E. M., Hofstede, F., & Wedel, M. (1999). A cross-national investigation into the individual and national cultural antecedents of consumer innovativeness. *Journal of Marketing, 63*, 55–69.

Steinmann, D. (1976). The effects of cognitive feedback and task complexity in multiple-cue probability learning. *Organizational Behavior and Human Performance, 15*, 168–179.

Stewart, T. R. (2001). The lens model equation. In K. R. Hammond & T. R. Stewart (Eds.), *The essential Brunswik: Beginnings, explications, applications* (pp. 357–362). Oxford University Press.

Stewart, T. R., Roebber, P. J., & Bosart, L. F. (1997). The importance of the task in analyzing expert judgment. *Organizational Behavior and Human Decision Processes, 69*(3), 205–219.

Stremersch, S., & Tellis, G. J. (2004). Understanding and managing international growth of new products. *International Journal of Research in Marketing, 21*, 421–438.

Su, X. (2008). Bounded rationality in newsvendor models. *Manufacturing & Service Operations Management, 10*, 566–589.

Sun, Z. L., Choi, T. M., Au, K. F., & Yu, Y. (2008). Sales forecasting using extreme learning machine with applications in fashion retailing. *Decision Support Systems, 46*(1), 411–419.

Taylor, J. W., & Buizza, R. (2003). Using weather ensemble predictions in electricity demand forecasting. *International Journal of Forecasting, 19*(1), 57–70.

Thomassey, S., Happiette, M., & Castelain, J. M. (2005). A global forecasting support system adapted to textile distribution. *International Journal of Production Economics, 96*(1), 81–95.

Tucker, L. R. (1964). A suggested alternative formulation in the developments by Hursch, Hammond, and Hursch, and by Hammond, Hursch, and Todd. *Psychological Review, 71*(6), 528–530.

Van den Bulte, C., & Stremersch, S. (2004). Social contagion and income heterogeneity in new product diffusion: A meta-analytic test. *Marketing Science, 23*(4), 530–544.

Vogel, H. L. (2007). *Entertainment industry economics: A guide for financial analysis* (7th ed.). Cambridge University Press.

Weick, K. (1988). Enacted sensemaking in crisis situation. *Journal of Management Studies, 25*(4), 305–317.

Wong, W. K., & Guo, Z. X. (2010). A hybrid intelligent model for medium-term sales forecasting in fashion retail supply chains using extreme learning machine and harmony search algorithm. *International Journal of Production Economics, 128*(2), 614–624.

Wood, R. E. (1986). Task complexity: Definition of the construct. *Organizational Behavior and Human Decision Processes, 37*, 60–82.

Xia, M., Zhang, Y., Weng, L., & Ye, X. (2012). Fashion retailing forecasting based on extreme learning machine with adaptive metrics of inputs. *Knowledge-Based Systems, 36*, 253–259.

Yu, Y., Choi, T. M., & Hui, C. L. (2011). An intelligent fast sales forecasting model for fashion products. *Expert Systems with Applications, 38*(6), 7373–7379.

Zuckerman, E. W., & Kim, T. Y. (2003). The critical trade-off: Identity assignment and box office success in the feature film industry. *Industrial and Corporate Change, 12*(1), 27–67.

Chapter 5
Judgmental Interventions and Behavioral Change

Fotios Petropoulos and Konstantinos Nikopoulos

Keywords Forecasting · Judgment · Adjustment · Statistical forecast · Combination · Rational behavior

1 Background

Since 1400 BC and the establishment of Pythia, the Oracle of Delphi, it was made clear that humans depend on predictions before they make important decisions. Kings and leaders would visit Pythia and, based on the prophecy, would decide whether or not to go into a war or proceed with a politically-directed marriage. Nowadays, forecasting plays a major role for an array of functions and processes in organisations. Accurate forecasting and minimisation of the uncertainty translate to improved operations and planning, significant savings in costs, and increased customer satisfaction. While forecasts are barely ever completely accurate, their sole existence enables humans to be more prepared for the future and offers a certain degree of comfort and safety.

Applying forecasting in practice is supported by dedicated software, usually referred to as "forecasting support systems". Such systems integrate an array of univariate and multivariate statistical forecasting methods and are able to automatically produce forecasts for a large number of series. However, they also feature interfaces that allow the application of managerial judgment on any stage of the forecasting process, from data cleaning and preprocessing, to selecting models and their parameters, adjusting the produced statistical forecasts, and even overriding the formal process altogether towards producing pure judgmental forecasts. In this paper, we focus on the case of judgmental adjustments (interventions) on the

F. Petropoulos (✉)
School of Management, University of Bath, Bath, UK
e-mail: fp359@bath.ac.uk

K. Nikopoulos
Business School, Durham University, Durham, UK

M. Seifert (ed.), *Judgment in Predictive Analytics*, International Series in Operations Research & Management Science 343,
https://doi.org/10.1007/978-3-031-30085-1_5

formally produced statistical forecasts. For a review of the literature results on pure judgmental forecasts, the reader is referred to (Lawrence et al., 2006) but also (Petropoulos et al., 2021, Sect. 2.10); the case of judgmental model selection has been examined by Petropoulos et al. (2018), Han et al. (2019), and De Baets and Harvey (2020).

Why are expert forecasters adjusting the formally produced statistical forecasts? One reason could be that the experts have information that is not available to the quantitative methods, or they know that the methods are deficient in properly incorporating this information. Such information could refer to the forthcoming effects of a promotional campaign or other one-off events. Adjusting the forecasts to include the anticipated impact of such events is likely to be beneficial, if the missing information is important and reliable (Goodwin & Fildes, 1999). Another reason for adjusting the statistically produced forecasts could be the lack of trust to the forecasting support system. Indeed, the users of forecasting support systems usually lack formal training on statistical algorithms and forecasting methods, which renders such methods black boxes. It is important that experts understand the value and limitations of statistical methods before performing judgmental adjustments (Donihue, 1993). In other cases, forecasts are judgmentally adjusted as the statistical forecasts are not "politically acceptable", depicted, for instance, downwards trends for the sales of the flagship product of a company. Finally, adjustments may simply be made so that the experts take the ownership of the forecasts while justifying the importance of their own roles. Anecdotal evidence on the reasoning behind performing adjustments, recorded via using a drop-down function of a forecasting support system, suggests that experts from a company were almost always selecting an option labelled "NGR". When they were asked what "NGR" stands for, they replied "No good reason"!

While some decades ago judgmental adjustments were discouraged (for example, see Carbone et al., 1983; Armstrong, 1985), Mathews and Diamantopoulos (1986, 1989, 1990) showed through a series of empirical studies that judgmental adjustments to statistical forecasts (or "forecast manipulation" as they called it) could be beneficial for the overall performance and could reduce the forecast error of the original statistical forecast. The benefits of judgmental adjustments were confirmed by other researchers (for example, see McNees, 1990; Donihue, 1993; Vere & Griffith, 1995).

A seminal empirical study on the performance of judgmental adjustments was conducted by Fildes et al. (2009) based on secondary data from four companies. Their main findings include: (*i*) judgmental adjustments are very frequent in practice, (*ii*) optimism bias, attributed to the confusion of the forecast with the target or the decision, resulted in many positive ineffective adjustments, and (*iii*) small adjustments usually add little value and should be avoided. Syntetos et al. (2010) challenge the last finding by providing evidence that small improvements in forecast accuracy as a result of judgmental adjustments may translate to significant inventory savings. Trapero et al. (2013) further analyse the performance of experts' forecast adjustment in the presence of promotions and show that the forecasting error can be reduced when the size of the adjustment is not too large. Syntetos et al. (2009)

focused on judgmental adjustments for intermittent demand time series, and confirmed the result of Fildes et al. (2009) regarding the efficacy of negative adjustments compared to positive ones.

In parallel, another research team analysed the judgmental adjustments to statistical forecasts using a large dataset from a pharmaceutical company. Similar to Fildes et al. (2009), Franses and Legerstee (2009) find that experts tend to perform judgmental adjustments very frequently, while the adjustments are more often positive (upwards) than negative (positivism bias). Focusing on the performance of such adjustments, Franses and Legerstee (2010) showed that often expert-adjusted forecasts were worse than model-based forecasts and at best equally good. However, the provision of feedback is likely to improve experts' behaviour towards making smaller statistical adjustments while improving the accuracy of their forecasts (Legerstee & Franses, 2014). Franses and Legerstee (2011a, b) analyse the interactions of managerial adjustments and the forecasting horizon and conclude that while interventions are made at all horizons, the added-value of the adjustments has a positive relationship with the length of the horizon.

Petropoulos et al. (2016) draw parallels of the behaviour of demand planners with that of poker players. Similar to how poker players react irrationally after losing a big pot towards playing more loose and, as a consequence, suffering even bigger loses (Smith et al., 2009), the behaviour of expert forecasters who perform judgmental adjustments to statistical forecasts is affected by erroneous adjustments made in the very previous period. Large wrong-direction adjustments or very large overshoots are more likely to be followed by large adjustments that will lead again to large losses. Fildes et al. (2019) designed two behavioural experiments and showed that forecasters not only misused the quantitative and qualitative information available to them while performing judgmental adjustments, but also significantly underestimated the effects of the promotions by neglecting the explicitly reported base rates.

One result that has been confirmed by several studies on judgmental adjustments is the very good performance of the Blattberg and Hoch (1990) approach on combining models with managerial intuition using equal weights (50–50%). Fildes et al. (2009) applied this simple technique of 50% formal/statistical forecast and 50% judgmentally adjusted forecast on the positive adjustments in the cases where positive adjustments were made. They observed a significant increased in forecasting performance, which is attributed to partly tackling the issue of positivism bias. Franses and Legerstee (2011a) combined statistical and expert forecasts without distinguishing between positive and negative adjustments, and tried to find the optimal combination weights. They concluded that a simple, equal-weighted, combination is a robust choice. Petropoulos et al. (2016) extended the results of Franses and Legerstee (2011a) and showed that while a 50–50% combination is indeed beneficial across the board, it is particularly effective when applied after a big-loss adjustment in the very last period. What does a combination of the statistical and the judgmentally adjusted forecast means? It suggests that the judgmental adjustments made by the experts are consequently further adjusted towards the statistical forecast. In other words, the judgmental adjustment is damped. In the 50–50% case, the final forecast equals to the statistical forecast plus half of the judgmental adjustment made on that forecast.

As noted by Petropoulos et al. (2016), a "critical question for future research is to examine if the automatic adjustment of judgmental adjustments (through a combination of statistical and expert forecasts, or dampening of the experts' adjustments) would lead to a long-term change of forecasters' behaviour with regards to how they perform judgmental interventions". In this study, we aim to investigate, by means of a laboratory experiment, whether or not the behaviour of experts changes if they have knowledge that their adjustments are further adjusted. If a change in the behaviour is observed, then we are interested to see if such a change is rational. Understanding experts' reactions on further and mechanically adjusting their judgmental interventions will offer us insights towards the design of forecasting support systems so that such behavioural changes are minimised and the acceptance of the integration of humans and algorithms is maximised.

2 The Design of a Behavioral Experiment

The task requires the judgmental adjustment of a statistical forecast produce using a time series in the light of an one-off event that is about to occur in the very next period. In more detail, we generated 30 synthetic time series of 36 periods each that we assume that depict the historical demand of 30 products from Company A. We also assume that these time series are in monthly frequency, so their length covers a period of three years, which is relevant with forecasting in practice (see, for instance, the Demand Planning module of the SAP APO software). We further assume that during these three years of history there has been no special events recorded, such as promotions, that would significantly affect the demand. As such, our generated series consist of level and noise, which is independent and identically distributed.

For each of these 30 series, we produce statistical forecasts using the Simple Exponential Smoothing (SES) method, which is a method able to model and extrapolate the level in the data, filtering out the noise, but it cannot deal with trend, seasonality or the expected impact of future special events. The forecasting horizon is assumed to be one period ahead ($h = 1$).

We asked the participants in our experiment to take the role of the demand planner of Company A, and to judgmentally adjust the statistical forecast produced by SES for period 37 in the light of a reliable piece of information that will affect the demand and is not taken into account by SES. This information relates to a heavy promotional activity of the type "Buy one, get one free" (BOGOF) that is applied by either our company (Company A) or our main competitor (Company B). Participants were given this information for each of the time series separately; see the representative screenshots in Fig. 5.1. For each participant and each series, we randomly determined if the promotion is run by our company or the competitor's company. The participants are expected to adjust the statistical forecast upwards (downwards) when Company A (B) is running the promotion for the next period. The participants

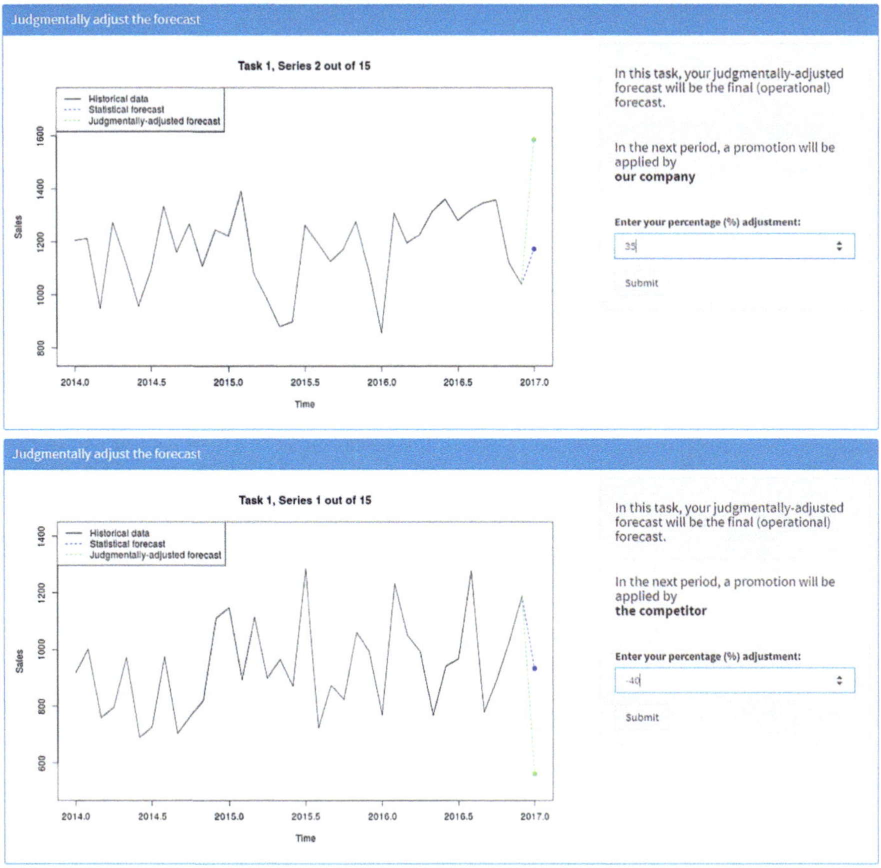

Fig. 5.1 Indicative screenshots of the experiment's user interface where the participant is informed regarding own promotion (top panel) or promotion by the main competitor (bottom panel)

input their adjustment for the current series as a percentage value through a numeric up-down control. Once a value has been input, the plot that displays the historical data (in black) and the statistical forecast (in blue) is updated to also depict the judgmentally-adjusted forecast (in green); see, also, Fig. 5.1. The participants may try different values of percentage adjustments they proceed to the next series.

We implemented these two conditions as previous studies have shown that experts tend to make larger positive adjustments compared to negative ones (see, for example, Fildes et al., 2009; Franses & Legerstee, 2009) due to positivism bias: "our promotion is going to bump up our sales considerably, however our competitors promotion will not negatively affect our sales to the same magnitude". It is important, however, to note that we do not explicitly provide a base rate with regards to the efficacy of promotions, as Fildes et al. (2019) did. So a rational behaviour would result, on average, in equal-sized adjustments for the two types of promotions (own

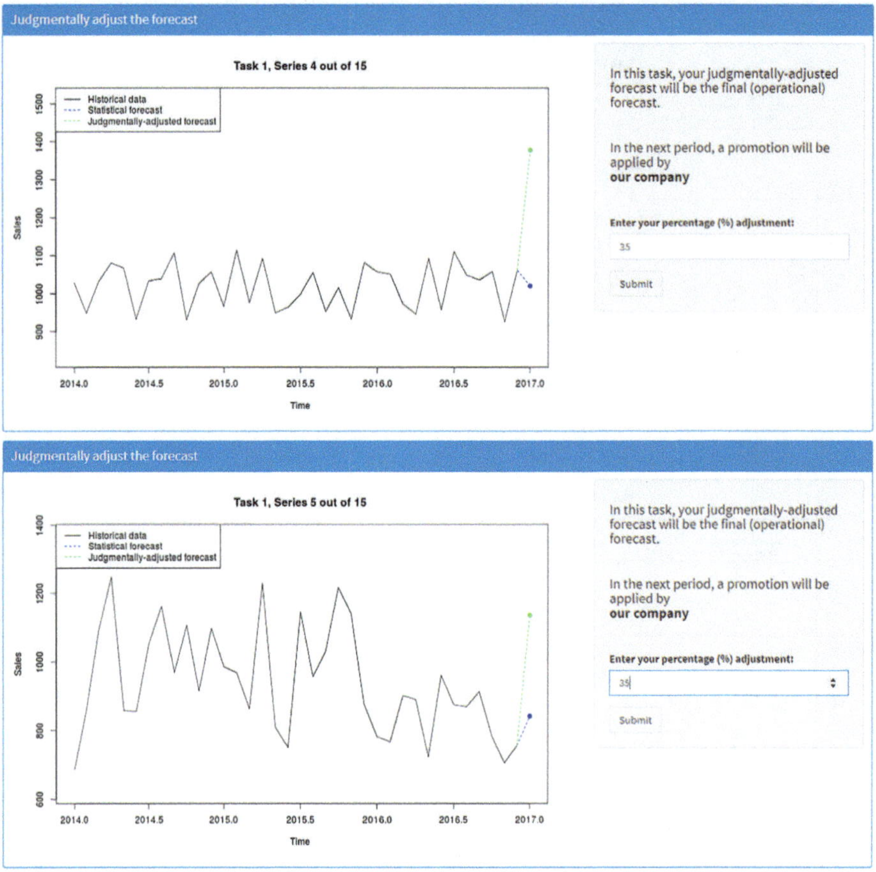

Fig. 5.2 Indicative screenshots of the experiment's user interface with examples of series with low (top panel) and high (bottom panel) levels of noise

or main competitor) as an uplift in sales following an own promotion should, on average, be the same to a decrease following a promotion from our main competitor.

Apart from controlling who is running the promotion, we also control for the level of noise in the series. We consider two levels of noise (low and high). Representative examples of series with low and high noise are presented in Fig. 5.2. Given that human judgment has been shown to be liable to biases and that experts are likely to confuse the signal with the noise (see, for example, Ian R. C. Eggleton, 1982; O'Connor et al., 1993), we would expect that the participants with make larger adjustments for series exhibiting high levels of noise compared to low ones even though a rational behaviour would not support that.

Finally, we provide to the participants explicit information regarding the calculation of the final forecast that will be considered for decision making (for example, ordering, replenishment, etc.). We consider two cases. The final forecast will be either the participants' judgmentally-adjusted forecast or the simple average (50–50%) of the participants' judgmentally-adjusted forecast and the statistical forecast.

For each participant, we randomly split the 30 series in two sets of 15 series each and assign each of these sets to one of the cases regarding the final forecast (solely the judgmentally-adjusted forecast or the combined statistical and judgmentally-adjusted forecast). Effectively, we create two sub-tasks (sub-task 1: the judgmentally-adjusted forecast is not further adjusted; sub-task 2: the judgmentally-adjusted forecast is further adjusted such that the adjustment is halved) that are presented in random order to each of the participants (sub-task 1 followed by sub-task 2, or vice versa). Let *MF*, *EF*, and *FF* denote the model forecast, expert (judgmentally-adjusted) forecast, and final forecast respectively, then the two sub-tasks can be defined as

$$\text{Sub}-\text{task1}: FF = EF$$

$$\text{Sub}-\text{task2}: FF = \frac{1}{2}(MF + EF) = MF + \frac{1}{2}(EF - MF)$$

where $EF - MF$ is the difference between the expert and model forecasts, or the judgmental adjustment. Note that sub-task 2 directly follows the proposition by Blattberg and Hoch (1990) and the empirical results of Fildes et al. (2009), Franses and Legerstee (2011a, b), and Petropoulos et al. (2016). Each participant must submit their judgmentally-adjusted forecasts for all the 15 series of one sub-task before moving to the other sub-task. Information specific to each sub-task is presented directly prior to the commencement of that sub-task and not at the beginning of the experiment. As such, and given the random order of presentation of the sub-tasks to each participant, if a change is observed with regards to how judgmental adjustments are made between the two tasks, then this change can be attributed to the information that is specific to each sub-task and relates to how the final forecast is calculated.

After the end of the two sub-tasks and the successful submission of judgmentally adjusted forecasts for all 30 synthetic time series, we presented participants with a short questionnaire. The first three questions focused on how the participants rate the performance of the statistical, judgmentally-adjusted (expert), and the performance of a 50–50% combination of the two. Their responses were captured in a 7-point likert scale, ranging from "extremely poor" to "excellent". In the next two questions, participants were asked to report their level of confidence (numerical input, from 0 to 100%) on whether their own individual, judgmental forecasts would be more accurate than the statistical forecasts or the combined statistical-judgmental forecasts using 50–50% weights. The next question asked for the participants' agreement (7-point likert scale) to the following statement:

> I made larger adjustments (to the given statistical forecast) for those cases where my judgmentally-adjusted forecasts would be used as final forecasts (as compared to the cases where 50-50 combination forecasts would be used as final forecasts).

In the final question, participants were asked to specify weights that if applied to the computer-produced statistical and their own judgmentally-adjusted forecast then the resulting weighted forecast would be more accurate than the combined forecast using the 50–50% weights. These weights were elicited through numerical-input controls that were interacting dynamically. Participants had the option not to respond to some questions.

The task was presented to two cohorts of students. The first cohort consisted of students enrolled in the Bachelor's level module "Forecasting Techniques" offered by the School of Electrical and Computer Engineering at the National Technical University of Athens (NTUA). The second cohort consisted of students at a Master's level taking the "Applied Business Projects" module offered by the Bangor Business School, Bangor University. The task was presented as an elective exercise to both cohorts. Prior to running the behavioural experiment, students in both cohorts were taught concepts surrounding forecasting and analytics. In total, we received 106 valid submissions, 70 from NTUA and 36 from Bangor University. 54 of the participants performed sub-task 1 before sub-task 2, while the rest 52 completed the two sub-tasks in reverse order.

3 Results

Given that we did not advise the participants regarding a representative uplift (or decrease) in the sales when a promotion is applied by our company (or our main competitor), we are interested in measuring the behavioural change participant separately by measuring the mean absolute adjustment for each sub-task (the expert forecast will not be further revised versus the expert forecast will be revised towards halving the adjustment). In particular, we define behavioural change as

$$\text{Behaviouralchange}(\%) = 100\left(\frac{MAA_2}{MAA_2} - 1\right),$$

in which MMA_1 and MMA_2 are the mean absolute adjustments for sub-tasks 1 and 2 respectively. A value close to zero suggests that the adjustment behaviour is not affected by the sub-task and the possible subsequent further adjustment of the expert forecast. A positive value indicates that participants apply larger adjustments when they are told that the final forecast will be the combination of the statistical and their judgmentally adjusted forecast. If a forecaster was to behave completely rational, aiming to negate the subsequent adjustment of their submissions, then their judgmental adjustment should be, on average, double the size for sub-task 2 compared to sub-task 1 which, in turn, would suggest a behavioural change of 100%. The measurement of behavioural change is scale-independent, so the respective values across multiple participants can be pooled together.

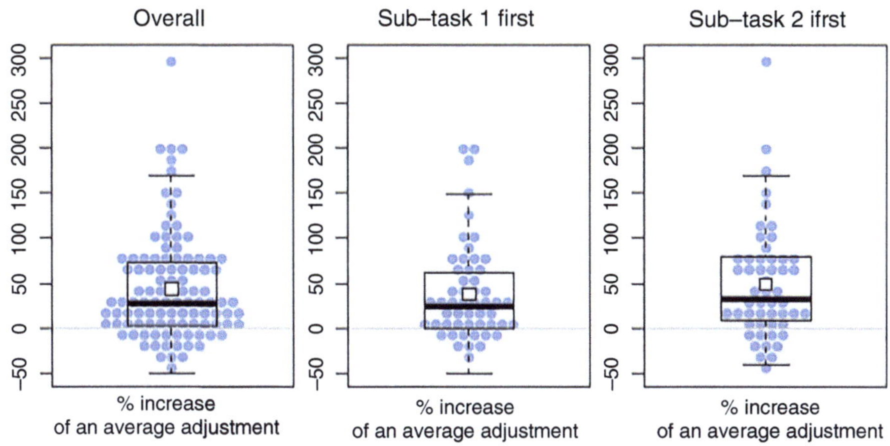

Fig. 5.3 Behavioural change when comparing the size of adjustments per individual between the two sub-tasks

Figure 5.3 presents the behavioural change values for each participant in our experiment (represented by blue dots) overlaid by box-plots. The first panel presents the values for all participants, while the second and third panel decompose the information with regards to the presentation order of the two sub-tasks (panel 2: sub-task 2 follows sub-task 1; panel 3: sub-task 1 follows sub-task 2). The gray horizontal lines indicate the value for no behavioural change, while the white squares represent the arithmetic mean of the values within each panel.

Focusing on the first panel of Fig. 5.3, we observe that there is a significant variation in the behavioural change across participants. 24% participants make, on average, smaller adjustments when they are told that their expert forecasts will be further revised, which is counter-intuitive. However, the majority of the participants (76%) increase their average adjustment. When this is happening, then the average increase is 61.6%, while 15% of the participants performed adjustments that are, on average, twice as large or more (i.e., behavioural change >100%). Across all participants, the mean (median) adjustment is 43.8% (27.7%).

In the last two panels of Fig. 5.3, we present separately the participants that first did sub-task 1 and then sub-task 2 (panel 2; 54 participants) or vice-versa (panel 3; 52 participants). We observe that the insights regarding the degree of average behavioural change hold, regardless the order of the tasks. When sub-task 1 was presented first, 74% of the participants changed their judgmental behaviour upwards between the two tasks with a mean change of 38.5%. In the opposite case (sub-task 2, followed by sub-task 1), 79% of the participants made lower adjustments in sub-task 1 compared to sub-task 2, on average by 49.3%. This suggests that participants (*i*) were clear regarding each sub-task and how their judgmental forecasts are used (if they are further adjusted) and (*ii*) showed trust towards the system, by significantly reducing (on average) the size of their adjustments when moving

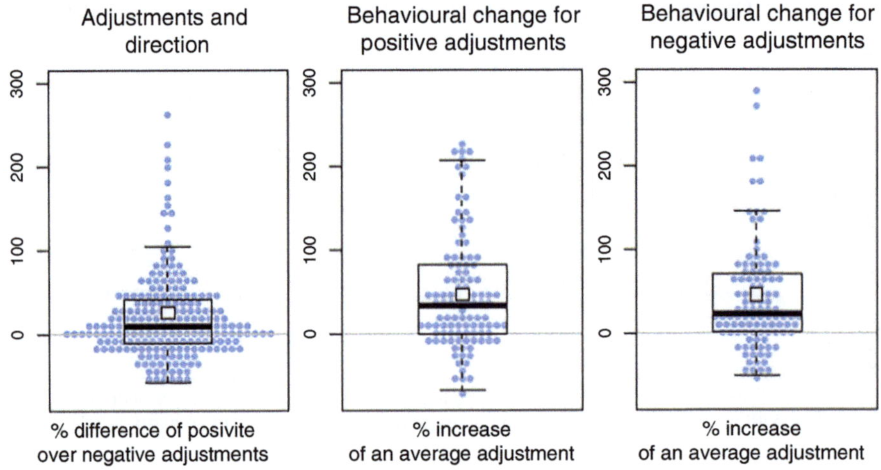

Fig. 5.4 The effect of the direction of adjustment

from sub-task 2 to sub-task 1 where they were told that their expert forecast will be used as the final forecast.

Next, we analyse the participants' adjustment behaviour with regards to the direction of the judgmental adjustment. Depending on the information provided to the participants regarding who is running the promotion in the next period (our company or our main competitor), the participants were limited by the system to perform positive or negative adjustments respectively. Now, we are interested to see whether or not the size of these adjustments were the same for the two cases of promotions (own or competitor's) and if the direction affects the behavioural change. The first panel of Fig. 5.4 presents for each participant and each sub-task (212 points) the difference in the average size of positive versus negative adjustments. We note a skewed distribution, with some participants performing positive adjustments that were on average more than double the size of the negative ones. Overall, in 63% of the sub-tasks, participants were performing larger positive than negative adjustments, with the mean difference of positive over negative adjustments being 26.1%. This tendency to make larger adjustments to the positive direction is well-linked with the positivism bias reported in the literature (Fildes et al., 2009). Does this bias affect the behavioural change between the two sub-tasks? Panels 2 and 3 of Fig. 5.4 offer some insights to this question. We observe that the behavioural change is similar for both positive (panel 2) and negative (panel 3) adjustments, with 74% and 78% of the cases showing an increase on the average adjustment when the participants were informed that their forecasts will be further adjusted (sub-task 2). The mean behavioural change was 47.6% and 46.1% respectively.

A similar analysis is performed for the level of noise (low or high) in the series presented to the participants in our experiment. Similar to the first panel of Fig. 5.4, the first panel of Fig. 5.5 presents the effects of the noise in the series on the size of

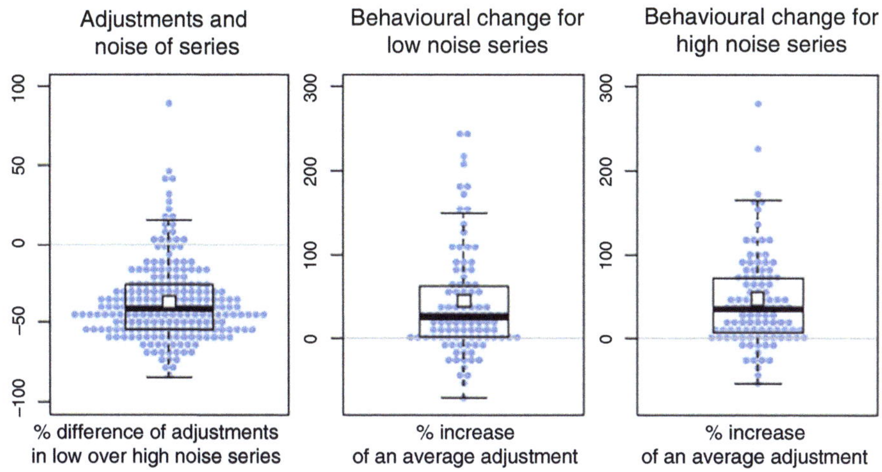

Fig. 5.5 The effect of the noise in the series

the judgmental adjustment for each participant and sub-task. We observe that the participants were performing much larger adjustments for the series with high noise. In fact, that was true in 92% of the cases, with the mean difference of adjustment in high over low noise series being 36.6%. Once again, this is supported from the literature: forecasters are liable in confusing the signal with the noise in the series (Eggleton, 1982); O'Connor et al., 1993). Does this affect the behavioural change regarding how the final forecast is calculated between the two sub-tasks? In short, no. The increase of an average adjustment from sub-task 1 to sub-task 2 is similar for the low and the high noise series (see panels 2 and 3 of Fig. 5.5). In 80% of the cases for the series with low noise (83% for the high noise series), we observe an increase on the average adjustment. The average increase is 45.5% and 47.7% for the low and high noise series respectively.

Next, we attempt to find associations between the participants' responses to the end-of-experiment questionnaire and their overall behaviour regarding judgmental adjustments. In the first three questions, we asked the participants their views on the performance of the three alternative options regarding the final forecast: the statistical forecast that was provided to them, their judgmentally-adjusted expert forecast, and the combined forecast. Their replies are summarised in 6. Interestingly, there is little to distinguish between the three options. If anything, it seems that the participants have less trust in the performance of their own forecast compared to the performance of the combined forecast. In an attempt to make sense of this result, we calculated the differences in their ranked responses (i.e., a participant's response regarding the performance of their forecast minus the statistical or the combined forecast) and tried to find associations with the size of the adjustments made or the degree of the behavioural change. All associations were weak and statistically insignificant (Fig. 5.6).

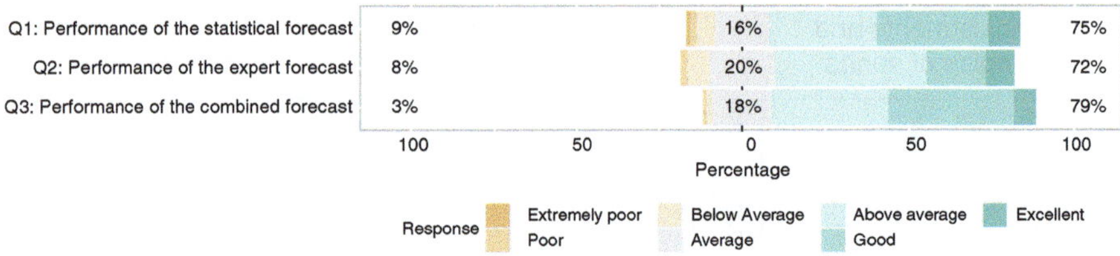

Fig. 5.6 The perceived performance of the statistical, expert, and combined forecast

In the next two questions, we asked the participants regarding their confidence that their judgmentally-adjusted forecast will be more accurate than the provided statistical forecast or the combined forecast. In contrast to the previous insights, the average participant reported a confidence level of 58% in that their forecast will outperform the provided statistical one, and 54% that will outperform the combination of the statistical and judgmental forecast. The levels of confidence further increased to 61% and 56% respectively when we focused on the participants with relatively large behavioural changes (at least + 50%; see also Fig. 5.3).

Another question focused on explicitly linking the bevahioural change of the participants with their reported behaviour. We asked them to report their agreement on the statement that in sub-task 2 they made larger adjustments than in the sub-task 1 as a result of the provided information that their final forecast in sub-task 2 would be further adjusted. While more participants agreed with that statement (45% versus 35%; with the remaining opting for the neutral response), the actual behavioural change in making larger judgmental adjustments in sub-task 2 is much more frequent (76%). This suggests that the participants either do not want to admit the change in their behaviour or are unaware of it.

Finally, we found only weak associations regarding the degree of behavioural change and the participants' reported "optimal" weight for their (judgmentally-adjusted) expert forecast to be used towards a weighted-combination between the statistical and the expert forecast. Moreover, the average (across participants) optimal weights for such a combined forecast were close to 50–50% with a slight preference for the statistical forecast, which was an unexpected result (50.75% versus 49.25%). When we focused only on the participants with positive values in behavioural change (i.e., larger adjustments in sub-task 2 compared to sub-task 1), then the average optimal weights were 51.9% and 48.1% in favour of the expert forecast.

4 Discussion

In the present study, we decided to use scientific—controlled to some extent—experiments in order to shed some light into the discussion about the potential behavioural change of experts. The context and task involved was a forecasting one and, in specific, whether (or to what extend if not at all) the behaviour of experts changes if they have a priori knowledge that their judgmental adjustments will be further adjusted. We would also like to take things one step and obtain some empirical evidence as to if a change in the behaviour is observed, whether or not this change is rational (where rationality obviously is hard to define anyway but is always an interesting aspect of the analysis).

To that end we observe that the experts' adjustments increase in size once they are informed that a subsequent adjustment will take place; in our experiment one that halves the expert adjustment. Participants in our experiment clearly try to mitigate for that further adjustment and retain the ownership of the final forecasts. But the former part of this last statement is the empirical fact, while the later part is merely a speculation, a guess. So, we do know that the experts in our experiment do change their behaviour if they have the knowledge that their forecasts/adjustment will be amended, but do we know that they do this because they want to retain the ownership of the forecast/adjustment? Not really; we can only speculate on the latter.

Even if we run a survey asking specifically this question, the fact that we ask the question in a specific way biases the process anyway: the very selection of words and order does matter. This is no news. It is a standard limitation that comes with any behavioural study. One can never know exactly 'why' she is finding 'what' she is finding. So we can definitely speculate, we can explain to some extent, and we should accept the empirical finding with the caveat of the small samples we are using in this instance. However, that should not be a major issue as we are after insights in this study rather than statistical significance.

We attempted to ask a direct question (in order to limit our speculation) and try get a direct scoring of a specific statement:

> I made larger adjustments (to the given statistical forecast) for those cases where my judgmentally-adjusted forecasts would be used as final forecasts (as compared to the cases where 50-50 combination forecasts would be used as final forecasts).

Unfortunately, this created even more room for speculation rather than reducing the ambiguity of our results. While 45% of the participants agreed with that statement, the actual behavioural change in making larger judgmental adjustments when they know that their adjustment will be halved was more frequent (76%). This suggests that the participants either do not want to admit the change in their behaviour or are unaware of it, which of course is our interpretation and speculation on this finding and could be the root for future studies in this topic. Unfortunately, and to the best of our knowledge, is very hard to depict why experts say one thing and do another.

There are clear implications from our study for theory, practice, and implementation, even when considering the small samples and the controlled environment:

- On the theoretical end, the clear implication is for decision theory and behavioural operations management/behavioural operations research. We do note—even with limited empirical evidence—that a behavioural change does takes place, and that is not prescribed by the respective theory so far to the best of our knowledge, although it may be anticipated—and as such we do contribute to this body of theory.

- On the practical end, the clear implication is for practitioners that in the light of information that their forecasts/adjustments will be manipulated, counter measures should be taken in advance in order to account for the forthcoming act upon their input.
- On the implementation end, the clear implication for software houses and respective vendors is to include facilities in forecasting support systems that would enable counter balancing of amplified adjustments from experts due to prior knowledge (from them) that their input would be further manipulated. This could be a dynamic feature where the level of counterbalance is calculated gradually or a set feature where, given the final adjustment formula, the respective counter adjustment is performed. Understanding experts' behaviors and reactions and automatically adjusting their judgmental interventions will offer forecasting and foresight support systems designers an invaluable tool in their arsenal.

5 Conclusions

Our research corroborates to a vast literature from psychology and decision sciences—more recently referred to as Behavioural Operations Research/ Behavioural Operations Management (BOR/BOM) on judgmental interventions and behavioural change. Previous empirical evidence studies have amalgamated the view that simple combinations of statistical forecasts and judgmentally-revised forecasts (expert forecasts) are beneficial, and can lead to increased forecasting accuracy. Nevertheless, it can be argued that if the expert is a-priori aware of this heuristic—or assuming that at some point in the process will find out that such a manipulation takes place—that would eventually lead to a long-term change of experts'/forecasters' behaviour. The degree of the change is, however, a typical behaviour reaction and quite hard to estimate in advance.

In this chapter, through a controlled experiment, we tried to have a first go at this phenomenon and the respective estimation of such a behaviour change. The adjustments of experts, under two treatments, with and without a 50–50% combination of system-expert forecast, were recorded and analysed. We believe that we provide sufficient empirical evidence in that the experts' adjustments increase in size once they are informed that a subsequent adjustment will take place. Our interpretation of this main finding (as elaborated in the previous section) assumes that the experts/ participants in our experiment do mitigate for our manipulation and respective adjustment, in an attempt to retain the ownership of the final forecasts.

We observed that 24% of experts make, on average, smaller adjustments when they are told that their expert forecasts will be further revised, which is counter-intuitive. However, the majority of the experts (76%) increase their average adjustment with an average increase of 61.6%. It is also worth noting that a good 15% of the experts overreacted and performed adjustments that were, on average, twice as large or more. Across all participants, the mean (median) adjustment is 43.8% (27.7%) which is along our initial expectations. The difference between the median and the mean, with the median being smaller, highlights the existence of outliers, the ones overreacting as per our previous elaboration. Also in accordance with prior literature we also found that the experts were making larger positive than negative adjustments, with the mean difference of positive over negative adjustments being 26.1%. Finally, we also observed that the experts performed much larger adjustments for the series with high noise.

As any other study in behavioural operations management and research, ours too comes with a series of limitations, that nevertheless we believe that do not limit the insights from our empirical findings. First of all, our study is an experiment, but one done is a carefully controlled environment, rather than in a classroom, where many external factors (starting from many trivial ones as in the day of the week, weather, semester, etc.) can affect the course of an experiment. Secondly, we are not certain about the level of expertise of our participants and as if these were novices, semi-experts or full experts and that can affect the outcome of an experiment (for example, Nikolopoulos et al., 2015). For a study with our experimental setup, out participants would be classified most commonly as semi-experts. Finally, in our experiment we asked the participants to take the role of the demand planner of Company A, and to judgmentally adjust statistical forecasts produced by SES for a following period 37 in the light of a reliable piece of information that will affect the demand and is not taken into account by SES. This information relates to a heavy promotional activity of the type "Buy one, get one free" (BOGOF) that is applied by either our company (Company A) or our main competitor (Company B). However, we are not sure the extent of the knowledge of our participants neither in the role of a demand planner nor in the type of the promotional activity (or the forecasting methods per se), and this can bring in some biases and limitations too.

We strongly believe that we can generalise our findings in other contexts and in other types of forecasting events and promotional activities. We find our results intuitively appealing and easy to be accepted from practitioners and, as such, we have no reason to believe that our insights would not apply in broader setting with the caveat of the level of expertise that was elaborated in the previous paragraph. Regardless, as in any other behavioural study, corroboration is key. We would like to see other researchers extending on our study and performing more experiments towards accounting for diverse levels of participants' expertise, diverse forecasting tasks, and diverse pieces of information. It would be also interesting to explore the impact of the level of forecasting and domain expertise as well as the very nature of the interface of the forecasting support systems used in the process.

References

Armstrong, J. S. (1985). *Long-range forecasting: From crystal ball to computer*. Wiley.

Blattberg, R. C., & Hoch, S. J. (1990). Database models and managerial intuition: 50% model + 50% manager. *Management Science, 36*(8), 887–899.

Carbone, R., Andersen, A., Corriveau, Y., & Corson, P. P. (1983). Comparing for different time series methods the value of technical expertise individualized analysis, and judgmental adjustment. *Management Science, 29*(5), 559–566.
De Baets, S., & Harvey, N. (2020). Using judgment to select and adjust forecasts from statistical models. *European Journal of Operational Research, 284*(3), 882–895.
Donihue, M. R. (1993). Evaluating the role judgment plays in forecast accuracy. *Journal of Forecasting, 12*(2), 81–92.
Eggleton, I. R. C. (1982). Intuitive time-series extrapolation. *Journal of Accounting Research, 20*(1), 68–102.
Fildes, R., Goodwin, P., Lawrence, M., & Nikolopoulos, K. (2009). Effective forecasting and judgmental adjustments: An empirical evaluation and strategies for improvement in supply-chain planning. *International Journal of Forecasting, 25*(1), 3–23.
Fildes, R., Goodwin, P., & Önkal, D. (2019). Use and misuse of information in supply chain forecasting of promotion effects. *International Journal of Forecasting, 35*(1), 144–156.
Franses, P. H., & Legerstee, R. (2009). Properties of expert adjustments on model-based SKU-level forecasts. *International Journal of Forecasting, 25*(1), 35–47.
Franses, P. H., & Legerstee, R. (2010). Do experts' adjustments on model-based SKU-level forecasts improve forecast quality? *Journal of Forecasting, 29*(3), 331–340.
Franses, P. H., & Legerstee, R. (2011a). Combining SKU-level sales forecasts from models and experts. *Expert Systems with Applications, 38*(3), 2365–2370.
Franses, P. H., & Legerstee, R. (2011b). Experts' adjustment to model-based SKU-level forecasts: Does the forecast horizon matter? *The Journal of the Operational Research Society, 62*(3), 537–543.
Goodwin, P., & Fildes, R. (1999). Judgmental forecasts of time series affected by special events: Does providing a statistical forecast improve accuracy? *Journal of Behavioural Decision Making, 12*(1), 37–53.
Han, W., Wang, X., Petropoulos, F., & Wang, J. (2019). Brain imaging and forecasting: Insights from judgmental model selection. *Omega, 87*, 1–9.
Lawrence, M., Goodwin, P., O'Connor, M., & Önkal, D. (2006). Judgmental forecasting: A review of progress over the last 25 years. *International Journal of Forecasting, 22*(3), 493–518.
Legerstee, R., & Franses, P. H. (2014). Do experts' SKU forecasts improve after feedback? *Journal of Forecasting, 33*(1), 69–79.
Mathews, B. P., & Diamantopoulos, A. (1986). Managerial intervention in forecasting. An empirical investigation of forecast manipulation. *International Journal of Research in Marketing, 3*(1), 3–10.
Mathews, B. P., & Diamantopoulos, A. (1989). Judgemental revision of sales forecasts: A longitudinal extension. *Journal of Forecasting*.
Mathews, B. P., & Diamantopoulos, A. (1990). Judgemental revision of sales forecasts: Effectiveness of forecast selection. *Journal of Forecasting, 9*(4), 407–415.
McNees, S. K. (1990). The role of judgment in macroeconomic forecasting accuracy. *International Journal of Forecasting, 6*(3), 287–299.
Nikolopoulos, K., Litsa, A., Petropoulos, F., Bougioukos, V., & Khammash, M. (2015). Relative performance of methods for forecasting special events. *Journal of Business Research, 68*(8), 1785–1791.
O'Connor, M., Remus, W., & Griggs, K. (1993). Judgemental forecasting in times of change. *International Journal of Forecasting, 9*(2), 163–172.
Petropoulos, F., Fildes, R., & Goodwin, P. (2016). Do 'big losses' in judgmental adjustments to statistical forecasts affect experts' behaviour? *European Journal of Operational Research, 249*(3), 842–852.
Petropoulos, F., Kourentzes, N., Nikolopoulos, K., & Siemsen, E. (2018). Judgmental selection of forecasting models. *Journal of Operations Management, 60*, 34–46.
Petropoulos, F., Apiletti, D., Assimakopoulos, V., Babai, M. Z., Barrow, D. K., Ben Taieb, S., Bergmeir, C., Bessa, R. J., Bijak, J., Boylan, J. E., Browell, J., Carnevale, C., Castle, J. L.,

Cirillo, P., Clements, M. P., Cordeiro, C., Oliveira, F. L. C., De Baets, S., Dokumentov, A., Ellison, J., Fiszeder, P., Franses, P. H., Frazier, D. T., Gilliland, M., Sinan Gönül, M., Goodwin, P., Grossi, L., Grushka-Cockayne, Y., Guidolin, M., Guidolin, M., Gunter, U., Guo, X., Guseo, R., Harvey, N., Hendry, D. F., Hollyman, R., Januschowski, T., Jeon, J., Jose, V. R. R., Kang, Y., Koehler, A. B., Kolassa, S., Kourentzes, N., Leva, S., Li, F., Litsiou, K., Makridakis, S., Martin, G. M., Martinez, A. B., Meeran, S., Modis, T., Nikolopoulos, K., Önkal, D., Paccagnini, A., Panagiotelis, A., Panapakidis, I., Pavia, J. M., Pedio, M., Pedregal, D. J., Pinson, P., Ramos, P., Rapach, D. E., James Reade, J., Rostami-Tabar, B., Rubaszek, M., Sermpinis, G., Shang, H. L., Spiliotis, E., et al. (2021). Forecasting: Theory and practice. *arXiv* https://doi.org/10.48550/arXiv2012.03854

Smith, G., Levere, M., & Kurtzman, R. (2009). Poker player behavior after big wins and big losses. *Management Science, 55*(9), 1547–1555.

Syntetos, A. A., Nikolopoulos, K., Boylan, J. E., Fildes, R., & Goodwin, P. (2009). The effects of integrating management judgement into intermittent demand forecasts. *International Journal of Production Economics, 118*(1), 72–81.

Syntetos, A. A., Nikolopoulos, K., & Boylan, J. E. (2010). Judging the judges through accuracy-implication metrics: The case of inventory forecasting. *International Journal of Forecasting, 26*(1), 134–143.

Trapero, J. R., Pedregal, D. J., Fildes, R., & Kourentzes, N. (2013). Analysis of judgmental adjustments in the presence of promotions. *International Journal of Forecasting, 29*(2), 234–243.

Vere, D. T., & Griffith, G. R. (1995). Modifying quantitative forecasts of livestock production using expert judgments: An application to the australian lamb industry. *Journal of Forecasting, 14*(5), 453–464.

Part II
Judgment in Collective Forecasting

Chapter 6
Talent Spotting in Crowd Prediction

Pavel Atanasov and Mark Himmelstein

Keywords Forecasting · Prediction · Crowdsourcing · Skill assessment

1 Introduction

Since Francis Galton's classic demonstration (1907), wisdom-of-crowds research has largely focused on methods for eliciting and aggregating estimates, while treating the skill of individual forecasters as exogenous. For example, Mannes et al. (2014) define the wisdom-of-crowds effect as the tendency for the average estimate to outperform the average individual forecaster. Davis-Stober et al. (2014) generalize this definition to include any linear combination of estimates and randomly chosen forecaster as the comparison point.

In this chapter, we summarize a complementary line of research that has thrived over the last decade—the search for skilled forecasters. The general idea is that accounting for individual forecasting skill is valuable in maximizing crowd accuracy. Research on superforecasting at the Good Judgment Project (GJP, Mellers et al., 2015a, b) has demonstrated identifying and cultivating highly skilled forecasters is a crucial lever in maximizing crowd wisdom. More recent work has shown that the skill of the forecasters making up the crowd may be more important to aggregate accuracy than the choice of elicitation or aggregation methods (Atanasov et al., 2022b). Many aggregation methods are flexible enough to incorporate performance weights (Atanasov et al., 2017; Hanea et al., 2021).

These superforecasters were famously identified using a single measure: performance rank at the end of each forecasting season, which generally lasted approximately 9 months and featured over 100 questions. Performance was measured using

P. Atanasov (✉)
Pytho LLC, Brooklyn, NY, USA
e-mail: pavel@pytho.io

M. Himmelstein
Department of Psychology, Fordham University, Bronx, NY, USA

M. Seifert (ed.), *Judgment in Predictive Analytics*, International Series in Operations Research & Management Science 343,
https://doi.org/10.1007/978-3-031-30085-1_6

Brier scores in prediction polls and market earnings in prediction markets. Given sufficient resources, a tournament designer—one who is tasked with collecting and scoring predictions—would be well-advised to follow this strategy: start with thousands of forecasters, pose 100 or more questions that resolve in the subsequent 6–9 months, and after all outcomes are known, pick the top 2% performers.

However, not all settings lend themselves to such an extensive pursuit of forecasting talent. For example, a forecasting tournament may feature questions that do not resolve for many years. Alternatively, tournament designers may lack the resources to wait several months or pose 100+ imminently resolvable questions to provide sufficiently reliable performance scores. In this chapter, we describe methods for identifying skilled forecasters in such time- or resource-limited environments.

We seek to make three main contributions to the research literature. First, we take stock of skill identification research, with a focus on ideas from the last decade, and propose an organizing schema for the various skill predictor measures.[1] It consists of five categories: accuracy-related, intersubjective, behavioral, dispositional, and expertise-based measures. Second, we provide a quantitative summary of effect sizes in pre-existing studies, expressed as correlation coefficients between predictor measures and performance outcome measures. Third, we address measurement challenges inherent to cross-study comparisons by conducting a new analysis of GJP data in which we recreate a subset of the predictor measures across the five categories, following the analytical framework originally developed in Atanasov et al. (2020b).

This chapter does not aim to develop a unified theory of what makes a great forecaster. Our main goal is descriptive: to summarize existing ideas, measures and evidence. However, we hope that this review will prove helpful developing deeper theoretical understanding of drivers of forecasting skill. The relationships we examine are correlational, not causal, so the recommendations stemming from this work are primarily relevant to tournament organizers—who collect and organize forecasts—and less relevant to forecasters working to improve their craft.

This chapter relates to several research strands within the judgment and decision making literature. Our outcome of interest, predictive skill, is conceptually related to decision making competence (Bruine de Bruin et al., 2007) and may share common correlates. Separately, research based on the Brunswikian lens model has examined the accuracy of expert predictive judgments, using regression-style models as benchmarks (Blattberg & Hoch, 1990; Stewart et al., 1997; Seifert et al., 2015). Our focus here is on relative forecaster skill. The ideas tested here may serve as a starting point for the task of identifying forecasters who outperform or add value to model-based estimates.

At a practical level, improved understanding of talent spotting measures can help tournament designers in at least three ways. First, skill measures can be used in

[1] We refer to measures correlating with skill as predictors or correlates. To avoid confusion, we refer to individuals engaged in forecasting tasks as forecasters.

forecast weighting aggregation schemas, as discussed in this volume (Collins et al., 2022). Second, skill measures allow tournament designers place forecasters into smaller, selective, high-performing crowds (Mannes et al., 2014; Goldstein et al., 2014), such as superforecasting teams (Tetlock & Gardner 2016). Lastly, accurate forecasters tend to also excel at other quantitative and logical challenges (Mellers et al., 2017), so talent identified in forecasting tasks may be beneficially deployed in other analytically challenging contexts.

1.1 Definition of Skill

When we use the term *talent spotting,* we do not mean to suggest that excelling in forecasting tournaments is mostly a matter of inborn talent. Rather, we use talent as a synonym of *consistently demonstrated forecasting skill*: the tendency to display consistently strong relative performance on forecasting tasks. Consistency in this context means that that performance is assessed across many questions, which reduces the importance of luck. In prediction polls, the method that is of primary focus here, accuracy, is usually measured using proper scoring rules such as the Brier scoring rule (Brier, 1950). Scores are usually modified to account for the varying difficulty of questions. The modifications range in complexity, from z-score normalization to item-response theory (IRT) modeling. In prediction markets, the core performance measure is market earnings. Performance is generally compared against peers—forecasters recruited in similar ways and assigned to equivalent experimental conditions.

Put simply, this chapter is focused on answering the question: "Who is good at prediction?" We note the subtle difference between this performance-focused question, and related research that focuses on the question of *expertise* in prediction or foresight (Mauksch et al., 2020). We treat expertise and forecasting skill as two conceptually and empirically distinct concepts (Burgman et al., 2011). Namely, we use expertise measures as correlates of forecasting skill, not as outcome measures. Experts can be identified by reviewing resumes, while skilled forecasters are those who demonstrate strong performance in settings where accuracy is rigorously tracked, such as forecasting tournaments. The extent to which expertise relates to forecasting skill is an empirical question.

1.2 Five Categories of Skill Correlates

We classify all measures in the literature in five categories: accuracy-related, intersubjective, behavioral, dispositional, and expertise-related. The order of the categories presented here roughly corresponds to the strength of their relationship to forecasting skill.

Accuracy-related measures all measures that rely on ground-truth resolutions (e.g., determination of whether or not a predicted event actually occurred). Accuracy is calculated based on the distance between a forecaster's estimates and ground-truth resolutions. Such measures fall in the general rubric of correspondence. The simplest versions rely directly on proper scoring rules (Gneiting & Raftery, 2007) such as quadratic Brier scores (Brier, 1950), and logarithmic scores (Good, 1952). Cross-forecaster comparison involves skill-based variants of proper scores, or standardized proper scores (e.g., Mellers et al., 2014). Item-response theory (IRT) models offer a more sophisticated approach, yielding estimates of two parameters: question discrimination ability and forecaster skill (Bo et al., 2017; Himmelstein et al., 2021; Merkle et al., 2016). Calibration and discrimination examine different facets of accuracy and can be obtained using Brier score decomposition (Murphy & Winkler, 1987).

Other measures in this category do not focus on accuracy directly but still rely on resolution data. Augenblick and Rabin (2021) develop a measure of Bayesian updating that gauges if time-series forecasts are characterized by insufficient or excessive volatility. Finally, Budescu and Chen (2015) proposed the use of contribution scores, which measure the extent to which including a person's forecast in an aggregate improves or reduces the aggregate accuracy.

Intersubjective measures do utilize ground-truth resolution data, which makes them suitable for settings in which outcomes are unverifiable or verification is delayed. Instead, individual forecasts verified based on their relation to consensus estimates, i.e., aggregate responses by peers. Measures like *proper proxy scoring rules* (Witkowski et al., 2017) can be applied to simple probability forecasts without requiring additional reports. In the canonical implementation, forecasts are scored by their squared distance from the aggregate (consensus) forecasts obtained from the same crowd, but other proper measures (e.g. logarithmic, spherical) can be used instead of squared distance. Surrogate scoring rules are conceptually similar, but rely on a model of forecast generation (Liu et al., 2020). Similarity-based measures (Kurvers et al., 2019) discretize probabilistic forecasts and calculate proportional agreement.

Other intersubjective measures depend on additional reports submitted by forecasters. For example, a forecaster may be asked to report her estimate of peer responses (e.g., the proportion of peers who select a given option, or the mean response across all peers). These methods include peer prediction (Miller et al., 2005), the Bayesian Truth Serum (Prelec, 2004), the Robust Bayesian Truth Serum (Witkowski & Parkes, 2012), and minimal pivoting (Palley & Soll, 2019). We do not discuss results from these elicitation mechanisms due to the additional report requirement.

Behavioral predictors measure what forecasters do on a forecasting platform. We distinguish among six sub-categories: activity, belief updating, extremity, coherence, rationale properties and question selection. Activity measures indicate how engaged forecasters are with the task, and include the number of forecasts, number of questions predicted, number of logins, time spent on forecasting platform, and news links clicked. It is generally expected that more active forecasters will perform better.

Belief updating measures describe how forecasters update their forecasts over time. Atanasov et al. (2020b) distinguish between three measures: frequency (how often updates occur per question), magnitude (how large is the average update in absolute terms) and confirmation propensity (how often forecaster re-enter their most recent forecast). Rationale-based measures rely on text analysis of the rationales that forecasters write on the platform. We also treat probabilistic extremity (how close a given probability forecast is to 0 or 1) as a behavioral measure of confidence, which we distinguish from self-reported expertise assessments used to assess calibration (see section on expertise below). Probabilistic coherence scores reflect the extent to which actual forecasts differ from logically and probabilistically coherent sets (Fan et al., 2019; Karvetski et al., 2013). Forecasters' choices about which questions to answer and which ones to skip may also be used as skill signals (Bennett & Steyvers, 2022).

Dispositional predictors are generally collected before or after the forecasting tournament, and involve psychometric tests. These aim to measure stable individual differences in fluid intelligence, thinking styles, and personality. Measures that closely relate to fluid intelligence include numeracy, cognitive reflection, matrix completion, number series completion, verbal and analytical aptitude. Thinking styles measures capture concepts such as active open-mindedness, need for cognition, need for closure and fox-hedgehog tendencies. Personality-type measures include the Big 5: conscientiousness, openness to experience, neuroticism, extraversion and agreeableness.

Expertise-based measures relate to the forecasters' knowledge and experience in the subject matter domain. *Demonstrated* knowledge is often measured using multiple-choice tests. Mellers et al. (2015a, henceforth Mellers et al. 2015b) describes knowledge scores as measures of crystallized intelligence. Such tests can also assess meta-knowledge, i.e., calibration, which relate the confidence expressed versus the rate of accurate responses (e.g., see the classical method, Cooke, 1991; Aspinall, 2010). *Biographical* expertise measures can generally be found on forecasters' resumes. These include education level, field of study, professional activities, publications and media appearances. Many of these measures were first described in Tetlock's (2005) book *Expert Political Judgment*. *Self-reported* expertise *measures* gauge how confident forecasters feel about their knowledge on a topic or about their predictive skills more generally.

This chapter consists of two studies. In Study 1, we review existing literature with the goal of providing a broad overview of skill identification measures. We first summarize all ideas in more detail, following the five-category structure, then report the correlation coefficients between prediction measures and accuracy outcomes. We do not provide a formal meta-analysis, mainly because the wide range of research designs makes such estimates tricky to aggregate or compare. Study 2 aims to address this comparability issue and provide more in-depth coverage: we reconstruct a subset of measures across each category and assess their correlations with accuracy, both in-sample and out-of-sample, using the same data and a uniform analytical framework originally developed and described in Atanasov et al. (2020b).

2 Study 1

2.1 Study 1: Methods

2.1.1 Literature Search

Articles of interest featured at least one of two elements: (a) new descriptions of predictive skill identification methods and measures, and (b) new empirical analyses featuring new or previously described measures. Relevant articles were identified in a four-step process.

First, we identified an initial set articles which we had read, reviewed or co-authored over the past decade. Second, we conducted literature searches, featuring search terms in two categories: (a) forecaster, forecasting, prediction, prediction, foresight, tournament; (b) talent, skill, performance, accuracy, earnings. Additional search terms included the award numbers for IARPA's Aggregative Contingent Estimation (ACE) and Hybrid Forecasting Competition (HFC) forecasting tournaments. Third, relevant articles that referenced or were referenced in the set compiled in the first two sets were added. Finally, we added several sources identified by peers. All in all, we identified over 40 individual measures from over 20 manuscripts from the above sources.

2.1.2 Outcome Variables

The core outcome variable in most studies was based on the Brier score (Brier, 1950). Although other proper scoring rules were mentioned in the literature, in practice, nearly all studies featured a version of the quadratic Brier scoring rule. We define one variant, mean standardized mean of daily Brier scores (MSMDB) in detail as it is used in both Study 1 and Study 2.

For any given forecast on a given day, the Brier score is the squared difference between probabilistic forecast and the ground truth, coded as 1 if event in question does occur, and 0 otherwise.

$$DB_{f,q,d} = 2\left(p_{f,q,d} - y_q\right)^2 \tag{6.1}$$

The daily Brier (DB) score for forecaster f, on question q on date d *is* twice the squared distance between the probability forecast p and the ground-truth outcome y (coded as 1 if event occurs, 0 otherwise). This result is a score that ranges from 0 (perfect accuracy) to 2 (reverse clairvoyance), with a 50% binary forecast earning a DB of 0.5. Mean daily Brier score is obtained by averaging Daily Brier scores across days within a question.

$$MDB_{f,q} = \frac{\sum_{d=1}^{D_q} DB_{f,q,d}}{D} \tag{6.2}$$

Standardized MDB (SMDB) is calculated as the difference between the forecaster's Mean Daily Brier score and the Mean of Mean Daily Brier scores across forecasters in a given condition, divided by the standard deviation of MDBs across these forecasters.

$$SMDB_{f,q} = \frac{MDB_{f,q} - \overline{MDB}_q}{SD\left(MDB_{f,q}\right)} \tag{6.3}$$

Accuracy across subsets of questions, s, SMDB scores can be averaged into a Mean SMDB (MSMDB) score. Variants of MSMDB are used in Mellers et al. (2014) and Atanasov et al. (2020b).

$$MSMDB_{f,s} = \frac{\sum_{q=1}^{Q_f} SMDB_{f,q}}{D} \tag{6.4}$$

Normalized Brier Score: $NBS_{f,\ q,\ d}$ refers to the *normalized accuracy* of the forecast made by a forecaster for a given question on a given date. It is a transformation of the Brier Score (or, put another way, a linking function) to make $\epsilon_{f,\ q,\ d}$ approximately normally distributed when used as an outcome measure in models which rely the assumption of normally distributed residuals. This variant is used in Himmelstein et al. (2021) and Merkle et al. (2016), as well as in the calculation of IRT scores in Study 2. In the original formulation, higher scores denote better accuracy. We reverse-code NBS to maintain consistency, so that all accuracy measures denote worse accuracy for higher values.[2]

$$\boldsymbol{NB_{f,q,d} = probit\left(1 - \sqrt{\frac{DB_{f,q,d}}{2}}\right)} \tag{6.5}$$

Delta Brier Score is based on the difference between the Brier score of the consensus estimate on a given question at a given time, and a forecaster's individual estimate for this question at this time. This version was used in Karvetski et al. (2021). It is reverse-coded in the current analysis, so that higher values denote worse accuracy, consistent with other outcome measures in this chapter.

[2]Normalization doesn't account for question difficulties on its own, just transforms the distribution. So, when used as criterion variables, normalized scores are then standardized: $SMNB_{f,q} = \frac{MNB_{f,q} - \overline{MNB}_q}{SD\left(MNB_{f,q}\right)}$.

2.1.3 Predictors of Skill

Study 1 provides an overview of measures and features that predict the skill level of forecasters. In order to avoid repetition between Study 1 and Study 2, we describe all measures here. All measures are summarized in Table 6.1 and detailed below, following the five-category structure.

2.1.3.1 Accuracy-Related

Raw Brier Score, Standardized Brier Score, Normalized Brier Score and Delta Brier Score measures also serve as outcome variables and are defined above. See Eqs. (6.1)–(6.5).

Item Response Theory Models: Item Response Theory (IRT) is a psychometric method for estimating latent traits, often latent ability or skill levels based on an objective measure, such as a standardized test or assessment (Embretson & Reise, 2013). Standardized tests, such as the SAT or GRE, are transformed based on an IRT estimation procedure.

The essential logic of IRT is that items on a psychological assessment instrument are not all created equal. Each item can carry unique diagnostic information people who answer it. For example, a very easy math problem may not be able to discriminate well between someone of moderate or high math ability, since either one would be very likely to get the item correct. However, it might be very well suited for discriminating between two different people of relatively low math ability, who would each have some chance of both getting the item right or wrong. Item Response Theory takes advantage of this by simultaneously estimating item-specific parameters that identify an item's unique diagnostic properties, as well as person-specific parameters that represent estimated ability levels.

Recent research has found that item response theory methods can be used to estimate the latent skill of forecasters based on the accuracy of their individual forecasts (Bo et al., 2017; Himmelstein et al., 2021; Merkle et al., 2016). This approach operates under the assumption that different forecasting problems convey different information about forecasters based on how accurate their forecasts are. Like standardized Brier scores, IRT assessment accounts for differences in question difficulty. It also allows questions to vary in discrimination—achieving a good raw score on some questions may be very informative about how good an individual forecaster is, while scores on other questions may not yield strong signal. There are additional features of the IRT approach that are especially appealing for assessing forecasting skill. Most crucially, IRT models are flexible enough to adjust for potential confounders. We describe one version of the model, which accounts for the role of time, in Appendix.

Calibration and *Discrimination* measures are facets of Brier score decomposition originally defined by Murphy and Winkler (1987). We use the operationalization in the context of individual forecasters in GJP, as described in Atanasov et al. (2020b).

Table 6.1 Description of forecasting skill identification measures

Measure	Brief description	Reference (Lead Author Year)
1. Accuracy-related		
Raw Brier Score	Strictly proper scoring rule, the squared distance between probability forecasts and ground-truth outcomes	Brier (1950)[a], Mellers et al. (2014)
Log Score	Log(p), where p is the probability estimate placed on the correct outcome	Good (1952)
Standardized Brier Score	Z-score transformed version of the raw Brier score; adjusts for question difficulty	Mellers et al. (2014), Atanasov et al., (2020b)
Normalized Brier Score	Transformed Brier score, see Appendix.	Himmelstein et al., (2021)
Calibration	Correspondence between predicted probabilities and observed base rates, Brier score decomposition component	Murphy & Winkler, 1987[a], Mellers et al. (2014)
Discrimination	Confidence of correct vs. incorrect forecasts, a.k.a. resolution, sharpness, Brier score decomposition component	Murphy & Winkler (1987)[a], Mellers et al. (2014)
Item Response Theory Models	Model-based estimate of forecaster skill, accounting for differences among questions	Himmelstein et al., (2021), Bo et al., (2017), Merkle et al., (2016)
Delta Brier	Difference in accuracy between an individual forecast and a contemporaneous consensus forecast	Karvetski et al., (2021)
Contribution Score	Difference in aggregate accuracy when a given individual is included vs. excluded from the aggregate	Budescu & Chen (2015)[a]
Excess volatility	Comparison of "measures of movement and uncertainty reduction given a Bayesian's changing beliefs over time." final beliefs correspond to the ground-truth outcome	Augenblick & Rabin (2021)[a]
2. Intersubjective		
Proper Proxy Scoring	Proxy scores are based on the distance between individual and consensus estimates; the latter are assumed to be unbiased	Witkowski et al., (2017)[a]
Surrogate scoring	Uses "noisy ground truth to evaluate quality of elicited information." Unlike proxy scoring, the noisy ground truth variable here is assumed to be biased	Liu et al., 2020[a]
Decision similarity	"The average percentage agreement of [the binarized forecasts of] this individual with all other N − 1 individuals."	Kurvers et al., (2019)[a]
Reciprocal Scoring	Forecasters are asked to estimate the consensus forecast from a large group of accurate forecasters; scored are based on the squared distance from consensus.	Karger et al., (2021)[a]

(continued)

Table 6.1 (continued)

Measure	Brief description	Reference (Lead Author Year)
Bayesian Truth Serum	Forecasters are asked to report both their true beliefs and their estimate of consensus belief. "The expected score [is a] measure of how much endorsing an opinion shifts others' posterior beliefs about the population distribution."	Prelec (2004)[a], Witkowski & Parkes, (2012)[a]
3. Behavioral		
3A. Activity		
Number of forecasts	Total number of forecasts entered	Mellers et al. (2015a)
Questions answered	Total number of questions with at least one forecast	Mellers et al. (2015a)
Number of sessions	Number of times a forecaster initiated a web session by logging in to a forecasting platform	Atanasov et al., (2020b)
Time on platform	Number of sessions multiplied by median session duration	Mellers et al. (2015a), Atanasov et al., (2020b)
News article clicks	Number of times forecasters clicked on unique news articles served in the forecasting platform	Atanasov et al., (2020b)
Training completion	Optional training completion, binary indicator on whether or not a forecaster completed an optional training module	Joseph & Atanasov (2019)
3B. Belief updating		
Update frequency	Number of forecasts per question, log-transformed	Mellers et al. (2015a)[a]
Update magnitude	Average absolute distance between subsequent forecasts, excluding confirmations	Atanasov et al., (2020b)[a]
Update confirmation propensity	Proportion of forecasts that actively confirm preceding forecasts	Atanasov et al., (2020b)[a]
3C. Other features		
Incoherence metric of Euclidean distance between observed responses and the closest coherent responses		Predd et al., (2008)[a], Karvetski et al., (2013), Collins et al., (2021)
Absolute distance from ignorance priors, normalized for number of answer options		Tannenbaum et al., (2017)
Rate of skipping impossible questions, i.e., questions with no correct answers		Bennett & Steyvers, (2022)[a]
3D. Rationale properties		
Average number of words or characters per rationale		Many
Flesch reading score uses features such as word and sentence length to determine the grade level proficiency needed to understand text		Zong et al., (2020)
Structural topic models discover sets of words that tend to occur together		Horowitz et al., (2019)
Integrative complexity, focus on the past, focus on the future, figures of speech		Karvetski et al., (2021), Zong et al., (2020)

(continued)

Table 6.1 (continued)

Measure	Brief description	Reference (Lead Author Year)
4. Dispositional		
4A. Fluid intelligence		
Number series	Correctness of open responses to ten questions involving number series completion	Dieckmann et al., (2017)[a], Himmelstein et al., (2021)
Numeracy	Berlin numerary: Computer adaptive score based on number of correct responses on up to three mathematical problems; others: % correct answers on mathematical problems	Cokely et al., (2012),[a] Lipkus et al., (2001)[a], Peters et al., (2006)[a]
Cognitive reflection	Original test included three mathematical questions, for which the obvious answers are incorrect. Extensions featured extra questions following this model.	Frederick (2005)[a], Toplak et al., (2014), Mellers et al. (2015a)
Inductive pattern recognition	Raven's progressive matrices test involves choosing one of six possible images to complete a series	Bors & Stokes (1998), Arthur et al., (1999)
Analytical intelligence	Shipley's analytical intelligence scale of Shipley–2 abstraction test	Shipley et al., (2009)[a]
Fluid intelligence	Equally-weighted combination of standardized scores from the available measures above	Mellers et al. (2015a), Atanasov et al., (2020b)
4B. Thinking styles		
Actively open minded thinking	Self-reported scale assessing the tendency to actively seek disconfirming information and keep an open mind	Baron (2000)[a], Haran et al., (2013), Mellers et al. (2015a)
Need for cognition	A self-reported tendency to structure relevant situations in meaningful, integrated ways	Cacioppo & Petty, (1982)[a]
Hedgehog-Fox	Hedgehogs see the world through a single big idea, while foxes use many perspectives. Multi-item self-report scale	Tetlock (2005)[a]
4C. Personality		
Conscientiousness	Personality trait reflecting the tendency to be organized, responsible, organized, hard-working and goal-directed	Costa & McCrae (2008)
5. Expertise		
5A. Demonstrated knowledge		
Knowledge test accuracy	Number of correct responses on binary or multiple choice questions about politics	Mellers et al. (2015a)[a], Himmelstein et al., (2021)[a]
Knowledge calibration	Difference between a forecaster's average confidence (subjective probability that answers are correct) and the proportion of correct responses on a knowledge test	Mellers et al. (2015a)
Classical method	The score includes calibration and information (discrimination) components, based on forecasters' confidence interval estimates for continuous values	Cooke (1991)[a], Aspinall (2010)

(continued)

Table 6.1 (continued)

Measure	Brief description	Reference (Lead Author Year)
5B. Biographical		
Fame	Frequency of engagement in policy advising, consulting and/or media appearances	Tetlock (2005)
Education	Advanced degree	Tetlock (2005)
h-index	Bibliographic measure of manuscript and citation counts	Benjamin et al., (2017), Atanasov et al., (2020a)
5C. Self-rated		
Self-rated expertise	Self-rating on scale from 1-not at all expert to 5-extremely expert	Mellers et al. (2015a)

Note: [a]Denotes that a source where the measure was first defined or operationalized, rather than just tested

Contribution Scores: There are many ways to define predictive skill. Most ground-truth-based approaches, such as proper scores and IRT assessments, involve assessing the accuracy of individual forecasts. A complementary approach is to ask how much individual forecasters contribute to the overall wisdom of the crowd. Framed differently: if you remove a given forecaster from a given crowd, how much does the crowd gain or lose in predictive accuracy? This is known as the *contribution*-based approach to predictive skill assessment (Budescu & Chen, 2015; Chen et al., 2016).

The contribution-based approach has a property that is appealing and absent from other ground-truth-based approaches to talent evaluation. Consider 20 forecasters who are equally skilled—each is likely to have the same amount of error in each of their forecasts as the those of the others. The first 19 all make an identical forecast for a given problem, while the 20th makes one that is very different. Because the first 19 contain redundant information, that information may be given more emphasis than the information from the 20th just by virtue of repetition, even though the 20th forecaster may have access to signals that are very useful.

This example highlights that even independent forecasters are often relying on redundant information to make their judgments (Broomell & Budescu, 2009; Palley & Soll, 2019). Having a strong consensus may indicate an informative signal, or it may be that the information shared between judges is creating shared bias, and the crowd wisdom would be improved with greater diversity (Davis-Stober et al., 2014). Assessing the contribution of individual analysts to the aggregate crowd judgment is a way to tease out which analysts are providing redundant information, and which are providing more unique information. For more details on calculating contribution scores, see Appendix.

Excess Volatility: This measure was originally defined by Augenblick and Rabin (2021) and is based on a comparison of "measures of movement and uncertainty reduction given a Bayesian's changing beliefs over time." Put simply, the more extreme the first judgment in a time-series, the smaller the subsequent updates should be. Question resolution is treated as a movement to $p = 1$ for the correct answer, and $p = 0$ for all other answer options. Thus, forecasters whose last reported estimates tend to be inaccurate would earn higher volatility scores than those who make the identically sized updates but report more accurate final estimates. Augenblick and Rabin (2021) operationalized the measure in the GJP context and reported that most forecasters consistently exhibited excess volatility, i.e., larger-than-optimal cumulative movements, given the forecasters' starting points. In the original formulation, negative scores denote insufficient volatility while positive scores denote excess volatility. In a sensitivity analysis, we use absolute deviations from optimal volatility, so forecasters straying far from the Bayesian standards in either direction receive higher scores.

2.1.3.2 Intersubjective

Proper Proxy Scoring Rules: Proxy scores are based on the distance between individual forecaster estimates and relevant consensus estimates (Witkowski et al., 2017). The proxy scoring variant utilizing squared distance can be defined as follows:

$$\boldsymbol{DPrS_{f,q,d} = 2\left(p_{f,q,d} - c_{q,d}\right)^2} \tag{6.6}$$

In the current formulation the daily proxy score DPrS for forecaster q on question q on date d is calculated as the squared distance between the probability forecast $p_{f,q,d}$ and a consensus forecast on this question at that time, $c_{q,d}$. The consensus is constructed as the aggregate of individual estimates elicited from the same group of forecasters. In our new analysis (Study 2), we use the aggregation algorithm used in Atanasov et al. (2017) to produce consensus estimates. It features subsetting of the 72% most recent forecasts, higher weights placed on more frequent updaters on a given question, and an extremizing constant of $a = 1.5$. These parameters were not optimized to produce maximally accurate estimates or serve as an optimal basis for proxy score calculation. Thus, the current analysis is likely conservative, as optimized algorithms for constructing consensus estimates may improve the performance of proxy scores.

The original application by Witkowski et al. (2017) is forecast aggregation, and the measure is validated in the GJP context, where forecasters receive feedback in terms of objective Brier scores. The underlying idea is that wisdom-of-crowds consensus estimates are more accurate than most individuals, so forecasters whose independent estimates heave closer to the consensus are likely to be accurate. The main assumption is that consensus estimates are unbiased. The original definition does not pose constraints on the relative timing of individual and consensus forecasts or who makes up the peer group. In our analyses for Study 2, we compare contemporaneous individual and consensus forecasts that are based on the same group (condition) of forecasters. Consensus estimates may be improved by relaxing the contemporaneity constraint, or by sourcing consensus estimates from a group of forecasters with superior track records, e.g., superforecasters. Neither of those modifications were employed here, which again makes our analyses of proxy scores' skill-spotting performance conservative.

A related variation, the expected Brier Score (EBS), is the average Brier score across each possible outcome weighted by the probability the crowd assigns to those outcomes (Himmelstein et al., 2023b). Formally,

$$\boldsymbol{EBS}_{f,q,d} \sum_{e=1}^{E} c_{f,q,d,e} \boldsymbol{DB}_{f,q,d,e}$$

Where $c_{f,\ q,\ d,\ e}$ is the probability assigned by the crowd to event e and $DB_{f,\ q,\ d,\ e}$ is the Brier Score the forecast would obtain if event e is realized as the ground truth.

Surrogate Scoring Rules: Surrogate scoring (Liu et al., 2020) is based on the similar underlying idea that consensus estimates are useful as departure points. Surrogate scoring models, however, build in the assumption that consensus forecasts are biased, and use "noisy ground truth to evaluate quality of elicited information" (p. 854). In practice, this makes surrogate scoring somewhat more complex to apply, as it involves the additional step of modeling the bias of the consensus crowd.

Decision Similarity: In a probabilistic elicitation context, decision similarity is assessed as "The average percentage agreement of [the binarized forecasts of] this individual with all other N − 1 individuals." (Kurvers et al., 2019; p. 2). Binarization involves transforming forecasts above 50% to 1 (i.e., 100%), and forecasts below 50% to 0. Binarized forecasts are then compared to the combined estimates made by all other forecasters. The authors use the measure in the context of skill identification and weighting, and do not test the incentive aspects of this schema. The measure is originally developed for non-probability contexts, where forecasters submit simple yes/no reports. The information loss stemming from binarizing forecasts make this measure suboptimal in the context of Study 2.

Reciprocal Scoring: This method incentivized forecasters to estimate the consensus forecast from a group of peers or a separate group of historically-accurate forecasters. Reciprocal scoring was defined and tested by Karger et al. (2021) mainly as an incentive schema, but the authors discussed how reciprocal scores may also serve as a skill identification or weighting measure. Forecasters in the reciprocal scoring only reported one set of estimates, as opposed to separate reports of personal vs. consensus beliefs.

Bayesian Truth Serum: This method was originally developed by Prelec (2004) and also applies to both resolvable and unresolvable questions. Respondents are asked about their own best guess about the true answer, as well as their estimate of the crowd's average answer. Responses are aggregated using the Surprisingly Popular algorithm, which boosts the likelihood of responses that are listed as correct more often than expected, based on the forecasters' consensus estimates. The method has shown to produce superior accuracy on questions where the obvious answer is incorrect. Witkowski and Parkes (2012) develop a version of this mechanism that applies to small crowds without common prior beliefs. Reciprocal scoring and Bayesian Truth Serum are not analyzed in Study 2, as they require additional reports from forecasters that are not available in the full GJP dataset.

2.1.3.3 Behavioral

Behavioral measures are generally sourced in the course of normal forecasting activities. Unlike accuracy-related features, they do not rely on question resolutions, and unlike intersubjective features, they do not involve comparisons of individual and consensus estimates.

Activity: Such measures assess forecaster task engagement and vary based on the specific features available in a forecasting platform. Activities measured on the GJP platforms included: the total number of forecasts submitted over the course of a season, the number of unique questions a forecaster answers (by reporting a probability estimate), the number of unique sessions on the forecasting platform, the per-session average or the total time spent on the forecasting platform, and the number of clicks on news articles served by the platform (Mellers et al., 2015a; Atanasov et al., 2020b).

Belief Updating: At first, updating measures simply measured forecast frequency: the mean number of estimates that a forecaster places on a given question (Mellers et al., 2015b). This measure was used as to determine forecast aggregation weights, where forecasters who submitted a larger number of estimates on a given question received higher aggregation weights (Atanasov et al., 2017). Later treatments distinguished among three separate aspects of belief updating (Atanasov et al., 2020b). First, *update frequency* is defined as the number of unique forecasts per question, which exclude forecast confirmations; if a forecaster submits two identical estimates on the same question in immediate succession, the latter is not counted. Frequency is log-transformed to reduce skew. Second, *update magnitude* is defined as the mean absolute distance among non-confirmatory estimates for a forecaster on a question. Third, *confirmation propensity* is the average proportion of all forecasts that confirm immediately preceding estimates. The original operationalization utilized forecasts on the first answer option on a question. In the current version, forecasts for all answer options are used to calculate update magnitude and confirmation propensity.

Probabilistic Coherence: Karvetski et al. (2013) define an incoherence metric as the "Euclidean distance between observed responses and the closest coherent responses." Examples of incoherent forecasts include ones for which the total probability across all answer options sums up to more or less than 100%, or forecasts that violate the Bayes rule, e.g., feature a combination of conditional and unconditional estimates that cannot be reconciled. In GJP, the platform interface forced forecast values to sum to 100%, and conditional forecasts were not elicited in ways that enable coherence assessment. In settings where forecasting activities do not enable coherence assessments, trait coherence can also be measured separately using an assessment tool specifically designed to identify analysts whose responses tend to be coherent (Ho, 2020; Budescu et al., 2021).

Probabilistic Extremity: The measure is based on the absolute distance from ignorance priors, normalized for the number of answer options. For a binary question, a 50%/50% would yield an extremity score of 0, while a forecast of 0%/100% would yield the highest possible extremity score. In Study 2, extremity is assessed exclusively based on the first estimate submitted by a forecaster on a question. Highly attentive forecasters tend to update toward the extremes as uncertainty is reduced over time, so aggregating extremity across all forecasts would yield a measure that partly reflects belief updating tendencies. Using only the first forecasts is meant to distinguish confidence from belief updating. Tannenbaum et al. (2017) used a closely related measure to assess how forecasters predict on questions that vary in levels of perceived epistemic versus aleatory uncertainty.

Rationale Properties: In addition to eliciting quantitative forecasts, tournament platforms also provide space for text-based rationales where forecasters can explain the reasoning and evidence underlying their predictions. In some conditions, forecasters work as members of a team (Mellers et al., 2014), and sharing rationales can help team members coordinate, challenge one another, and otherwise contribute to team accuracy. Outside of the team context, the incentives for rationales are less clear: forecasters may be motivated to post detailed rationales in order to improve the overall predictions of the crowd, their own reputation in the forecasting community. In certain contexts, such as the Hybrid Forecasting Competition (Morstatter et al., 2019), rationales may also be analyzed to determine payment.

In the GJP independent elicitation condition that is the focus of Study 2, no specific incentives are provided for writing rationales, so it is possible that forecasters in that condition wrote rationales mostly as notes for their own use. Because of the sparsity of rationales in this condition, Study 2 does not feature linguistic rationale properties.

Linguistic properties of rationales vary in complexity. The simplest analyses focus on the rationale length, measured by the number of words or number of characters. More sophisticated natural-language processing techniques (NLP) include bag-of-words and topic modeling, which analyze which words and phrases tend to co-occur together (Horowitz et al., 2019; Zong et al., 2020).[3] NLP techniques have also been used to measure latent psychological factors, such as forecasters' tendency to consider base rates, to focus on the future versus the past, and to engage in complex thought patterns (Karvetski et al., 2021). The practice of gauging complexity of thought by analyzing written text predates automated NLP techniques (e.g., Suedfeld & Tetlock, 1977).

2.1.3.4 Dispositional

In both the ACE and HFC tournaments, a variety of dispositional variables were determined to be valid correlates of forecasting accuracy (ACE: Mellers et al. 2015b; HFC: Himmelstein et al., 2021). A major practical benefit of these results is that dispositional traits are a dimension of talent spotting that can be assessed *a priori*. Unlike ground-truth based and behavioral information, or even

[3]The authors were members of the SAGE team in the Hybrid Forecasting Competition. Linguistic properties of rationales were among the features used in aggregation weighting algorithms. The SAGE team the achieved highest accuracy in 2020, the last season of the tournament.

intersubjective approaches, no information about actual forecasting behavior, let alone ground-truth resolutions, are required to assess dispositional information. As a result, dispositional data can give talent-spotters a head start in picking out likely high performers before any forecasting starts (Himmelstein et al., 2021).

Dispositional information can be assessed with a battery of surveys. These can be broken into two classes, objective and subjective, sometimes referred to as performance-based and self-report-based, or as cognitive and non-cognitive in the psychometric literature (Bandalos, 2018). Objective surveys are akin to tests: they include math and reasoning problems with objectively correct answers, and can be scored based on how many responses were correct. Subjective surveys include self-reports that reflect how people view themselves.

Fluid Intelligence and Related Measures

Numeracy is considered a measure of statistical reasoning ability and risk literacy. It is most often measured with the four-item Berlin Numeracy scale (Cokely et al., 2012). It was employed during both ACE and HFC.

The *Cognitive Reflection Test (CRT)* is a measure of people's ability to reason reflectively in the presence of intuitively appealing, but incorrect answers. The original three-item measure (Frederick, 2005) has been expanded into longer versions, often containing 6–8 items (Baron et al., 2015; Toplak et al., 2014). Two versions of the CRT were administered during ACE (Mellers et al., 2015a). An extended version was administered during two HFC seasons (Himmelstein et al., 2021).

Matrix Reasoning tasks have long been staples of cognitive assessments, such as IQ tests. Matrix reasoning tasks are rooted in visual pattern matching. A series of shapes are displayed which contain a pattern of some sort, with one figure in the series left blank. Participants then must determine which among several choices of shapes would fit the pattern. A classic matrix reasoning scale, Raven's progressive matrices (Bors & Stokes, 1998), was administered during ACE (Mellers et al., 2015a). A newer matrix reasoning task, based on randomly computer-generated problems (Matzen et al., 2010), was administered during HFC Season 1 (Himmelstein et al., 2021).

Number Series is a more recently developed nine item scale, which has received less standalone psychometric validation than some of the others. The number series task is similar in structure to matrix reasoning, except featuring numerical patterns instead of visual patterns (Dieckmann et al., 2017). People are shown a series of numbers which follow a particular pattern, with one number missing, and must determine the missing value. The scale was administered in both HFC seasons (Himmelstein et al., 2021).

Thinking Style Measures are usually based on forecasters' self-report responses about the ways in which they think, behave and process information. The following four measures are included here:

Actively Open-Minded Thinking measures willingness to reason and accept information that is contrary to one's beliefs is necessary to forecast objectively.

See Baron (2000), Stanovich & West (1997). There are several versions of the scale. In the ACE study, a 7-item version was used (Haran et al., 2013, Mellers et al., 2015a).

Foxes and *hedgehogs* are defined as two poles of intellectual heterogeneity in Tetlock (2005). Hedgehogs represent people who tend to be highly specific in their expertise, while foxes tend to be more eclectic. Tetlock (2005) also found that foxes tended to be less overconfident in their predictions. In ACE, participants were asked a single self-report item about whether they would classify themselves as foxes or hedgehogs (Mellers et al. 2015a), as well as a 10-item scale; the latter is used in Study 2.

Need for Closure, which is conceptually opposed to open mindedness, is considered a hindrance to forecasting talent. People who have a higher need for closure will tend to more easily accept conclusions that conform with their preconceptions, while people with less need for closure will tend to seek counterfactual information. In ACE, an 11-item need for closure scale (Webster & Kruglanski, 1994) was included as a potential correlate of forecasting skill in ACE (Mellers et al., 2015a).

Need for Cognition measures people's willingness to engage in effortful reasoning behavior (Cacioppo & Petty, 1982), and was included in HFC Season 1 (Himmelstein et al., 2021).

Personality: Conscientiousness is a facet in the Big-Five personality inventory (Costa & McCrae, 2008), which describes a person's tendency to be organized, responsible, hard-working and goal-directed. Among the five facets, conscientiousness stands out as the one with most "consistent relations with all job performance criteria for all occupational groups" (Barrick & Mount, 1991). This result motivated us to include conscientiousness in Study 2, despite the lack of studies reporting its relation to predictive accuracy.

2.1.3.5 Expertise-Related

Expertise-related measures focus on forecasters' level of expertise with potential relevance to a given domain. By our definition, an expert is someone who demonstrates knowledge in a topic, has relevant educational or professional experiences, or considers themselves an expert. However, as originally noted by Tetlock (2005), an expert is not necessarily more accurate on a given topic than a non-expert. Again, we do not treat expertise and forecasting skill as synonymous with one another.

Demonstrated Expertise is usually assessed using subject-matter knowledge questionnaires. Mellers et al. (2015a) report on political knowledge questionnaires used in GJP, while Himmelstein et al. (2021) describes similar measures used in HFC. Expertise was measured as the proportion of correct responses, and probabilistic knowledge calibration. For example, a forecaster who places an average confidence of 80% that their answers are correct, but has an actual accuracy rate of 70% is considered as overconfident, while someone with the same 70% accuracy rate who places an average confidence of 60% is considered underconfident. Cooke (1991) originally developed a related so-called "classical method", which involves elicitation of confidence intervals for continuous quantities. Individuals are then scored based on their knowledge calibration and resolution/sharpness. Forecasters in GJP did not provide estimates on continuous quantities, so measures based on the classical method are not used in Study 2.

Biographical measures can be assessed based on information forecasters would put in their resumes, such as educational level, educational specialty, professional experience, publication record and media mentions. Tetlock (2005) also uses several related measures of professional fame, such as the frequency with which individuals engage in media appearances, government or private sector consulting.

Self-Rated Expertise is assessed by asking the forecaster if they consider themselves an expert on the subject matter related to an individual forecasting question or a set of forecasting questions. In GJP, expertise was elicited on a 5-point scale, from 1-Not at all Expert to 5-Extremely Expert.

2.2 *Study 1: Results*

We identified 89 correlation measures across 16 manuscripts. Manuscripts that did not report correlations between predictors and accuracy measures are not discussed here. We organize the results following the five-category structure outlined above. To keep results consistent, we reverse-code outcome measures for which high scores denote better accuracy. After the reversal, higher scores for all measures denote larger errors, and thus lower accuracy. The median absolute correlation coefficient among all measures was $r = 0.23$; among all non-accuracy-related measures, it was $r = 0.20$. Correlation coefficients are tabulated in Table 6.2 and summarized visually in Fig. 6.1.

2.2.1 Accuracy-Related

As expected, accuracy-related (predictor) measures were most closely correlated with other accuracy (outcome) measures. The correlation coefficients exhibited variation across studies. On the high end, Atanasov et al. (2020b) used data and set-up very similar to that in Study 2 and found a correlation of $r = 0.75$ between standardized mean daily Brier scores (MSMDB) on one set of question and the same measure on another. Their sample included n = 515 forecasters across 4 seasons of the Good Judgment Project who answered a mean of 43 questions (SD = 35, Median = 32) per forecaster. The sample was split randomly in two question subsets, yielding a mean of 21.5 questions per forecaster for each subset.

Atanasov et al. (2020b) calculated cross-sample correlation differently: they tracked GJP forecasters across seasons, assessing the correlation of end-of-season leaderboard ranks between Seasons 2 & 3 (S1), and Seasons 3 & 4 (S2). In prediction polls, leaderboard rankings were based on Brier scores. The study also tracked performance rankings in prediction markets, which were based on end-of-season

Table 6.2 Correlations with outcome measures, where higher values denote larger errors, i.e., worse accuracy. Pearson's r coefficients reported, unless otherwise noted

Outcome variable	Predictor	Correlation Coefficient	Source
1. Accuracy-related			
Normalized Brier	Normalized, out-of-sample	0.54	Himmelstein et al., (2021), S2
Normalized Brier	IRT, out-of-sample	0.53	Himmelstein et al., (2021), S2
Normalized Brier	Normalized, out-of-sample	0.36	Himmelstein et al., (2021), S2
Normalized Brier	IRT, out-of-sample	0.30	Himmelstein et al., (2021), S2
Standardized Brier	Standardized Brier, out-of-sample	0.75	Atanasov et al., (2020b)
Brier-score rank	Brier-score rank, out-of-sample	0.37	Atanasov et al., (2022b), S1
Brier-score rank	Brier-score rank, out-of-sample	0.44	Atanasov et al., (2022b), S2
Market earnings rank	Market earnings rank, out-of-sample	0.25	Atanasov et al., (2022b), S1
Market earnings rank	Market earnings rank, out-of-sample	0.18	Atanasov et al., (2022b), S2
Long-term calibration	Short-term calibration	0.53	Tetlock (2005)
Long-term discrimination[a]	Short-term discrimination[a]	0.44	Tetlock (2005)
Calibration, out-of-specialty area	Calibration, in specialty area	0.39	Tetlock (2005)
Discrimination, out of specialty area[a]	Discrimination, in specialty area	0.31	Tetlock (2005)
2. Intersubjective			
Brier	Mean distance proxy score	0.66	
Brier	Mean distance proxy score, out-of-sample	0.44	
Brier	Mean expected Brier score	0.66	
Brier	Mean expected Brier score, out-of-sample	0.43	
Accuracy, balanced[a]	Similarity	−0.56[b]	Kurvers et al., (2019), S1
Accuracy, balanced[a]	Similarity	−0.83[b]	Kurvers et al., (2019), S2
Accuracy, % correct[a]	Similarity	−0.84[b]	Kurvers et al., (2019), S3
Accuracy, % correct[a]	Similarity	−0.84[b]	Kurvers et al., (2019), S4
3. Behavioral			
Standardized Brier	Number of questions attempted	0.07	Mellers et al. (2015a)
Standardized Brier	Deliberation time	−0.30	Mellers et al. (2015a)

(continued)

Table 6.2 (continued)

Outcome variable	Predictor	Correlation Coefficient	Source
Standardized Brier	Optional training completion	−0.21	Joseph & Atanasov (2019)
Standardized Brier	Frequency	−0.32	Atanasov et al., (2020b)
Standardized Brier	Magnitude	0.49	Atanasov et al., (2020b)
Standardized Brier	Confirmation	0.03	Atanasov et al., (2020b)
Standardized Brier	Frequency, out-of-sample	−0.32	Atanasov et al., (2020b)
Standardized Brier	Magnitude, out-of-sample	0.45	Atanasov et al., (2020b)
Standardized Brier	Confirmation, out-of-sample	0.03	Atanasov et al., (2020b)
Standardized Brier	Frequency	−0.49	Mellers et al. (2015a)
Delta Brier[a]	Rationale word count	−0.12	Karvetski et al., (2021), S1
Delta Brier[a]	Rationale comparison class	−0.18	Karvetski et al., (2021), S1
Delta Brier[a]	Rationale integrative complexity	−0.17	Karvetski et al., (2021), S1
Delta Brier[a]	Rationale dialectical complexity	−0.20	Karvetski et al., (2021), S1
Delta Brier[a]	Rationale elaborative complexity	−0.11	Karvetski et al., (2021), S1
Delta Brier[a]	Rationale tentativeness	−0.17	Karvetski et al., (2021), S1
Delta Brier[a]	Rationale focus on the past	−0.13	Karvetski et al., (2021), S1
Delta Brier[a]	Rationale focus on the future	0.09	Karvetski et al., (2021), S1
Delta Brier[a]	Rationale word count	−0.22	Karvetski et al., (2021), S2
Delta Brier[a]	Rationale comparison class IC	−0.32	Karvetski et al., (2021), S2
Delta Brier[a]	Rationale integrative complexity	−0.28	Karvetski et al., (2021), S2
Delta Brier[a]	Rationale dialectical complexity	−0.28	Karvetski et al., (2021), S2
Delta Brier[a]	Rationale elaborative complexity	−0.23	Karvetski et al., (2021), S2
Delta Brier[a]	Rationale source count	−0.25	Karvetski et al., (2021), S2
Delta Brier[a]	Rationale use of quotes	−0.12	Karvetski et al., (2021), S2
Delta Brier[a]	Rationale focus on the future	0.21	Karvetski et al., (2021), S2
Overconfidence	Balance in rationales	−0.37	Tetlock (2005)
Correspondence error	Coherence error	0.68	Tsai & Kirlik, (2012)
Brier	Coherence forecasting scale (9-item)	−0.39	Budescu et al., (2021)
Brier	Coherence forecasting scale (18-item)	−0.50	

(continued)

Table 6.2 (continued)

Outcome variable	Predictor	Correlation Coefficient	Source
Brier	Impossible question criterion	−0.46	Bennett & Steyvers, (2022)
4. Dispositional			
Normalized Brier[a]	Number series	−0.15	Himmelstein et al., (2021), S1
Normalized Brier[a]	Number series	−0.30	Himmelstein et al., (2021), S2
Brier	Number series	−0.34	Budescu et al., (2021)
Normalized Brier[a]	Berlin numeracy	−0.15	Himmelstein et al., (2021), S1
Normalized Brier[a]	Berlin numeracy	−0.28	Himmelstein et al., (2021), S2
Brier	Berlin numeracy	−0.28	Budescu et al., (2021)
Standardized Brier	Numeracy	−0.09	Mellers et al. (2015a)
Brier	Subjective numeracy	−0.20	Budescu et al., (2021)
Normalized Brier[a]	Cognitive reflection test	−0.20	Himmelstein et al., (2021), S1
Brier	Cognitive reflection test	−0.28	Budescu et al., (2021)
Standardized Brier	Cognitive reflection test	−0.15	Mellers et al. (2015a)
Standardized Brier	Extended CRT	−0.14	Mellers et al. (2015a)
Normalized Brier[a]	Matrix reasoning	−0.15	Himmelstein et al., (2021), S1
Standardized Brier	Raven's progressive matrices	−0.23	Mellers et al. (2015a)
Normalized Brier[a]	Actively open minded thinking	−0.15	Himmelstein et al., (2021), S1
Normalized Brier[a]	Need for cognition	−0.15	Himmelstein et al., (2021), S1
Normalized Brier[a]	Actively open minded thinking	0.00	Himmelstein et al., (2021), S1
Standardized Brier	Actively open minded thinking	−0.10	Mellers et al. (2015a)
Standardized Brier	Need for closure	−0.03	Mellers et al. (2015a)
Calibration[a]	Fox-hedgehog	−0.35	Tetlock (2005)
Standardized Brier	Fox-hedgehog	0.09	Mellers et al. (2015a)
5. Expertise-related			
Normalized Brier[a]	Political knowledge % correct	−0.11	Himmelstein et al., (2021), S1
Normalized Brier[a]	Political knowledge, overconfidence)	0.15	Himmelstein et al., (2021), S1
Normalized Brier[a]	Political knowledge, % correct	−0.10	Himmelstein et al., (2021), S2
Normalized Brier[a]	Political knowledge, overconfidence)	0.14	Himmelstein et al., (2021), S2

(continued)

Table 6.2 (continued)

Outcome variable	Predictor	Correlation Coefficient	Source
Standardized Brier	Political knowledge, % correct	−0.18	Mellers et al. (2015a), Measure 1
Standardized Brier	Political knowledge, % correct	−0.20	Mellers et al. (2015a), Measure 2
Normalized Brier[a]	Education	−0.13	Himmelstein et al., (2021), S1
Normalized Brier[a]	Education	−0.03	Himmelstein et al., (2021), S2
Standardized Brier	Co-investigator (Y = 1, N = 0)	0.09	Atanasov et al., (2020a)
Standardized Brier	h-index	0.00	Atanasov et al., (2020a)
Brier score	h-index	−0.15	Benjamin et al., (2017)
Overconfidence	Fame/in-demand	0.33	Tetlock (2005)
Brier score	Confidence	0.20	Benjamin et al., (2017)
Calibration[a]	Self-rated relevance of expertise	−0.09	Tetlock (2005)

Notes: [a]Measures with asterisks were reverse-coded to maintain consistency. Positive correlation coefficients denote that higher levels of predictor variables are associated with larger error, i.e., worse predictive performance. Predictor measures were reverse-coded in cases where predictors and outcome measures were the same (e.g., discrimination)
[b]Denotes a Spearman's rank correlation coefficient

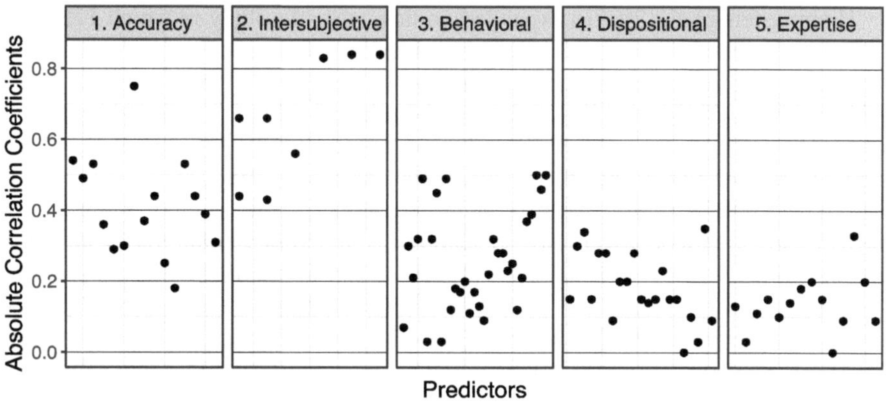

Fig. 6.1 Visual summary of absolute correlation coefficients for the five categories of predictors. Studies vary widely in design, outcome and predictor variables, so the figure aims to provide a general overview, not a detailed, self-sufficient summary of evidence. Average correlations are not reported. Horizontal axis coordinates for datapoints are random

market earnings. In prediction polls, leaderboard rank correlations were $r = 0.37$ between Seasons 2 and 3; $r = 0.44$ between Seasons 3 & 4. In prediction markets, rank correlations were lower: $r = 0.25$ between Seasons 2 and 3; $r = 0.18$ between Seasons 3 & 4. These results suggest that prediction polls' rankings based on Brier scores tend to be more reliable over time than earnings-based rankings in markets.

Himmelstein et al. (2021) used data from the Hybrid Forecasting Competition. Season 1 data was for 326 forecasters who were recruited openly on the web and who made at least 5 forecasts. Correlations were assessed between two sets of 94 questions, and the resulting cross-sample reliability was $r = 0.36$ for normalized Brier scores, slightly lower for IRT. Season 2 (uses data from n = 547 forecasters recruited through Amazon's Mechanical Turk. Cross-sample reliability was $r = 0.54$ for normalized Brier scores, again slightly lower for IRT. Tetlock (2005) reported correlations of calibration and discrimination measures in vs. outside of forecaster specialty area, and on long vs. short-term questions.

2.2.2 Intersubjective

Witkowski et al. (2017) introduced proxy scoring rules, which provide scores for individual forecasters based the distance between a forecaster's measure and the group's consensus. Instead of correlations, the validation takes the approach of pairing forecasters, comparing their proxy scores on one set of questions against their accuracy in a validation set. When the training set consists of 30 questions, the forecaster with the better proxy score achieves better accuracy scores in the validation set approximately in 65% of the comparisons. Thus, there appears to be a significant association between proxy and accuracy scores across question samples. The original analysis did not include correlation coefficients.

Himmelstein et al. (2023b) expanded on this work in a study in which 175 forecasters made predictions about 11 events related to politics, economics, and public health. Each forecaster forecasted each event five times at three-week intervals leading up to event resolution. Across all forecasts and time points forecasters' mean daily proxy scores (MDPrS) and mean expected Brier scores (MEBS) were significantly correlated with their actual MDB scores, r(EBS, MDB) = 0.66 and r(MDPrS, MDB) = 0.66. (Because each forecaster forecasted each question during each wave, Brier scores were not standardized.) To cross validate the results, the authors also split the 11 questions up into all 462 possible combinations of separate samples of 5 and 6 questions. They estimated cross-correlations between MDPrS and MEBS with MDB. The average out-of-sample correlations were r(MEBS, MDB) = 0.43 and r(MDPrS, MDS) = 0.44. The authors note that intersubjective scores were slightly more effective at discriminating poor performers than strong performers.

Liu et al. (2020) show that Surrogate Scoring Rules show slightly higher correlations with Brier and logarithmic scores in-sample than Proxy Scores across 14 datasets. Surrogate Scoring Rule also performed slightly better than Proxy Scoring in selecting top forecasters. However, correlation coefficients were not reported. Kurvers et al. (2019) show that similarity scores strongly related to accuracy in-sample, with Spearman rank correlations ranging between $r_s = 0.56$ to and $r_s = 0.84$, the latter of which was based on GJP data. These data are included with the caveat that Spearman rank and Pearson correlation coefficients are not directly comparable.

2.2.3 Behavioral

Activity measures varied in their correlation with accuracy: Mellers et al. (2015a) reported that answering more questions was associated with worse SMDB accuracy (r = 0.07); while spending more deliberation time on the platform tended to correlate with better accuracy (r = −0.30). Using data from HFC, Joseph and Atanasov (2019) documented that when forecasters were given the option to review training materials, those who chose to complete training performed better than those who did not (Cohen's $d = 0.42$ for the full sample, converted to Pearson $r = -0.21$). Based on additional analyses and experimental data, they argued that the association between training and accuracy is primarily causal (training improves accuracy), and to a lesser extent a matter of self-selection (better forecasters choosing to engage in training).

Belief updating: More frequent updating corresponded to lower (better) MSMDB. The number of forecasts per question (update frequency) was moderately correlated with accuracy both in-sample and out-of-sample (r = −0.32 for both (Atanasov et al., 2020b). Mellers et al. (2015a) documented a stronger association between frequency and accuracy (r = −0.49). While both papers were based on GJP data, they used data from different seasons and slightly different selection criteria.

Update magnitude was among the strongest correlates of MSMDB accuracy measures, both in-sample (r = 0.49) and out-of-sample (r = 0.45), whereas incremental updaters tended to be more accurate than large-step updaters. Confirmation propensity was weakly correlated to accuracy on its own (r = 0.03, both in- and out-of-sample), but improved fit in multiple regression models that included frequency and magnitude. See Atanasov et al. (2020b).

Rationale Text Features: In his study of expert political judgment, Tetlock (2005) showed that forecasters who tended to produce forecast rationales with more balance (e.g., using terms like 'however' and 'on the other hand') tended to be less overconfident (r = 0.36). Similarly, a positive correlation was observed between integratively complex thought protocols and calibration (r = 0.31).

Karvetski et al. (2021) documented the relationship between linguistic properties of forecast rationales and accuracy in more recent forecasting tournaments, including ACE and the global forecasting challenge. In their study, the outcome is a Delta-Brier measure where higher scores denote better accuracy. We reverse-coded this measure, to maintain consistency with other Brier-based measures we review. In our reports, higher scores denote worse accuracy. Even simple word-count measures showed weak but consistent correlations ($r = -0.12$ to $r = -0.22$), as long-rationale writers tended to be more accurate. Frequent mention of reference classes was among the strongest correlates of better accuracy scores ($r = -0.18$ to $r = -0.32$), as were measures of integrative ($r = -0.17$ to $r = -0.23$) and dialectic complexity ($r = -0.20$ to $r = -0.28$). Notably, using words and terms about the past related to better scores ($r = -0.13$), while using more terms about the future correlated with worse accuracy ($r = 0.09$ to $r = 0.21$).

In a separate analysis based on data from Good Judgment Open, Zong et al. (2020) also found that statements about uncertainty, and ones focused on the past relate to better accuracy, while statements focused on the future correspond with worse accuracy. Zong et al. (2020) also documented that absolute sentiment strength (positive or negative) tended to relate to worse accuracy. Usage of cardinal numbers and nouns predicted better accuracy, while frequency of verb usage predicted worse accuracy. In a second study, the authors analyze earnings forecast statements. The results regarding focus on the past and future were replicated; uncertainty terms correlated with worse accuracy. Zong et al., did not report correlation coefficients, only significance tests.

Coherence: Because coherence can be difficult to distill from the experimental designs common to forecasting research (i.e., forecasting tournaments), Ho (2020) developed an independent coherence assessment: the *coherence forecasting scale* (CFS). Himmelstein et al. (2023b) ran a performance test in which 75 forecasters each forecasted 11 questions at five different time points. The longer, 18-item version of the CFS was correlated with MDB, $r = -0.50$, while a shortened 9-item version exhibited slightly lower correlation, $r = -0.39$ (Budescu et al., 2021).

2.2.4 Dispositional

Fluid Intelligence: Out of all psychometric measures, fluid intelligence measures maintained the strongest and most consistent relationship with accuracy. Correlations for individual test measures generally fall in the range of $r = 0.15$ to $r = 0.30$. For example, the correlation of cognitive reflection test (CRT) with standardized Brier is $r = -0.15$ in Mellers et al. (2015a), while Himmelstein et al. (2021) document a correlation with normalized scores of $r = -0.20$, and Budescu et al. (2021) reported a correlation with raw Brier scores of $r = -0.28$. Similar patterns were observed for Number Series, Berlin Numeracy, and Matrix Reasoning. While each measure is conceptually distinct, individual measures appear to load well on a single underlying factor that Mellers et al. (2015a) called fluid intelligence.

Cognitive Styles: In contrast to fluid intelligence measures, which are objective (performance-based), thinking style measures rely on forecasters to self-reflect and report on their own proclivities. Among these measures, actively open-minded thinking was the only one that has been significantly linked with better accuracy: Mellers et al. (2015a) reported a small but significant correlation ($r = -0.10$) with standardized Brier scores; Himmelstein replicated this result in HFC Season 1 ($r = -0.15$), but not in Season 2 ($r = 0.00$). Tetlock (2005) documented a strong correlation between expert political forecasters' fox-hedgehog scores and their calibration scores (r = 0.30), whereas foxier experts tended to be better calibrated. In the context of open forecasting tournaments, however, Mellers et al. (2015a) did not find a significant correlation between fox-hedgehog scores and standardized Brier scores (r = 0.09). Mellers et al., also showed that Need for Closure measure did not corelate to accuracy scores (r = 0.03).

2.2.5 Expertise-Related

Demonstrated Expertise: these measures focus on subject matter knowledge, e.g., political knowledge tests measuring how much forecasters know about topics covered in geopolitical forecasting tournaments. Mellers et al. (2015a) reported on two political knowledge measures collected across two seasons of GJP, where scores are based on the proportion of correct responses to multiple-choice questions. The reported correlations were $r = -0.18$ and $r = -0.20$. Using a similar measure (percentage correct responses to political knowledge questions), Himmelstein et al. (2021) reported somewhat lower correlations with normalized accuracy (r = −0.10 to r = −0.11). In addition to raw accuracy, Himmelstein et al., also computed calibration scores, where perfect calibration denotes that a forecaster's average confidence, expressed as probability (e.g., 60%), equals the proportion of correct responses. These calibration scores significantly correlated with normalized accuracy in both volunteer ($r = 0.14$) and MTurk ($r = 0.15$) samples.

Biographical: One measure of general expertise is education level. Himmelstein et al. (2021) show that in a volunteer sample (HFC Season 1), higher educational attainment significantly predicted normalized accuracy (r = −0.13), but that pattern did not hold among forecasters recruited on Mechanical Turk (r = −0.03, Season 2). In the life-sciences context, Atanasov et al. (2020b) showed trial co-investigators—physicians who worked on a specific trial—were slightly but not significantly less accurate than independent observers ($r = 0.09$) in predicting efficacy outcomes. This study also showed no correlation between bibliographic measure of research impact (h-index) and accuracy ($r = 0.00$). In a similar context Benjamin et al. (2017) reported a low but significant correlation between h-index and brier score accuracy (r = −0.15).

In the context of expert political judgment, Tetlock (2005) showed that experts' degree of fame, as measured by the experts' ratings of "how often they advised policy makers, consulted with government or business, and were solicited by the media for interviews." This fame measure correlated with overconfidence (r = 0.33), whereas more famous experts tended to be more overconfident. A similar correlation with overconfidence (r = 0.26) was observed for an alternative measure of fame, based on the number of media mentions. Finally, a self-rating of media contact frequency (0–never to 7–every week) had a low correlation with calibration (r = −0.12).

Self-Rated Expertise: We did not find published results of confidence or expertise self-ratings in the recent forecasting tournaments literature. In EPJ, Tetlock (2005) collected self-ratings of forecasters' relevance of expertise and found that the correlation between these ratings and calibration was not statistically significant ($r = 0.09$).

2.3 *Study 1 Discussion*

The summary of measures is a reasonable starting point, and the correlation coefficients summarized in Table 6.2 and Fig. 6.1 provide a general sense of how skill identification measures relate to accuracy in the individual studies we summarize. However, these coefficients are not directly comparable across studies. This is partly because individual studies vary in the types of stimuli (forecasting questions) and forecaster samples. It is possible that some questions are better than others at measuring underlying skill. In fact, that possibility is a central motivation for developing IRT models. Separately, correlation coefficients may vary across samples. As a hypothetical example, imagine that a tournament only accepts forecasters with IQ scores above 140. Such a tournament will likely yield low correlations between IQ and accuracy. While it is possible to adjust statistically for such restricted-range effects (e.g., Bland & Altman, 2011), any such adjustments often rely on information that is unavailable to tournament designers, e.g., because some measures are not normed for a given population. Studies also used a variety of different outcome variables for defining accuracy.

Even within a study, where question types and forecaster samples are held constant, estimated correlation coefficients will increase with the *number* of forecasting questions used to estimate the accuracy outcome measure. Increasing the number of questions boosts the outcome measure's reliability, and thus its potential correlation with other measures. For example, a study that assesses accuracy on 100 questions will yield a larger correlation coefficient between, say, IQ and accuracy, relative to a study that uses the same IQ measure, but assesses accuracy based on a subset of 30 forecasting questions. In Study 2, we aim to address these measurement challenges by directly comparing a subset of skill identification measures within the same analytical framework.

3 Study 2

3.1 *Study 2: Methods*

3.1.1 Good Judgment Project Data

All Study 2 analyses are based on the data and analytical framework described in Atanasov et al. (2020b). We provide a brief overview. The ACE tournament featured 481 forecasting questions over four seasons. Questions lasted approximately 3 months on average (Median = 81 days, M = 110 days, SD = 93). Our sample consists of N = 515 participants (forecasters) who made at least two forecasts on at least 10 forecasting questions over one or more seasons. These forecasters worked independently, not as members of forecasting teams. These forecasters made at least one forecast on over one hundred questions on average (M = 113, SD = 73), and made at least two forecasts on forty-three questions (M = 43, SD = 35). Forecasters made an average of two forecasts per question (M = 2.0, SD = 1.6).

Forecasters were scored based on mean daily Brier scores, as described above. Instead of standardizing scores, performance across questions of varying difficulty was equalized through imputation. First, once a forecaster placed an estimate, their forecasts were carried over across days until an update. Second, a forecaster placed an estimate after the first day that a question was open, their scores were imputed as the median daily Brier score across all their peers in a given condition. Third, if a forecaster skipped a question altogether, they received the median overall Brier score for their condition on a question. Overall, Brier scores were displayed on a leaderboard, which featured only peers within a condition. The top 2% forecasters in each condition were invited to work as superforecasters in the following season.

3.1.2 Cross-Validation and Outcome Variable Definition

For each forecaster included in the analysis, we randomly divided all questions they answered in two random question subsets. Let's call then A and B. Each subset consists of approximately half of the questions on which each forecaster placed an estimate. The subsets are randomly split for each forecaster, so that even if two forecasters answered the same set of questions, these questions will most likely be split differently When skill predictors in one subset are correlated with MSMDB (see Eq. 6.4) accuracy in the same subset (A and A, or B and B), we refer these as in-sample. When skill predictors in one question subset are correlated with forecasting accuracy in the other, we refer to these analyses as out-of-sample.

3.1.3 Predictor Selection

We used three main criteria to select predictors for further testing in Study 2: importance, data availability and fidelity. First, with regard to importance, we focused on skill predictors that demonstrated significant associations with accuracy in the literature, in either univariate or multivariate analyses. Second, data for some measures was either unavailable or insufficient. Insufficiency was the reason for excluding linguistic rationale data, as forecasters in our sample made independent forecasts and were not incentivized to write detailed rationales. Data availability also eliminated measures that require additional forecast reports from forecasters, such as Bayesian Truth Serum.

Third, fidelity concerns centered on our ability to reproduce the measures in the context of the current study. These involved contribution scores, as well as intersubjective measures such as surrogate scores. For all of these, our initial examination led to the assessment that the measures would be difficult to reproduce, as small details in decisions about the adaptation of the methods to our analytical framework may have large impacts on results. The fidelity criterion is admittedly subjective, and we see the benefits of including these measures in future research.

3.1.4 Statistical Tests

The core univariate analyses focus on the Person's r correlation coefficient between each predictor and mean standardized Brier scores (MSMDB). The univariate correlation analyses provide a useful starting point for examining the value of various predictors of skill. To provide useful recommendations for skill spotting in forecasting tournaments with limited resources, we need to go a step further. Namely, we need to understand which measures add the most value in the presence of others.

To address this need, we fit a series regularized LASSO regression models. Regularization involves penalizing complexity in model building, so that predictors are only included with non-zero coefficients if the improvements in fit overcome the penalty. The models we report follow ten-fold cross validation; these models prioritize sparsity, and are based on "the value of λ that gives the most regularized model such that the cross-validated error is within one standard error of the minimum" (Hastie et al., 2021, p. 5). All predictors were standardized before entry into the model, to distributions with mean zero and standard deviation of one. We report two runs for each model, one for each subset of questions. We include at least one predictor measure from each category: accuracy-related (out-of-sample mean standardized Brier scores), intersubjective (proxy scores), behavioral (update frequency, update magnitude, forecast extremity), dispositional (fluid intelligence composite scores, AOMT), expertise (knowledge test scores, advanced degree, self-assessed expertise ratings).

3.2 *Study 2: Results*

3.2.1 Correlational Analyses

We first report the univariate correlation coefficients with our core measure of accuracy: Mean Standardized Mean Daily Brier (MSMDB) scores. Results are organized according to the categories describe above. For question-specific measures, we report in-sample and out-of-sample correlations with accuracy. In-sample correlations are calculated for predictors and outcomes (MSMDB) assessed on the same set of questions.

We also report cross-sample reliability of each measure, also in the form of Pearson r coefficient. For the full sample of $n = 515$ forecasters, absolute values above $r = 0.10$ are statistically significant at $\alpha = 0.05$, for a two-tailed test, and those

Table 6.3 Correlations with accuracy (MSMDB) and reliability for predictors in Study 2

Predictor	Correlation with accuracy		Cross-sample reliability
	In-sample	Out-of-sample	
Standardized Brier (SB)	NA	NA	0.74
Debiased Brier	0.96	0.69	0.69
SB, first forecast	0.85	0.62	0.68
SB, last forecast	0.84	0.68	0.83
IRT forecaster score	0.30	0.22	0.89
Calibration	0.51	0.36	0.67
Discrimination	0.71	0.57	0.74
Excess volatility	0.40	0.30	0.85
Proper proxy, all forecasts	0.69	0.60	0.81
Proper proxy, first forecast	0.57	0.52	0.77
Number of questions	0.25	0.25	NA
Update magnitude, abs. Distance	0.51	0.45	0.75
Update frequency	−0.31	−0.32	0.98
Confirmation propensity	0.03	0.03	0.86
Extremity, first forecast	0.19	0.15	0.91
Fluid IQ composite all	NA	0.27	NA
Fluid IQ composite free	NA	0.28	NA
Political knowledge score	NA	−0.10	NA
AOMT	NA	0.10	NA
Fox-hedgehog scale	NA	0.08	NA
Conscientiousness	NA	0.13	NA
Education (advanced degree = 1)	NA	0.01	NA
Expertise self-ratings	0.00	0.00	0.97

above $r = 0.12$, are statistically significant at $\alpha = .01$. To avoid repetition, we do not report p-values. We do report sample size only for predictors that are not available in the full sample. The cross-sampling procedure differs somewhat from Atanasov et al. (2020b), as the results reported here are based on two sampling iterations. Thus, the results reported here occasionally differ slightly (by $r = 0.01$ or less). Correlations with MSMDB and reliability coefficients are reported in Table 6.3. Cross-correlations among predictors are presented in Appendix Table 6.6.

3.2.1.1 Accuracy-Related Measures

The cross-sample reliability of standardized Brier scores (MSMDB) was $r = 0.74$. Notably, standardized Brier scores for the last estimate a forecaster made on a question had higher cross-sample reliability ($r = 0.83$) than those for first forecast ($r = 0.68$). This may be because last-forecast accuracy relates to updating effort, a reliable individual difference. IRT estimates exhibited a low correlation with

MSMDB ($r = 0.30$), suggesting that the two are distinct measures of skill. Notably, IRT estimates demonstrated high cross-sample reliability (r = 0.89).

In terms of Brier score decomposition, discrimination was more strongly correlated with overall MSMDB than calibration. Discrimination and MSMDB were strongly negatively correlated both in-sample (r = −0.71) and out-of-sample (r = −0.57), while calibration error and MSMDB were positively correlated in-sample (r = 0.51) and out-of-sample (r = 0.36), as expected. Discrimination (r = 0.74) and calibration error (r = 0.67) exhibited similar levels of cross-sample reliability. Overall, forecasters' discrimination scores were more strongly related to accuracy than their calibration scores. This result was consistent with a pattern where most forecasters are relatively well-calibrated, and the best forecasters mostly distinguish themselves through superior discrimination.

The Augenblick-Rabin measure of volatility exhibited high cross-sample reliability ($r = 0.85$) but was only moderately correlated with accuracy in-sample ($r = 0.40$), and out-of-sample ($r = 0.30$), where forecasters who produced time-series with more excess volatility tended to be less accurate. The core version of this measure was coded such that the results indicated that a forecaster exhibiting insufficient volatility would be expected to be more accurate than one exhibiting optimal levels of volatility, who in turn would be expected to be more accurate than one producing excessively volatile forecast series. As a sensitivity analysis, we calculated a different version of this measure in which we calculated absolute deviations from optimal volatility levels at the forecaster level, treating errors of excess volatility as equivalent to errors of insufficient volatility. Curiously, this absolute-distance-from-optimal-volatility measure had lower correlations with accuracy, both in-sample ($r = 0.29$), and out-of-sample ($r = 0.23$).

3.2.1.2 Intersubjective Measures

Proper proxy scores calculated based on all forecasts by a person on a question were highly correlated with MSMDB, both in-sample ($r = 0.69$) and out-of-sample ($r = 0.60$), whereas forecasters who tended to place independent estimates closer to the consensus were generally more accurate than those who strayed from the consensus. Even when proxy scores were calculated only based on the first forecast made by a forecaster on a question, the correlations remained very high in-sample ($r = 0.57$) and out-of-sample ($r = 0.52$). First-forecast proxy scores are useful as they are available as soon as a question is posed and several forecasters have placed their initial estimates. Proxy scores exhibited cross-sample reliability similar to that of accuracy: $r = 0.81$ for all-forecast proxy scores, and r = 0.77 for first-forecast proxy scores. Among predictors that could be calculated without the need for ground-truth question resolutions, proper proxy scores yielded the highest out-of-sample correlations with MSMDB. These results highlight the promise of intersubjective measures in talent spotting, especially in settings where forecaster selection decisions must take place before questions resolutions are known.[4]

3.2.1.3 Behavioral Measures

Behavioral measures comprised the widest and most diverse category in the literature. We distinguished four sub-categories: general activity measures, belief updating, probabilistic confidence, linguistic properties of forecast rationales and coherence. The GJP user interface forced within-forecast coherence, so such a measure is not included here. Our analysis follows Atanasov et al. (2020b) in focusing on independently elicited forecasts, where forecasters had no incentive to write detailed rationales, so we do not include linguistic rationale properties.

Among activity measures, the number of questions a forecaster attempted was a predictor of worse performance, correlating with higher standardized Brier scores ($r = 0.25$) both in-sample and out-of-sample. In other words, forecasters who answered more questions registered worse accuracy. Cross-sample reliability of number of questions was not assessed as the sample was constructed by splitting questions into equal categories. In contrast, the number of questions with forecast updates was weakly correlated with better accuracy ($r = -0.10$).

Among belief updating measures, update frequency (the number of non-confirmatory forecasts per question), was the measure with the highest cross-sample reliability ($r = 0.98$), showing that individual differences in how often forecasters update are stable across questions. Update frequency was moderately correlated with accuracy both in-sample and out-of-sample ($r = -0.31$ and $r = -0.32$). Absolute update magnitude between forecast updates was also reliable ($r = 0.75$), and relatively highly correlated with MSMDB both in-sample ($r = 0.51$) and out-of-sample ($r = 0.45$). The positive signs denote that that forecasters who updated in small-step increments tended to register better (lower) accuracy scores. Confirmation propensity was highly reliable ($r = 0.82$), but it had a low correlation with MSMDB: $r = 0.03$ in-sample and $r = 0.03$ out-of-sample.

Probability extremity, the absolute distance between forecasts and the ignorance prior, was assessed based on the first estimate for a forecaster on a question. Extremity was negatively correlated with MSMDB, both in-sample ($r = -0.19$), and out-of-sample ($r = -0.15$), denoting that forecasters who tended to make more extreme (confident) probabilistic estimates tended to have better accuracy scores. Examination of correlations across predictors provides an interpretation for this result: forecasters' who exhibited higher probabilistic confidence tended to earn better discrimination scores ($r = 0.32$), but did not earn significantly worse calibration-error scores ($r = 0.05$).

[4]We do not offer complete coverage of intersubjective measures, including surrogate scores and similarity measures, but given our current results, further empirical investigation seems worthwhile.

3.2.1.4 Dispositional Measures

The strongest psychometric predictor of MSMDB accuracy was a fluid intelligence. Our composite measure was calculated as an equal-weight combination of available standardized scores on Berlin Numeracy, Cognitive Reflection, Raven's Progressive Matrices and Shipley's Analytical Intelligence test (Cronbach's alpha = 0.62). This Fluid IQ measure was negatively correlated with MSMDB (n = 409, $r = -0.27$). The first two measures (Berlin Numeracy and CRT) are freely available, while the last two are commercially available for a fee. A combination of the freely available measures yielded lower in reliability (Cronbach's alpha = 0.43), but similar correlations with MSMDB (n = 408, $r = -0.28$). Thus, it does not appear that the available-for-purchase fluid intelligence measures add value in terms of predicting MSMDB measures of accuracy. Actively open minded thinking (AOMT) measure had moderately low internal reliability (Cronbach's alpha = 0.64). AOMT scores yielded a marginally significant correlation with SMDB (n = 379, $r = -0.10$).

Fox-hedgehog measure scores (Cronbach's alpha = 0.31) were positively but not significantly correlated with SMDB (n = 311, $r = 0.08$). The positive sign indicates that forecasters who rate themselves as hedgehogs tend to have worse accuracy. Conscientiousness measure was high in internal reliability (Cronbach's alpha = 0.81) and scores were positively correlated with SMDB (n = 311, $r = 0.13$), indicating that forecasters who rated themselves as more conscientious tended to perform worse.

3.2.1.5 Expertise Measures

Demonstrated Expertise: Political Knowledge (PK) test scores were reliable across GJP Seasons 1, 2 and 3 (Cronbach's alpha = 0.75). The combined PK test scores were marginally correlated with SMDB (n = 409, r = −0.10). Zooming in on individual tests, PK scores from Season 2 (n = 263, r = −0.24) yielded somewhat higher correlations with SMDB than did PK scores in Season 1 (n = 281, r = −0.18) and Season 3 (n = 323, r = −0.10). The forecaster sample of PK test completers differs across seasons, making cross-season comparisons less direct. For each of Season 1 and Season 2, the overall PK scores (measured as the number of correct responses) were as good or better predictors of accuracy than calibration and discrimination measures based on the same tests.

Biographical: The most general measure of demonstrated expertise was education, coded as binary variable indicating whether the forecaster had obtained an advanced (post-Bachelor) degree or not. This binary indicator was uncorrelated to SMDB (r = 0.01).

Self-Rated Expertise: Self-ratings of forecasters' own relevance of expertise in the question domain were highly reliable across question sets, indicating that some forecasters tended to exhibit consistently higher confidence in their own expertise than others ($r = 0.97$). However, expertise self-ratings were completely uncorrelated with accuracy (SMDB) in-sample (n = 404, $r = 0.00$), and out-of-sample (n = 404, $r = 0.00$).

3.2.2 Multivariate LASSO Models

We constructed a set of LASSO models with only out-of-sample predictors and without any accuracy-related measures. Such models mirror a setting in which forecasters have registered dozens of predictions, but no accuracy data are available yet. LASSO models tend to produce zero coefficients for some predictors, meaning that they do not improve fit enough to overcome the overfitting penalty. We show results for two model runs, A and B. In model run A, predictors are calculated for question subset 1, and then used to predict accuracy on subset 2. In model run B, the direction is reversed. This pattern is equivalent to two-fold cross validation and both sets of coefficients are shown to provide a look into the variability of model fits across questions sub-samples. All predictors are Z-score transformed (i.e., standardized) to enable more direct coefficient comparison. Table 6.4 reports the coefficients of the final model results. Predictors with non-zero coefficients are estimated to improve fit enough to offset the overfitting penalty, which does not necessarily mean that the coefficients would be statistically significant in a conventional ordinary-least squares model.

In Table 6.4, column A, we report the coefficients in the final model specification in order of decreasing absolute value: first-forecast proxy scores (b = 0.137), first-forecast extremity (b = −0.060), update frequency (b = −0.031), update magnitude (b = 0.004), fluid intelligence composite score (b = 0.004). All other coefficients were zero, as the predictors did not improve fit enough to overcome the LASSO penalty. In the converse run B, all coefficients were somewhat similar, except absolute update magnitude which was notably larger (b = 0.040).

Several notable patterns emerged. First, intersubjective proxy scores made the strongest out-of-sample predictor of accuracy in our set. Second, forecast extremity was the second strongest predictor, a result that would not be obvious from examining univariate correlations. Third, belief updating measures remained relevant. Coefficients for frequency were more consistent than those of magnitude. The most likely explanation for this pattern is that proxy scores were highly correlated with update magnitude ($r = 0.66$), but weakly correlated with update frequency ($r = -0.10$). See Appendix Table 6.6.

In Table 6.4, Columns C and D, we report the results models including out-of-sample accuracy (MSMDB) and excess volatility, both of which depend on resolution data, as well as out-of-sample and in-sample measures that do not rely on resolution data. These model specifications mirror a setting in which the tournament has been running for long enough to accuracy data on approximately one half of questions. More informally, these specifications follow an exploratory approach where we err on the side of over-inclusion of predictors, and rely on the regularization to reduce the risk of overfitting.

Table 6.4 LASSO regression models predicting SMDB accuracy measures. Non-zero coefficients do not imply statistical significance in an OLS model

	Out-of-sample, no accuracy data		In and out-of-sample, accuracy data	
Predictors	A	B	C	D
Intercept	−0.072	−0.082	−0.069	−0.080
Out-of-sample				
Accuracy, full MSMDB	NA	NA	0.056	0.081
Accuracy, first forecast Brier	NA	NA	0	0
Accuracy, last forecast Brier	NA	NA	0.059	0.029
IRT skill	NA	NA	0	−0.009
Excess volatility	NA	NA	0	0
Proxy score, first forecast	0.137	0.098	0	0
Update magnitude	0.004	0.040	0	0
Update frequency	−0.031	−0.033	0	−0.016
Confirmation propensity	0	0	0	0
Extremity, first forecast	−0.060	−0.042	0	0
Political knowledge score	0	0	0	0
Fluid IQ score	−0.004	−0.012	0	0
AOMT	0	0	0	0
Education, advanced degree	0	0	0	0
Expertise self-rating	0	0	0	0
In-sample				
Proxy score, first forecast	NA	NA	0.109	0.116
Update frequency	NA	NA	0	0
Update magnitude	NA	NA	0.017	0.014
Extremity, first forecast	NA	NA	−0.051	−0.042
All other predictors[a]	NA	NA	0	0

Note: [a]Only in-sample predictors with non-zero coefficients are shown. Other in-sample predictors are omitted due to space considerations

In the model run reported in column C, the only out-of-sample predictors with non-zero coefficients were overall MSMDB ($b = 0.056$) and last-forecast standardized Brier score (b = 0.059). Among in-sample predictors, the largest absolute coefficients were for first-forecast proxy scores ($b = 0.109$), first-forecast extremity ($b = -0.051$), followed by update magnitude ($b = 0.017$). The second sampling run, reported in column D, produced similar results, with one notable exception: out-of-sample IRT forecaster parameter had a non-zero regression coefficient ($b = -0.009$).

3.3 *Study 2: Discussion*

In summary, even when out-of-sample accuracy data on dozens of questions was available, in-sample intersubjective and behavioral measures still added value in identifying skilled forecasters. More specifically, forecasters whose independent initial estimates were both relatively close to the consensus (yielding better proxy scores) and were relatively extreme, as well as those who updated in frequent, small steps, tended to be most accurate. At that point, none of the other predictors such as psychometric scores, other behavioral measures or self-reported confidence provided enough marginal value in improving fit to warrant inclusion into the model.

4 General Discussion

4.1 Research Synthesis

Our main objective was to summarize the existing evidence on measures for identifying skilled forecasters who tend to perform consistently better than their peers. Our review catalogued over 40 measures in a growing body of research from a wide range of academic fields, including psychology, judgment and decision making, decision science, political science, economics and computer science. The wide range of ideas, measures and naming conventions poses challenges to summarizing all in one place, but makes this summary more useful in enabling learning and synergy across disciplinary boundaries.

While not the result of a formal meta-analysis, the median absolute correlation coefficient among non-accuracy-related measures ($r = 0.20$) provides a rough but useful baseline for researchers conducting power analyses for studies about new skill-identification measures. More importantly, the current research helps us confirm or update views about the strongest correlates of prediction skill. Among the five categories, accuracy-related measures were, unsurprisingly, most highly correlated to the outcome measures, which were also based on accuracy. Put simply: predictive accuracy is reliable. Posing dozens of rigorously resolvable questions and scoring individuals on their accuracy on those questions remains the undisputed gold standard in skill spotting.

The results of Study 2 provide an upper limit of cross-sample reliability for accuracy measures of approximately r = 0.74 across random sub-samples of questions. As Atanasov et al. (2022b) noted, test-retest reliability across *seasons* tends to be lower, at approximately r = 0.45. In other words, skill assessments become less reliable with time (see Himmelstein et al., 2023a, this volume, for an in-depth discussion of temporally driven issues in judgmental forecasting). While relative accuracy appears to be consistent across questions and over time, the limits to reliability also relate to our expectations of the predictive fit of any measures, whether they are based on accuracy or not. It is difficult to predict the future values of any measure better than by using past values of the same measure.

The relatively low correlation IRT-model based skill estimates with of our accuracy measure highlights the importance of specific details in measurement definition, such as imputation, time-trends and transformations. In open tournaments, where forecasters generally answer a small proportion of available questions, simpler measures, such as standardized Brier scores may be most practical. IRT model skill estimates may be most useful in settings where most forecasters answer the majority questions, avoiding sparse-matrix data issues. These models also show potential in adjusting for potential confounders, such as timing effects, and understanding the diagnostic properties of different types of forecasting questions.

In many real-world settings, gold-standard accuracy data are not available. Among the other categories, intersubjective measures demonstrated the strongest correlation withs accuracy. In Study 2, proxy scores based on the forecasters' initial estimates on questions provided stronger predictor fit that any other non-accuracy measure. Given this result, we see the study of intersubjective measures as an especially promising avenue for future research. Additional research may focus on improving intersubjective measures by maximizing the accuracy of consensus estimates that are used as proxies. For example, tournament designers must choose which forecasters are included in the consensus (e.g., superforecasters or less selective crowds), how consensus is updated over time, and how individual estimates are aggregated. It appears likely that more accurate consensus estimates will make for more effective proxies, but more research is needed to examine potential edge cases.

Promising applications of intersubjective measures include skill identification and incentive provision (Karger et al., 2022). At the same time, as Himmelstein et al. (2023b) point out, intersubjective measures that relate an individual's estimates to the consensus may be limited in their utility in spotting accurate forecasters with unique views. Intersubjective measures may be most helpful in identifying a small group of individuals whose aggregate estimates tend to be as accurate as those generated by a larger crowd. This is a useful property. To spot outstanding forecasters, intersubjective measures may need to be complemented by others.

Our analysis underscores the importance of behavioral measures. Building on Atanasov et al. (2020b), we showed that update frequency and magnitude add value in identifying accurate forecasters, even in the presence of accuracy-related and intersubjective measures. Frequent, small-increment updaters tend to generate accurate predictions across questions. Probabilistic extremity also appears useful in spotting accurate forecasters, especially as a complement to intersubjective measures. This finding may be specific to the construction of our proxy estimates, e.g., if we had applied stronger extremization in the aggregation algorithm producing the proxy estimates, forecaster extremity may have added less or no value. Our Study 2 analysis focused on independent forecasters who were not incentivized to write detailed rationales, so we did not analyze linguistic features of rationales. However, strong results across multiple previous studies (Horowitz et al., 2019; Zong et al., 2020; Karvetski et al., 2021) show that that such can be very helpful in spotting consistently accurate forecasters in settings where inter-forecaster communication is encouraged.

Among dispositional measures, performance-based scores related to forecasters' fluid intelligence were by far the most useful in assessing forecaster skills. As we showed, combinations of freely available measures can provide a useful starting point for spotting consistently accurate forecasters; fluid intelligence measures' correlations with accuracy ranged up to $r = 0.3$. Thinking-styles measures, generally based on self-reports, registered relatively low correlations with forecasting skill. The measure with the highest correlation was actively-open minded thinking, and even for that, the range of correlations was between $r = 0.10$ and 0.15.

Other thinking-style measures yielded low and generally not statistically significant correlations. One notable example is the fox-hedgehog scale. Tetlock's (2005) seminal research on expert political judgment highlighted a version of this measure as a key correlate of accuracy among geopolitical experts in his multi-decade research study. The result that foxy forecasters tend to be better than their hedgehog-like peers is well known among researchers and forecasters. However, this finding did not replicate in the 2011–2015 ACE tournament. More specifically, Mellers et al. (2015a, p. 7) included fox-hedgehog scale, along with need for closure and actively open-minded thinking and concluded that: "Only one of the measures, actively open-minded thinking, was significantly related to standardized Brier score accuracy." Our current analysis, which included two additional years of GJP data, replicated this null relationship.

Popular science accounts of crowd prediction are still catching up to this evidence. Epstein (2019), for example, noted that Tetlock and Mellers' approach in GJP was to "identify a small group of the foxiest forecasters."[5] Foxiness was not actually used for the selection of superforecasters, nor in the weighting schemas for tournament-winning aggregation algorithms. While it is plausible that the measure is still useful in identifying relatively accurate subject matter experts, it is not predictive in an open forecasting tournament environment. This measure may also serve as an example of a broader concern about self-report measures: when a measure becomes well-known, it loses some of its predictive validity as survey respondents learn which responses will make them look good.[6]

One classic result from Tetlock (2005) that appears valid in our context is the notion that biographical measures of expertise are not effective at identifying consistently accurate forecasters. The literature review in Study 1 included several such measures education level and h-index, and most studies did not show strong correlations between biographical expertise measures and skill. In Study 2, we showed that forecaster self-reports about their own expertise were completely uncorrelated with accuracy.

Our results underscore a methodological challenge to researchers: seek ways to assess forecaster tendencies through their behaviors, and rely less on their self-reports. For example, if you seek confident forecasters, track the extremity of their

[5]We have notified Epstein of this. As a result, he shared plans to edit the sentence in future editions of Range.

[6]Readers who have been exposed to research on forecaster skill identification through general media or popular science outlets may find some of our findings surprising. For example, a recent admittedly non-scientific poll of 30 twitter users by one of us (Atanasov) revealed that the plurality (40%) of respondents thought active open mindedness was more strongly correlated with accuracy than update magnitude, fluid intelligence or subject matter knowledge scores. Fewer than 20% correctly guessed that the closest correlate of accuracy was update magnitude.

estimates and ignore their expertise self-ratings. If you seek open-minded forecasters, pay more attention to the frequency of their updates than to their responses on open-mindedness questionnaires. These two examples are consistent with our results. The challenge lies in creatively constructing behavioral measures suitable to new contexts.

4.2 Use Cases

We illustrate the real-world use of skill spotting measures with two vignettes, summarized in Table 6.4. Both involve forecasting tournaments consisting of hundreds of participants with dozens of questions. In the first vignette, the tournament takes place within a large corporation. All participants are employees of the firm. The questions focus on outcomes relevant to the firm, such as product launch dates, sales, popularity of product features (Cowgill & Zitzewitz, 2015), or clinical trial development milestones (Atanasov et al., 2022a). Questions resolve within weeks or months. The company runs the tournament to inform its strategy and operations but also to uncover analytical talent.

Given the short-term questions, accuracy-related measures become relevant quickly, and can thus add much value. At the start of the tournament, intersubjective and behavioral measures can be very helpful in assessing aggregation weights to individual forecasters. Most dispositional measures will likely be of little utility, as human-resource regulations may constrain the use of IQ-related tests, while self-report measures tend to have low predictive validity. Expertise information may be available from forecasters' biography and record at the company, but such information is not very useful in uncovering skilled forecasters. The most useful expertise measures will likely be knowledge tests with a calibration component, which tend to have moderate predictive validity.

The second vignette involves a public tournament focused on existential risk. The tournament is open for anyone to participate. Questions range in duration from several months to one hundred years. Due to the long-time horizon of most questions, accuracy-related measures do not provide sufficient skill signals early on. Intersubjective measures may prove especially useful here in terms of skill identification, as well as means of providing feedback and incentives to forecasters (Karger et al., 2022; Beard et al. 2020). Behavioral measures can also add value in the short to medium-run, mostly as inputs to aggregation weights. In open tournaments, the range of allowable dispositional measures expands, as fluid intelligence measures can be included, subject to IRB approval. Such measures may even be used as initial screening tools e.g., if there are thousands of interested forecasters but sufficient resources to administer or pay only a subset. Dispositional measures can also provide signals for aggregation algorithms, addressing the "cold start" problem. Over time, as data from intersubjective and behavioral measures accumulates, the relative value of dispositional measures will likely diminish. Expertise measures are

Table 6.5 Predicted value-added for each category of measures in two application scenarios

Skill identification measure category	Predicted value added Corporate tournament: short-term questions, small teams	Open tournament: long-term questions, large crowds
1. Accuracy-related	Highest	Low
2. Intersubjective	High	Highest
3. Behavioral	High	High
4. Dispositional	Moderate	Moderate
5A. Expertise: Knowledge tests	Moderate	Moderate
5B. Expertise: Others	Low	Low

again of limited usefulness, except for knowledge tests with a calibration component (Table 6.5).

4.3 *Limitations and Future Directions*

We must acknowledge several limitations of the current research. First, while we aim to provide comprehensive summary of measures and empirical relationships, it is possible that we have missed some measures, especially ones older than 10 years, as well as new measures in unpublished studies. Relatedly, most measures included in our Study 1 review could not be practically included in our own analysis (Study 2), because of data availability, contextual differences or sensitivity to key assumptions. More comprehensive follow-up studies simultaneously testing multiple ideas will likely be beneficial, and consistent with the recent trend of "megastudies" in behavioral science (Milkman et al., 2022).

Second, the main statistical test utilized in most empirical analyses, the Pearson correlation coefficient, is designed to capture linear relationships. As such, we did not attempt to capture any non-linearities. For example, our research does not allow us to assess if a specific measure is particularly well suited for distinguishing among skill levels near the bottom or near the top of the distribution. Item-response theory (IRT) models are designed for this purpose. Such models are most useful in data-rich environments, *i.e.,* cases where most forecasters have answered most questions, and resolution information is available. Short of that, follow-up research should address non-linearities by zooming in on forecaster sub-sets or using more advanced statistical techniques, such as quantile regression models.

Third, most of the evidence summarized here is based on forecasting tournaments in which forecasters are asked and incentivized to produced maximally accurate forecasts, with the prospect of ground-truth verification. Different patterns may emerge in settings where forecasters produce unincentivized predictions (Dana et al., 2019), or are held accountable for process rather than accuracy-related outcomes (Chang et al., 2017). Finally, this research relies heavily on data from forecasting tournaments focused on geopolitics and economics. The body of research focused on public health and life sciences outcomes is growing (Benjamin et al. 2017; Atanasov et al., 2020a, 2022a; McAndrew et al., 2022; Sell et al., 2021), but the evidence base on correlates of individual skill outside of geopolitics and economics remains relatively thin. Future research should examine if subject matter expertise is more or less closely related to forecasting skill in other domains.

4.4 Conclusion

Individual forecasting performance is largely a function of skill, as some forecasters perform consistently better than others. In the presence of plentiful historical accuracy information across dozens of questions, accuracy track records constitute the overwhelming gold-standard in talent spotting. In settings where such information is not available, however, we show that researchers have plenty of options for gauging predictive skill. Unfortunately, most measures that seem intuitively attractive at first sight are not very effective. Asking forecasters about their expertise, or about their thinking patterns is not useful in terms of predicting which individuals will prove consistently accurate. Examining their behaviors, such as belief updating patterns, as well as their psychometric scores related to fluid intelligence offer more promising avenues. Arguably the most impressive performance in our study was for registered intersubjective measures, which rely on comparisons between individual and consensus estimates. Such measures proved valid as predictors of relative accuracy. As our research focus moves away from large crowds of amateurs staring at oxen to smaller, more selective crowds, we need better maps to navigate through a peculiar terrain sown with broken expectations. This chapter aims to provide the most complete rendition of such a map.

Acknowledgments We thank Matthias Seifert, David Budescu, David Mandel, Stefan Herzog and Philip Tetlock for helpful suggestions. All remaining errors are our own. No project-specific funding was used for the completion of this chapter.

Appendix: Methodological Details of Selected Predictors

Item Response Theory Models

In forecasting, one such confounder is the timing in which forecasts are made. In forecasting tournaments, forecasters make many forecasts about the same problems at various time points. Those who forecast problems closer to their resolution date have an accuracy advantage which may be important to account for in assessing their talent level (for more detail, see Himmelstein et al., this issue). IRT models can be extended so that their diagnostic properties change relative to the time point at which a forecaster makes their forecast. One such model is given below (Himmelstein et al., 2021; Merkle et al., 2016).

$$NB_{f,q,d} = b_{0,q} + \left(b_{1,q} - b_{0,q}\right)e^{-b_2 t_{f,q,d}} + \lambda_q \theta_f + \epsilon_{f,q,d}$$

The three b parameters represent how an item's difficult changes as time passes: $b_{0,\ q}$ represents an item's maximum difficulty (as time to resolution goes to infinity), $b_{1,\ q}$ an item's minimum difficulty (immediately prior to resolution), and b_2 the shape of the curve between $b_{0,\ q}$ and $b_{1,\ q}$ based on how much time is remaining in the question at the time of the forecast ($t_{f,\ q,\ d}$). The other two parameters represent how well an item discriminates between forecasters of different skill levels (λ_q) and how skilled the individual forecasters are (θ_f). As the estimate of forecaster skill, talent spotters will typically be most interested in this θ_f parameter, which is conventionally scaled so that it is on a standard normal distribution, $\theta_f \sim N(0, 1)$, with scores of 0 indicating an average forecaster, -1 a forecaster that is 1 SD below average, and 1 a forecaster that is 1 SD above average.

One potential problem with this model is that, in some cases, the distribution of Brier Scores is not well behaved. This typically occurs in cases which have many binary questions, so that the Brier score is a direct function of the accuracy assigned to the correct option. In such cases, the distribution of Brier scores can be multi-modal, because forecasters will tend to input many extreme and round number probability estimates, such as 0, .5, and 1 (Bo et al., 2017; Budescu et al., 1988; Merkle et al., 2016; Wallsten et al., 1993). To accommodate such multi-modal distributions, one option is to discretize the distribution of Brier scores into bins and reconfigure the model into an ordinal response model. Such models, such as the graded response model (Samejima, 1969), have a long history in the IRT literature.

Merkle et al. (2016) and Bo et al. (2017) describe examples of ordinal IRT models for forecasting judgment. However, the former found that the continuous and ordinal versions of the model were highly correlated (r = .87) in their assessment of forecaster ability level, and that disagreements tended to be focused on poor performing forecasters (who tend to make large errors) than high performing forecasters.

Contribution Scores

To obtain contribution scores for individual forecasters, it is necessary to first define some aggregation method for all of their judgments for each question. The simplest, and most common form of aggregation would just be to obtain the mean of all probabilities for all events associated with a forecasting problem. The aggregate probability (*AP*) for each of the *c* events associated with a forecasting question across all forecasters would be

$$\boldsymbol{AP_{q,c}} = \frac{\sum_{f=1}^{F} \boldsymbol{p_{q,c,f}}}{\boldsymbol{F}}$$

And the aggregate Brier score (*AB*) would then be

$$\boldsymbol{AB_q} = \sum_{c=1}^{C} \left(\boldsymbol{AP_{q,c}} - \boldsymbol{y_q}\right)^2$$

Based on this aggregation approach, defining the contribution of individual forecasters to the aggregate is algebraically straightforward. We can define $AP_{q,\ c,\ -f}$ as the aggregate probability with an individual forecaster's judgment removed as

$$\boldsymbol{AP_{q,c,\,-f}} = \frac{(\boldsymbol{F})\left(\boldsymbol{AP_{q,c}}\right) - \boldsymbol{p_{q,c,f}}}{\boldsymbol{F} - \boldsymbol{1}}$$

And the aggregate Brier score with an individual forecaster's judgment removed as

$$AB_{q,\,-f} = \sum_{c=1}^{C} \left(AP_{q,c,\,-f} - y_q\right)^{2.}$$

Finally, we define a forecaster's average *contribution* to the accuracy of the aggregate crowd forecasts as

$$\boldsymbol{C_f} = \frac{\sum_{q=1}^{Q} \boldsymbol{AB_q} - \boldsymbol{AB_{q,\,-f}}}{\boldsymbol{Q}}$$

C_f is a representation of how much information a forecaster brings to the table, on average, that is both *unique* and *beneficial.* It is possible that a forecaster ranked very highly on individual accuracy might be ranked lower in terms of their contribution, because their forecasts tended to be very similar to the forecasts of others, and so they did less to move the needle when averaged into the crowd.

Both weighting members of the crowd by average contribution scores, as well as selecting positive or high performing contributors, have been demonstrated to improve the aggregate crowd judgment (Budescu & Chen, 2015; Chen et al., 2016). The approach is especially appealing because it can be extended into a model that is dynamic, in that it is able to update contribution scores for each member of a crowd as more information about their performance comes available; it requires relatively little information about past performance to reliably estimate high performing contributors; and it is cost effective, in that is able to select a relatively small group of high performing contributors who can produce an

Table 6.6 Correlation matrix for measures in Study 2. Pearson correlation coefficients reported. Below-diagonal values are assessed in-sample, above diagonal values are calculated out-of-sample. Diagonal values highlighted in gray are cross-sample reliability coefficients. The bottom five measures are not question specific, so out-of-sample correlation coefficients or cross-sample reliability coefficients are not relevant

Measure	Standard. Brier (SB)	Debiased Brier	SB, 1st Forecast	SB, Last Forecast	IRT Model	Excess Volatility	Proper Proxy, All Forecasts	Proper Proxy, 1st Forecast	N. of Questions	Update Magnitude, Abs. Dist.	Update Freq.	Confirm. Prop.	Extremity 1st Forecast	Mean Expert.
Standardized Brier (SB)	0.74	0.69	0.62	0.68	-0.22	0.30	0.58	0.49	0.25	-0.32	0.45	0.03	-0.15	0.00
Debiased Brier	0.96	0.69	0.58	0.64	-0.15	0.32	0.59	0.50	0.22	-0.29	0.45	0.03	-0.07	0.02
SB, 1st Forecast	0.85	0.82	0.68	0.49	-0.15	0.39	0.56	0.53	0.21	-0.09	0.44	0.09	-0.22	0.02
SB, Last Forecast	0.84	0.80	0.64	0.83	-0.29	0.18	0.51	0.38	0.22	-0.44	0.35	0.01	-0.17	-0.06
IRT Model	-0.30	-0.23	-0.24	-0.34	0.89	0.42	0.15	0.21	0.05	0.16	0.23	0.27	0.60	0.17
Excess Volatility	0.40	0.43	0.48	0.26	0.40	0.85	0.63	0.67	0.12	0.04	0.46	0.35	0.38	0.15
Proper Proxy, All Forecasts	0.69	0.69	0.68	0.60	0.11	0.55	0.81	0.75	0.22	-0.18	0.58	0.23	0.28	0.13
Proper Proxy, 1st Forecast	0.60	0.60	0.67	0.44	0.15	0.61	0.93	0.77	0.18	-0.10	0.58	0.25	0.31	0.15
Number of Questions	0.25	0.23	0.22	0.24	0.05	0.11	0.21	0.18	0.87	-0.15	0.29	-0.11	0.00	-0.10
Update Magnitude, Abs. Dist.	-0.31	-0.28	-0.07	-0.44	0.16	0.05	-0.17	-0.10	-0.16	0.96	-0.29	0.25	-0.03	0.05
Update Frequency	0.51	0.52	0.51	0.35	0.21	0.51	0.63	0.63	0.31	-0.29	0.75	0.00	0.22	0.02
Confirmation Propensity	0.03	0.02	0.09	0.02	0.27	0.34	0.23	0.25	-0.07	0.24	0.00	0.86	0.25	0.32
Extremity, 1st Forecast	-0.19	-0.11	-0.28	-0.20	0.63	0.38	0.27	0.30	0.01	-0.05	0.21	0.26	0.91	0.24
Mean Expertise	0.00	0.01	0.01	-0.06	0.18	0.15	0.13	0.14	-0.11	0.05	0.02	0.32	0.23	0.97
Non-Question Specific Measures														
Fluid IQ Composite All	-0.27	-0.24	-0.29	-0.22	0.03	-0.25	-0.23	-0.23	-0.05	-0.09	0.04	-0.33	0.07	-0.15
Political Knowledge Score	-0.10	-0.06	-0.01	-0.17	0.13	0.06	-0.07	-0.05	-0.08	-0.04	0.14	0.00	0.01	0.03
AOMT	-0.10	-0.08	-0.05	-0.13	0.08	0.08	-0.04	-0.01	-0.14	-0.01	0.08	-0.01	0.08	-0.03
Fox-Hedgehog Scale	0.08	0.07	0.05	0.09	-0.05	0.00	0.03	0.02	0.07	0.07	0.00	0.00	-0.01	-0.04
Conscientiousness	0.13	0.12	0.14	0.12	0.03	0.12	0.15	0.15	0.16	0.20	-0.10	0.02	-0.05	0.01

aggregate judgment that matches or exceeds the judgment of larger crowds in terms of accuracy (Chen et al., 2016).

The advent of contribution assessment was initially designed with a particular goal in mind: to improve the aggregate wisdom of the crowd (Budescu & Chen, 2015; Chen et al., 2016). One might challenge as slightly as a slightly narrower goal than pure talent spotting. It is clearly an effective tool for maximizing crowd wisdom, but is it a valid tool for assessing expertise? The answer appears to be yes. Chen et al. (2016) not only studied contribution scores as an aggregation tool but tested how well contribution scores perform at selecting forecasters known to have a skill advantage through various manipulations known to benefit expertise, such as explicit training and interactive collaboration.

References

Arthur, W., Jr., Tubre, T. C., Paul, D. S., & Sanchez-Ku, M. L. (1999). College-sample psychometric and normative data on a short form of the raven advanced progressive matrices test. *Journal of Psychoeducational Assessment, 17*(4), 354–361.

Aspinall, W. (2010). A route to more tractable expert advice. *Nature, 463*(7279), 294–295.

Atanasov, P., Rescober, P., Stone, E., Servan-Schreiber, E., Tetlock, P., Ungar, L., & Mellers, B. (2017). Distilling the wisdom of crowds: Prediction markets vs. prediction polls. *Management Science, 63*(3), 691–706.

Atanasov, P., Diamantaras, A., MacPherson, A., Vinarov, E., Benjamin, D. M., Shrier, I., Paul, F., Dirnagl, U., & Kimmelman, J. (2020a). Wisdom of the expert crowd prediction of response for 3 neurology randomized trials. *Neurology, 95*(5), e488–e498.

Atanasov, P., Witkowski, J., Ungar, L., Mellers, B., & Tetlock, P. (2020b). Small steps to accuracy: Incremental belief updaters are better forecasters. *Organizational Behavior and Human Decision Processes, 160*, 19–35.

Atanasov, P., Joseph, R., Feijoo, F., Marshall, M., & Siddiqui, S. (2022a). *Human forest vs. random forest in time-sensitive Covid-19 clinical trial prediction*. Working Paper.

Atanasov, P., Witkowski, J., Mellers, B., & Tetlock, P. (2022b) *Crowdsourced prediction systems: Markets, polls, and elite forecasters*. Working Paper.

Augenblick, N., & Rabin, M. (2021). Belief movement, uncertainty reduction, and rational updating. *The Quarterly Journal of Economics, 136*(2), 933–985.

Bandalos, D. L. (2018). *Measurement theory and applications for the social sciences*. Guilford Publications.

Baron, J. (2000). *Thinking and deciding*. Cambridge University Press.

Baron, J., Scott, S., Fincher, K., & Metz, S. E. (2015). Why does the cognitive reflection test (sometimes) predict utilitarian moral judgment (and other things)? *Journal of Applied Research in Memory and Cognition, 4*(3), 265–284.

Barrick, M. R., & Mount, M. K. (1991). The big five personality dimensions and job performance: A meta-analysis. *Personnel Psychology, 44*(1), 1–26.

Beard, S., Rowe, T., & Fox, J. (2020). An analysis and evaluation of methods currently used to quantify the likelihood of existential hazards. *Futures, 115*, 102469.

Benjamin, D., Mandel, D. R., & Kimmelman, J. (2017). Can cancer researchers accurately judge whether preclinical reports will reproduce? *PLoS Biology, 15*(6), e2002212.

Bennett, S., & Steyvers, M. (2022). Leveraging metacognitive ability to improve crowd accuracy via impossible questions. *Decision, 9*(1), 60–73.

Bland, J. M., & Altman, D. G. (2011). Correlation in restricted ranges of data. *BMJ: British Medical Journal, 342*.

Blattberg, R. C., & Hoch, S. J. (1990). Database models and managerial intuition: 50% model + 50% manager. *Management Science, 36*(8), 887–1009.

Bo, Y. E., Budescu, D. V., Lewis, C., Tetlock, P. E., & Mellers, B. (2017). An IRT forecasting model: Linking proper scoring rules to item response theory. *Judgment & Decision Making, 12*(2), 90–103.

Bors, D. A., & Stokes, T. L. (1998). Raven's advanced progressive matrices: Norms for first-year university students and the development of a short form. *Educational and Psychological Measurement, 58*(3), 382–398.

Brier, G. W. (1950). Verification of forecasts expressed in terms of probability. *Monthly Weather Review, 78*(1), 1–3.

Broomell, S. B., & Budescu, D. V. (2009). Why are experts correlated? Decomposing correlations between judges. *Psychometrika, 74*(3), 531–553.

Bruine de Bruin, W., Parker, A. M., & Fischhoff, B. (2007). Individual differences in adult decision-making competence. *Journal of Personality and Social Psychology, 92*(5), 938–956.

Budescu, D. V., Weinberg, S., & Wallsten, T. S. (1988). Decisions based on numerically and verbally expressed uncertainties. *Journal of Experimental Psychology: Human Perception and Performance, 14*(2), 281–294.

Budescu, D. V., & Chen, E. (2015). Identifying expertise to extract the wisdom of crowds. *Management Science, 61*(2), 267–280.

Budescu, D.V., Himmelstein, M & Ho, E. (2021, October) Boosting the wisdom of crowds with social forecasts and coherence measures. In *Presented at annual meeting of Society of Multivariate Experimental Psychology (SMEP).*

Burgman, M. A., McBride, M., Ashton, R., Speirs-Bridge, A., Flander, L., Wintle, B., Fider, F., Rumpff, L., & Twardy, C. (2011). Expert status and performance. *PLoS One, 6*(7), e22998.

Cacioppo, J. T., & Petty, R. E. (1982). The need for cognition. *Journal of Personality and Social Psychology, 42*(1), 116–131.

Chang, W., Atanasov, P., Patil, S., Mellers, B., & Tetlock, P. (2017). Accountability and adaptive performance: The long-term view. *Judgment and Decision making, 12*(6), 610–626.

Chen, E., Budescu, D. V., Lakshmikanth, S. K., Mellers, B. A., & Tetlock, P. E. (2016). Validating the contribution-weighted model: Robustness and cost-benefit analyses. *Decision Analysis, 13*(2), 128–152.

Cokely, E. T., Galesic, M., Schulz, E., Ghazal, S., & Garcia-Retamero, R. (2012). Measuring risk literacy: The Berlin numeracy test. *Judgment and Decision making, 7*(1), 25–47.

Collins, R. N., Mandel, D. R., Karvetski, C. W., Wu, C. M., & Nelson, J. D. (2021). The wisdom of the coherent: Improving correspondence with coherence-weighted aggregation. *Preprint available at PsyArXiv*. Retrieved from https://psyarxiv.com/fmnty/

Collins, R., Mandel, D., & Budescu, D. (2022). Performance-weighted aggregation: Ferreting out wisdom within the crowd. In M. Seifert (Ed.), *Judgment in predictive analytics*. Springer [Reference to be updated with page numbers].

Cooke, R. (1991). *Experts in uncertainty: Opinion and subjective probability in science*. Oxford University Press.

Costa, P. T., Jr., & McCrae, R. R. (2008). *The revised neo personality inventory (NEO-PI-R)*. Sage.

Cowgill, B., & Zitzewitz, E. (2015). Corporate prediction markets: Evidence from Google, Ford, and Firm X. *The Review of Economic Studies, 82*(4), 1309–1341.

Dana, J., Atanasov, P., Tetlock, P., & Mellers, B. (2019). Are markets more accurate than polls? The surprising informational value of "just asking". *Judgment and Decision making, 14*(2), 135–147.

Davis-Stober, C. P., Budescu, D. V., Dana, J., & Broomell, S. B. (2014). When is a crowd wise? *Decision, 1*(2), 79–101.

Dieckmann, N. F., Gregory, R., Peters, E., & Hartman, R. (2017). Seeing what you want to see: How imprecise uncertainty ranges enhance motivated reasoning. *Risk Analysis, 37*(3), 471–486.

Embretson, S. E., & Reise, S. P. (2013). *Item response theory*. Psychology Press.

Epstein, D. (2019). *Range: How generalists triumph in a specialized world*. Pan Macmillan.

Fan, Y., Budescu, D. V., Mandel, D., & Himmelstein, M. (2019). Improving accuracy by coherence weighting of direct and ratio probability judgments. *Decision Analysis, 16*, 197–217.

Frederick, S. (2005). Cognitive reflection and decision making. *Journal of Economic Perspectives, 19*(4), 25–42.

Galton, F. (1907). Vox populi (the wisdom of crowds). *Nature, 75*(7), 450–451.

Gneiting, T., & Raftery, A. E. (2007). Strictly proper scoring rules, prediction, and estimation. *Journal of the American Statistical Association, 102*(477), 359–378.

Goldstein, D. G., McAfee, R. P., & Suri, S. (2014, June). The wisdom of smaller, smarter crowds. In *Proceedings of the Fifteenth ACM Conference on Economics and Computation* (pp. 471–488).

Good, I. J. (1952). Rational decisions. *Journal of the Royal Statistical Society. Series B (Methodological), 1952*, 107–114.

Hanea, A. D., Wilkinson, D., McBride, M., Lyon, A., van Ravenzwaaij, D., Singleton Thorn, F., Gray, C., Mandel, D. R., Willcox, A., Gould, E., Smith, E., Mody, F., Bush, M., Fidler, F., Fraser, H., & Wintle, B. (2021). Mathematically aggregating experts' predictions of possible futures. *PLoS One, 16*(9), e0256919. https://doi.org/10.1371/journal.pone.0256919

Haran, U., Ritov, I., & Mellers, B. A. (2013). The role of actively open-minded thinking in information acquisition, accuracy, and calibration. *Judgment and Decision making, 8*(3), 188–201.

Hastie, T., Qian, J., & Tay, K. (2021). *An introduction to glmnet*. CRAN R Repository.

Himmelstein, M., Atanasov, P., & Budescu, D. V. (2021). Forecasting forecaster accuracy: Contributions of past performance and individual differences. *Judgment & Decision Making, 16*(2), 323–362.

Himmelstein, M., Budescu, D. V., & Han, Y. (2023a). The wisdom of timely crowds. In M. Seifert (Ed.), *Judgment in predictive analytics*. Springer.

Himmelstein, M., Budescu, D. V., & Ho, E. (2023b). The wisdom of many in few: Finding individuals who are as wise as the crowd. *Journal of Experimental Psychology: General.* Advance online publication.

Ho, E. H. (2020, June). *Developing and validating a method of coherence-based judgment aggregation*. Unpublished PhD Dissertation. Fordham University, Bronx NY.

Horowitz, M., Stewart, B. M., Tingley, D., Bishop, M., Resnick Samotin, L., Roberts, M., Chang, W., Mellers, B., & Tetlock, P. (2019). What makes foreign policy teams tick: Explaining variation in group performance at geopolitical forecasting. *The Journal of Politics, 81*(4), 1388–1404.

Joseph, R., & Atanasov, P. (2019). *Predictive training and accuracy: Self-selection and causal factors*. Working Paper, Presented at Collective Intelligence 2019.

Karger, E., Monrad, J., Mellers, B., & Tetlock, P. (2021). *Reciprocal scoring: A method for forecasting unanswerable questions*. Retrieved from SSRN

Karger, J., Atanasov, P., & Tetlock, P. (2022). *Improving judgments of existential risk: Better forecasts, questions, explanations, policies*. SSRN Working Paper.

Karvetski, C. W., Olson, K. C., Mandel, D. R., & Twardy, C. R. (2013). Probabilistic coherence weighting for optimizing expert forecasts. *Decision Analysis, 10*(4), 305–326.

Karvetski, C. W., Meinel, C., Maxwell, D. T., Lu, Y., Mellers, B. A., & Tetlock, P. E. (2021). What do forecasting rationales reveal about thinking patterns of top geopolitical forecasters? *International Journal of Forecasting, 38*(2), 688–704.

Kurvers, R. H., Herzog, S. M., Hertwig, R., Krause, J., Moussaid, M., Argenziano, G., Zalaudek, I., Carney, P. A., & Wolf, M. (2019). How to detect high-performing individuals and groups: Decision similarity predicts accuracy. *Science Advances, 5*(11), eaaw9011.

Lipkus, I. M., Samsa, G., & Rimer, B. K. (2001). General performance on a numeracy scale among highly educated samples. *Medical Decision Making, 21*(1), 37–44.

Liu, Y., Wang, J., & Chen, Y. (2020, July). Surrogate scoring rules. In *Proceedings of the 21st ACM Conference on Economics and Computation* (pp. 853–871).

Mannes, A. E., Soll, J. B., & Larrick, R. P. (2014). The wisdom of select crowds. *Journal of Personality and Social Psychology, 107*(2), 276.

Matzen, L. E., Benz, Z. O., Dixon, K. R., Posey, J., Kroger, J. K., & Speed, A. E. (2010). Recreating Raven's: Software for systematically generating large numbers of Raven-like matrix problems with normed properties. *Behavior Research Methods, 42*(2), 525–541.

Mauksch, S., Heiko, A., & Gordon, T. J. (2020). Who is an expert for foresight? A review of identification methods. *Technological Forecasting and Social Change, 154*, 119982.

McAndrew, T., Cambeiro, J., & Besiroglu, T. (2022). Aggregating human judgment probabilistic predictions of the safety, efficacy, and timing of a COVID-19 vaccine. *Vaccine, 40*(15), 2331–2341.

Mellers, B., Ungar, L., Baron, J., Ramos, J., Gurcay, B., Fincher, K., Scott, S. E., Moore, D., Atanasov, P., Swift, S. A., Murray, T., Stone, E., & Tetlock, P. E. (2014). Psychological strategies for winning a geopolitical forecasting tournament. *Psychological Science, 25*(5), 1106–1115.

Mellers, B., Stone, E., Atanasov, P., Rohrbaugh, N., Metz, S. E., Ungar, L., Bishop, M. M., Horowitz, M., Merkle, E., & Tetlock, P. (2015a). The psychology of intelligence analysis: Drivers of prediction accuracy in world politics. *Journal of Experimental Psychology: Applied, 21*(1), 1.

Mellers, B., Stone, E., Murray, T., Minster, A., Rohrbaugh, N., Bishop, M., Chen, E., Baker, J., Hou, Y., Horowitz, M., Ungar, L., & Tetlock, P. (2015b). Identifying and cultivating superforecasters as a method of improving probabilistic predictions. *Perspectives on Psychological Science, 10*(3), 267–281.

Mellers, B. A., Baker, J. D., Chen, E., Mandel, D. R., & Tetlock, P. E. (2017). How generalizable is good judgment? A multitask, multi-benchmark study. *Judgment and Decision making, 12*(4), 369–381.
Merkle, E. C., Steyvers, M., Mellers, B., & Tetlock, P. E. (2016). Item response models of probability judgments: Application to a geopolitical forecasting tournament. *Decision, 3*(1), 1–19.
Milkman, K. L., Gandhi, L., Patel, M. S., Graci, H. N., Gromet, D. M., Ho, H., Kay, J. S., Lee, T. W., Rothschild, J., Bogard, J. E., Brody, I., Chabris, C. F., & Chang, E. (2022). A 680,000-person megastudy of nudges to encourage vaccination in pharmacies. *Proceedings of the National Academy of Sciences, 119*(6), e2115126119.
Miller, N., Resnick, P., & Zeckhauser, R. (2005). Eliciting informative feedback: The peer-prediction method. *Management Science, 51*(9), 1359–1373.
Morstatter, F., Galstyan, A., Satyukov, G., Benjamin, D., Abeliuk, A., Mirtaheri, M., et al. (2019). SAGE: A hybrid geopolitical event forecasting system. *IJCAI, 1*, 6557–6559.
Murphy, A. H., & Winkler, R. L. (1987). A general framework for forecast verification. *Monthly Weather Review, 115*(7), 1330–1338.
Palley, A. B., & Soll, J. B. (2019). Extracting the wisdom of crowds when information is shared. *Management Science, 65*(5), 2291–2309.
Peters, E., Västfjäll, D., Slovic, P., Mertz, C. K., Mazzocco, K., & Dickert, S. (2006). Numeracy and decision making. *Psychological Science, 17*(5), 407–413.
Predd, J. B., Osherson, D. N., Kulkarni, S. R., & Poor, H. V. (2008). Aggregating probabilistic forecasts from incoherent and abstaining experts. *Decision Analysis, 5*(4), 177–189.
Prelec, D. (2004). A Bayesian truth serum for subjective data. *Science, 306*(5695), 462–466.
Samejima, F. (1969). Estimation of latent ability using a response pattern of graded scores. *Psychometrika Monograph, 34*, 1–97.
Seifert, M., Siemsen, E., Hadida, A. L., & Eisingerich, A. B. (2015). Effective judgmental forecasting in the context of fashion products. *Journal of Operations Management, 36*, 33–45.
Sell, T. K., Warmbrod, K. L., Watson, C., Trotochaud, M., Martin, E., Ravi, S. J., Balick, M., & Servan-Schreiber, E. (2021). Using prediction polling to harness collective intelligence for disease forecasting. *BMC Public Health, 21*(1), 1–9.
Shipley, W. C., Gruber, C. P., Martin, T. A., & Klein, A. M. (2009). *Shipley-2 manual*. Western Psychological Services.
Stanovich, K. E., & West, R. F. (1997). Reasoning independently of prior belief and individual differences in actively open-minded thinking. *Journal of Educational Psychology, 89*(2), 342–357.
Stewart, T. R., Roebber, P. J., & Bosart, L. F. (1997). The importance of the task in analyzing expert judgment. *Organizational Behavior and Human Decision Processes, 69*(3), 205–219.
Suedfeld, P., & Tetlock, P. (1977). Integrative complexity of communications in international crises. *Journal of Conflict Resolution, 21*(1), 169–184.
Tannenbaum, D., Fox, C. R., & Ülkümen, G. (2017). Judgment extremity and accuracy under epistemic vs. aleatory uncertainty. *Management Science, 63*(2), 497–518.
Tetlock, P. E. (2005). *Expert political judgment*. Princeton University Press.
Tetlock, P. E., & Gardner, D. (2016). *Superforecasting: The art and science of prediction*. Random House.
Toplak, M. E., West, R. F., & Stanovich, K. E. (2014). Assessing miserly information processing: An expansion of the cognitive reflection test. *Thinking & Reasoning, 20*(2), 147–168.
Tsai, J., & Kirlik, A. (2012). Coherence and correspondence competence: Implications for elicitation and aggregation of probabilistic forecasts of world events. In *Proceedings of the Human Factors and Ergonomics Society Annual Meeting* (Vol. 56, pp. 313–317). Sage.
Wallsten, T. S., Budescu, D. V., & Zwick, R. (1993). Comparing the calibration and coherence of numerical and verbal probability judgments. *Management Science, 39*(2), 176–190.

Webster, D. M., & Kruglanski, A. W. (1994). Individual differences in need for cognitive closure. *Journal of Personality and Social Psychology, 67*(6), 1049–1162.
Witkowski, J., & Parkes, D. (2012). A robust bayesian truth serum for small populations. *Proceedings of the AAAI Conference on Artificial Intelligence, 26*(1), 1492–1498.
Witkowski, J., Atanasov, P., Ungar, L., & Krause, A. (2017) Proper proxy scoring rules. In *Presented at AAAI-17: Thirty-First AAAI Conference on Artificial Intelligence.*
Zong, S., Ritter, A., & Hovy, E. (2020). Measuring forecasting skill from text. *arXiv preprint arXiv:2006.07425.*

Chapter 7
Performance-Weighted Aggregation: Ferreting Out Wisdom Within the Crowd

Robert N. Collins, David R. Mandel, and David V. Budescu

Keywords Judgment · Weighted aggregation · Accuracy · Correspondence

1 Introduction

Individuals, private organizations, and governments often consult a bevy of experts and nonexperts to inform their decision-making. In some cases, they seek probabilistic estimates on binary outcomes, such as, "Will the US be hit by a devastating hurricane this year?" In other cases, multiple categories may be of interest, such as, "Who will be the next Prime Minister in a four-candidate race?" In other cases, decision-makers require quantitative estimates, qualified by some degree of confidence or margin of error, such as, "With 90% confidence, what are your lower and upper bound estimates of how much will the sea levels rise within the next 50 years?" These questions can take various forms, for instance, calls for unconditional estimates (e.g., "Will Russia conduct military actions in Ukraine in the next 6 months?") or conditional estimates (e.g., if Russia conducts military actions in Ukraine in the next 6 months, what will NATO's response likely be?).[1]

Decision-makers face multiple challenges that go beyond the substantive questions they are dealing with. Principal among these is deciding which sources of information

The original version of this chapter was revised: The copyright holder name has been corrected. The correction to the chapter is available at: https://doi.org/10.1007/978-3-031-30085-1_12

[1] As a postscript, we note that these examples, which now appear dated, were incorporated roughly one month prior to Russia's invasion of Ukraine.

R. N. Collins (✉) · D. R. Mandel
Intelligence, Influence, and Collaboration Section, Defence Research & Development Canada, Toronto, ON, Canada

D. V. Budescu
Department of Psychology, Fordham University, Bronx, NY, USA

M. Seifert (ed.), *Judgment in Predictive Analytics*, International Series in Operations Research & Management Science 343,
https://doi.org/10.1007/978-3-031-30085-1_7

to follow. Often, decision-makers have access to judgments not from one but from a "crowd" (a collection) of judges. Typically, the crowd will contain individuals with differing knowledge and expertise on the relevant topic. Some may know the correct answer, others may know something useful about the topic, and still others may know nothing at all about the topic. How should decision-makers harness the crowd, exploiting the wisdom it offers? Should they combine the judgments of the entire crowd—experts and non-experts alike? Should they seek out the best expert within the crowd in the hope they will have the correct answer? Or should they combine the two strategies, weighting the judgments based on a "tell" regarding their informativeness on the given task? As we discuss in this chapter, there is mounting evidence that this hybrid approach known as *performance-weighted aggregation* holds much promise.

This chapter addresses key issues related to the use of performance-weighted aggregation strategies combining judgments. These methods assign unequal weights to judges based on a performance indicator believed to predict expertise and judgment quality. In this chapter, we discuss the theory behind aggregation and arguments for and against different weighting strategies. We present a framework for deciding when decision-makers and aggregators should prefer different strategies. We introduce the reader to practical methods and functions for combining opinions using performance-weighted aggregation. The second part of the chapter consists of an in-depth examination of research focused on identifying valid, reliable indicators of expertise that can serve as bases for implementing performance-weighted aggregation.

1.1 The Wisdom of Crowds

In any given context, the amount of useful information held by individuals can vary dramatically. Consequently, the quality of their judgments will vary in turn. Compared to a layperson, for instance, a weather specialist will predict rainfall amounts with greater accuracy, and an oncologist will diagnose tumors with greater accuracy than non-experts. These individuals qualify as true experts at their respective judgment tasks (Kahneman et al., 2021; Mannes et al., 2014). In many (bot not all) domains there exist true experts with highly specialized knowledge to which few laypeople have access. These highly trained and skilled individuals are also likely to be less noisy and biased than their peers. Consequently, they can be relied upon to produce accurate judgments within their field.

Unfortunately, many so-called experts still produce biased, noisy, and error-prone judgments (Kahneman et al., 2021). Although genuine experts do exist (e.g., weather forecasting; Wallsten & Budescu, 1983; strategic intelligence forecasting; Mandel & Barnes, 2014, 2018), many areas of judgment and decision-making have proven challenging to the production of consistent, high-quality judgments. In others, valid claims or measures of expertise are questionable. For example, wine experts cannot reliably distinguish between wines in a blind test (Goldstein et al., 2008). This suggests that wines do not systematically differ in objectively measurable qualities or, if they do, claims that experts are sensitive to these objective qualities are vastly overstated. In yet other cases, experts may exist, but reliable indicators of expertise are hard to detect and measure. Further complicating matters, it is easy to overgeneralize an expert's expertise. For instance, while oncologists will outperform others in diagnosing cancer, their discrimination in predicting the outcomes of clinical cancer trials is virtually at chance levels (Benjamin et al., 2021). Indeed, only a small minority of individuals can produce consistent, high-quality true judgments across a range of topics (Satopää et al., 2021; Tetlock & Gardner, 2015). In these cases, key predictors of performance were cognitive ability, political knowledge, and open-mindedness, rather than domain-specific knowledge.

Critics regard this approach as 'chasing the expert' or CTE for short (Larrick & Soll, 2006; Surowiecki, 2004). A superior approach, they argue, is to mathematically combine the independent opinions of large, diverse groups of individuals. These groups will include individuals with varying levels of expertise. A simple average of a group's judgment is often quite accurate, in many cases surpassing the accuracy of even the single best judge in the group. Famous classical examples include guessing the weight of a slaughtered and dressed ox (Galton, 1907), and students in a classroom guessing the number of jellybeans in a jar. Surowiecki (2004) labeled this effect the *wisdom of crowds* (*WOC*). Mathematically speaking, a crowd is wise if the average of judges beats the average judge (Larrick et al., 2011). Davis-Stober et al. (2014) offered a broader definition: a crowd is wise if a linear combination of the judges produces a higher quality judgment, on average, than a randomly selected judge from the crowd.

The simplest crowd aggregation method is to rely on measures of the central tendency of the group's judgments (Larrick et al., 2011). In this respect, the *WOC* works by exploiting the benefits of error cancellation. In most judgment tasks, individual judges will contribute differing levels of useful information, bias, and noise (Davis-Stober et al., 2014; Satopää et al., 2021). The distributions of this bias and noise matters. Some judges overestimate, while others underestimate. Furthermore, some random errors will skew the judgment higher, while others will skew the judgment lower. Regardless, exempting cases where there is a definitive true or false answer (Collins et al., in press; Karvetski et al., 2013), a group of judges is likely to 'bracket' the true value (Larrick & Soll, 2006). When this occurs, the bias and noise of the individual judges and judgments will cancel out.

Thus, the *WOC* effect amplifies judgment quality by combining the private information of its constituent judges while reducing noise and counterbalancing biases. The crowd can even comprise biased laypersons, so long as they supply unique, diverse information that aggregators can combine (Davis-Stober et al., 2014; Wallsten & Diederich, 2001). Decision-makers have exploited the WOC phenomenon to improve judgment in a variety of tasks from prediction markets and political forecasts (Silver, 2012) to policymaking (Hastie & Kameda, 2005). Critical to the effectiveness of the *WOC* effect is the cognitive diversity and independence[2] of the

[2]By independence we mean that judges operate independently of each other and cannot influence their peers' judgments. This does not imply that their judgements are uncorrelated. In fact, they can be highly correlate simply because all the judges rely on similar information when formulating their individual opinions (e.g., Broomell & Budescu, 2009).

judges who contribute. Correlated errors or biases (such as those due to 'group think') will quickly compound to influence the collective judgment of the group, requiring the solicitation of ever more opinions (Broomell & Budescu, 2009; Lorenz et al., 2011).

Unweighted averages often perform as well as more sophisticated methods of combining judgments (Armstrong, 2001; Clemen, 1989; Makridakis & Winkler, 1983). Critics of the *WOC* approach point out that the unweighted average under-weights the information offered by experts within the crowd. They argue that it reduces the potential benefits by focusing exclusively on noise reduction (Budescu & Chen, 2015). Rather, aggregators can amplify the signal-to-noise ratio using unequal weighting. Unequal weights can boost aggregated judgment quality by tapping into the contributions of experts within the crowd who have high-quality information and by reducing the contributions of low-performing non-experts. Accordingly, aggregators should weight contributions in pooled judgments based on an indicator that reliably tracks expertise.

Performance-weighting methods have proven useful for improving judgment accuracy in several tasks and distinct domains. These include sports betting (Predd et al., 2008), prediction markets for political events (Atanasov et al., 2017; Wang et al., 2011a, b), geopolitical forecasts (Budescu & Chen, 2015), general-knowledge questions (Fan et al., 2019; Karvetski et al., 2013), and Bayesian probability judgment tasks (Collins et al., in press; Karvetski et al., 2020; Mandel et al., 2018). The principal goal of aggregators who wish to employ performance-weighted aggregation is to identify indicators of *true* expertise, ferreting out signals of accurate judges within the crowd. Rather than answer the judgment task directly, aggregators who use performance-weighting seek to identify experts within the crowd and weight their judgments accordingly. Several strategies for identifying expertise have proven successful, such weighting judges according to: their past performance on similar judgment tasks (Budescu & Chen, 2015; Cooke et al., 1988), performance on related tests and psychometric inventories (Mellers et al., 2015; Satopää et al., 2021), or the internal consistency or coherence of one's judgments (Karvetski et al., 2013; Predd et al., 2008). Each approach provides a potential signal of expertise for aggregators to exploit and optimize the quality of aggregated judgments.

1.2 Judgment Quality: Defining and Identifying Expertise in the Crowd

There is naturally a wide stretch of judgment terrain that is value-laden or entirely subjective, and therefor not amenable to scoring in terms of objective accuracy measures. While these judgments may be important in our lives, they are not the focus of the present chapter. Rather, we focus exclusively on epistemic judgments about events whose outcomes can eventually be known or about hypotheses where the truth values can eventually be ascertained. In other words, we focus on cases where accuracy can be objectively scored.

The quality of a judgment with objective ground truth is synonymous with its *accuracy* or *correspondence*, or the degree to which a provided judgment matches the ground truth in the observed world. The nature of scoring may vary across different tasks. For example, consider error measures such as the mean absolute error (*MAE*; Willmott & Matsuura, 2005) or Brier score (*BS*, Brier, 1950). The *MAE* is the arithmetic average of absolute errors across *N* judgments, defined as

$$MAE = \frac{1}{N}\sum_{i=1}^{N} |y_i - x_i|,$$

and the *BS* is a mean squared error measure across *N* judgments defined as

$$BS = \frac{1}{N}\sum_{i=1}^{N} (y_i - x_i)^2,$$

where x is the judge's opinion and y is the resolved truth value. Note that N judgments can refer either to the number of judgments made or the number of judges. The former definition assesses the average error of one individual across multiple judgments. The latter definition assesses the average error of a group of judges. Although both are suitable for measuring error on a variety of tasks, *MAE* is typically used for measuring the error of point estimates of quantities, whereas *BS* is typically used for measuring the error of point probability estimates (Gneiting & Raftery, 2007; Han & Budescu, 2019; Predd et al., 2009). For both *MAE* and *BS*, lower scores indicate higher-quality judgments. A score of 0 indicates a perfect judgment.

Measures of error are decomposable into at least two components: *bias* and *noise* (Kahneman et al., 2016, 2021; Satopää et al., 2021). *Bias* is a judge's tendency to systematically over- or under-estimate judgments and is, in principle, predictable. Conversely, *noise* represents non-systematic error that is impossible to predict. The component of judgment not attributable to error, then, is *information.* Information points toward the true or correct answer. Perfect information, devoid of bias and noise, is sufficient to produce optimal, perfect, and correct judgments.

Using this framework, we can define the ideal *true* expert (or, more generally, *perfect judgment*). A true expert is someone who has complete, exhaustive, and correct *information* on a topic of judgment and who displays no bias or noise.[3] This ideal expert would have an error score no greater than the irreducible uncertainty of the system or event being judged and can serve as one benchmark for evaluating the

[3]To the extent that the aggregator can predict and correct for bias, aggregators can relax this definition and simply model the true expert based on the elicited judgment.

quality of individual judges and their aggregates. Another relevant benchmark is defined by an uninformed and naïve judge who contributes only *noise* and no *information* whatsoever, achieving an error score no better than random chance. The former individual will strictly improve the quality of the aggregated judgment, while the latter will have random effects. Depending on the task, both individuals may exist. Most individuals, including spurious experts, will lie somewhere in between these two extremes.

The ideal performance-weighting measure, to borrow language from *signal detection theory* literature (Peterson et al., 1954), acts as a *signal* of a judge's *information*. *Noise*, in this context, is *prediction error*. That is, error of the performance measure to correctly identify an expert, as opposed to the *noise* of the judgment itself. An ideal performance measure, therefore, has high *sensitivity*, correctly assigning high scores to experts, and high *specificity*, correctly assigning low scores to naïve individuals (Yerushalmy, 1947). In the absence of noise (i.e., a perfect correlation between the *signal* and *information*) and assuming perfect *sensitivity* and *specificity*, a perfect score on the *performance measure* would deterministically indicate a *true* expert. In the absence of signal (i.e., no correlation between *signal* and *information*), weighting according to such a performance measure will produce judgments that are, at best, equal to the performance of the unweighted average. However, the judgments will likely be strictly inferior to the unweighted average due to the introduction of sampling and prediction error.

Although *true* experts may exist for any number of judgment topics, a perfect diagnostic signal of that expertise does not. Fortunately, the description we provided above provides a framework for decision-makers aiming to optimize aggregated decision quality. The default approach of the savvy aggregator should be unweighted aggregation, relying on the cancellation of noise and bias to enhance judgment quality (Larrick et al., 2011; Surowiecki, 2004). Conversely, if a perfect diagnostic signal of expertise existed and was known, aggregators should use those signals to *chase the expert* (*CTE*). We propose that the ideal solution most often lies between these alternatives (Jaspersen, 2021). Thus, we suggest that savvy aggregators should prefer the unweighted *WOC* unless they have evidence that a particular performance measure is a valid and reliable measure of expertise. They should increasingly favor *CTE* as evidence mounts regarding the sensitivity, specificity, and reliability of the measures relation to expertise. Ideally, the performance metric should be closely related to the judgment topic at hand. In these two points, we agree with Armstrong (2001). In practice, aggregators should favor a middle-ground hypothesis, using some level of performance-weighted aggregation instead of the unweighted *WOC* or *CTE* model in most cases. At the very least, aggregators can balance these models in consideration with unweighted judgments. Aggregators must further balance their beliefs with the cost and difficulty of implementing the strategy successfully (Hanea et al., 2018; Hemming et al., 2020).

2 Judgment Aggregation Strategies

Broadly speaking, aggregation is any strategy or algorithm for selecting, combining, or collating information from multiple judgments in an opinion pool. This includes both extreme strategies we previously discussed—*WOC* and *CTE*—and everything in between. There are infinitely many aggregation strategies that an aggregator could use to combine judgments, each with different properties. Our focus in this chapter is on performance-weighted aggregation, so we will focus on simple mean and median aggregation functions and their performance-weighted counterparts. Aggregators may apply more sophisticated trimming or extremization procedures (e.g., Jose et al., 2013) alongside performance-weighting if desired. Furthermore, the aggregation of interval judgments (Cooke, 2014; Han & Budescu, 2019; Park & Budescu, 2015), multiplicative aggregation (Dietrich & List, 2017), or other sets of dependent judgments will often require more complex aggregation methods than what we discuss here. However, the merits of performance weighting should still apply, even if the implementation differs.

2.1 Mean Strategies

The mean is the arithmetic average of the opinion pool. Mean measures minimize the sum of squared differences between the aggregate and the judgments in an opinion pool. Formally, the mean or unweighted average model is

$$F(y_1, \ldots, y_i) = \frac{1}{N} \sum_{i=1}^{N} y_i,$$

where N represents the number of judgments and y represents individual judgments. This model will yield the arithmetic mean of the crowd's judgment and is the simplest aggregation model a practitioner could employ.

The simplest formula for performance-weighted aggregation, the *linear opinion pool* (sometimes called the LinOP, Clemen & Winkler, 1999), defined as

$$F(y_1, \ldots, y_i) = \sum_{i=1}^{N} \omega_i \times y_i,$$

where ω_i is a normalized non-negative weight (i.e., sums to 1) for individual judgments derived from some measure of performance.

Mean aggregation has several attractive properties. It satisfies the unanimity property: if all experts agree, the combined probability must also agree. Mean aggregation also satisfied the marginalization property: the combined judgment is the same whether one combines the experts' own marginal distributions or combines the joint distributions and then calculates the marginal distributions (Clemen & Winkler, 1999). Because judgments often bracket the true value, mean measures can often produce judgments that are better than even the single best deterministically selected judge (Afflerbach et al., 2021; Surowiecki, 2004). A potential weakness of mean aggregation—both unweighted and weighted—is its sensitivity to outliers, which can skew the aggregated result.

2.2 Median Strategies

The median is another measure of the central tendency of a pool of estimates. The median minimizes the sum of absolute deviations between the aggregate and all the individual judgments in the opinion pool. In practice, this means selecting the judgment that lies in the middle of an ordered list of all judgments in the opinion pool. Formally, we define the unweighted median model as

$$F(\mathrm{y}_1, \ldots, \mathrm{y}_i) = \textit{if N is odd}, \textit{median}(y) = y_{\frac{n+1}{2}}$$

$$F(\mathrm{y}_1, \ldots, \mathrm{y}_i) = \textit{if N is even}, \textit{median}(y) = \frac{y_{(N/2)} + y_{(N/2)+1}}{2.}$$

for the distinct ordered individual judgments $y_1, y_2, y_3, \ldots, y_N$, where N is the number of judgments in the opinion pool.

We can define a performance-weighted equivalent of median aggregation. For n distinct ordered judgments with non-negative normalized weights that satisfy

$$F(\mathrm{y}_1, \ldots, \mathrm{y}_i) = \sum_{i=1}^{N} \omega_i = 1,$$

the weighted median is the element y_k satisfying

$$F(\mathrm{y}_1, \ldots, \mathrm{y}_i) = \sum_{i=1}^{k-1} \omega_i \leq \frac{1}{2} \text{and} \sum_{i=k+1}^{N} \omega_i \leq \frac{1}{2}.$$

In other words, the weighted median is the judgment that corresponds to the ordinal position above and below which half of the summed total crowd weights reside. Median measures also respect the unanimity principle but are less vulnerable to the effects of extreme values in the opinion pool than mean aggregation. One drawback is the potential for information loss in skewed or asymmetric distributions of responses. Furthermore, the best possible accuracy of aggregated median judgments is that of the median judge (when n is odd) or the average of the two inner-most judges (*when n is even*), whereas a mean aggregation strategy has greater resolution to produce values in between any individual judgment in the pool.

Both mean and median strategies are effective methods for exploiting the WOC. Furthermore, performance-weighted implementations will often outperform their equal-weighted counterparts in each case. Interestingly, Galton's (1907) often-cited demonstration of the WOC effect used a median aggregation strategy. In this sense, median models represent the earliest WOC strategy applied. Furthermore, there is increasing evidence that median aggregation outperforms mean measures in terms of *BS* (Han & Budescu, 2019). Han and Budescu attributed this to the skewed distributions in most judgment tasks, which have longer and fatter tails in the direction of deficient performance. Thus, outlier sensitivity is often a key factor in aggregated judgment quality, a fact that becomes more salient with larger group sizes and increasingly extreme outliers (Han & Budescu, 2019). Trimming and outlier removal procedures can address this issue, potentially ameliorating this discrepancy.[4] However, the argument that median aggregation tends to outperform mean aggregation remains compelling. As with deciding whether to prefer an unweighted or weighted aggregation model, the aggregator's decision should reflect their belief about whether outliers are likely to be significant or dominant source of error.

2.3 *Weighting Functions*

There are infinitely many functions an aggregator could use to construct an aggregation weight from a performance measure, which we define as ρ. The only restriction is that, based on the aggregation formulas we provide, the weights derived from ρ must be non-negative. Here, we review some functions that aggregators can use to implement the different performance-weighting strategies discussed in the introduction. We break these functions down into three types of strategies: *Weight All*, which differentially weights all judgments in the opinion pool; *Select Crowd*, in which a subset of judgments is included but judgments in that subset are equally weighted (including the *select crowd* and *chase the expert* strategies); and *Hybrid Functions*, which select a subset of judgments to be weighted differentially. These functions differ primarily in terms of the discriminability or gradient of bias for indicators of expertise.

For simplicity's sake, we present these functions before normalization. In all cases, aggregators should normalize these values for use in aggregation functions. Furthermore, where task-specific performance measures are available, aggregators should substitute ρ_i, the judgment-specific performance measure, for ρ.

[4]Technically, the median is an (extremely) trimmed mean where one removes the lowest and highest 50% observations, so all trimmed means can be thought of as compromises between the mean and the median.

2.3.1 Weight All

The first class of weighting functions aggregates all of the judgments within the opinion pool. These functions assign differential weights to everyone within the opinion pool. In the specific case that $\rho \geq 0$ and positively and linearly correlates with the error function, the simplest weighting function is

$$\omega_i = \rho,$$

in which case our aggregation weight is simply equal to ρ. In practice, however, aggregators will often need to transform the performance metric ρ into a more suitable weight.

When ρ is negatively correlated with judgment quality (Karvetski et al., 2013; Fan et al., 2019), another potentially useful function is the standardized linear distance

$$\omega_i = \frac{\max(\rho) - \rho}{\max(\rho)},$$

which assigns the maximum weight of 1 to the judge with the lowest ρ and the minimum score of 0 to the judge with the highest ρ. In practice, this function guarantees the worst judge is ignored (i.e., assigned a weight of 0). The nullification of the contributions of the worst judge has different effects depending on the group. That is, eliminating the worst of four judges produces greater entropy and larger shifts in the aggregated average—relative to the unweighted average—than eliminating the worst of 40 judges. Note that, in the special case of a dyad ($N = 2$), this function is equivalent to simply choosing the *best expert*.

A more generalized, universal weighting function may be desirable, such as when an aggregator hypothesizes some relationship between a performance measure and judgment quality but is unsure of its precise nature. One solution is to simply rank the judges from best to worst such as the rank function used by Fan et al. (2019)

$$\omega_i = \frac{1}{rank(\rho)},$$

where larger values of ρ indicate worse performance and lower assigned weights. If higher values of ρ indicate greater performance, aggregators may substitute $-\rho$ in the denominator function.[5]

Another general-purpose function that aggregators may wish to consider is the exponential function (Wang et al., 2011a, b), defined as

[5] Ties are assigned the average of their collective ranks, i.e., three participants tied for first would receive a rank of $\frac{1+2+3}{3} = 2$.

$$\omega_i = e^{-\rho \times \beta},$$

where β is a tuning parameter chosen by the aggregator. This function assigns smaller weights in response to increasing values of ρ, but with decreasing gradient of penalties for increasingly deficient performance. The function is particularly useful as the aggregator can tune β to control the penalties associated with increasing values of the performance measure.

2.3.2 Select Crowd

The second class includes weighting functions that select or subset judgments within the opinion pool. These functions harness the *wisdom of select crowds* (Mannes et al., 2014). They may be appropriate if ρ is a reliable binary predictor or inflection point of an indicator of expertise. The simplest form of a *select crowd* strategy is the step function, which we express as

$$\omega_i = \begin{cases} 1; if \rho \in A \\ 0; if \rho \notin A \end{cases},$$

where *A* is a defined number or set of numbers. Judges who provide a value within the set will receive a weight of 1 if ρ is a member of *A* and a weight of 0 if ρ is not a member of *A*. The result is an equal weighting of the contributions of the former and nullifying the contributions of the latter. For example, we could define the set

$$\{A\} \mid \{A\} > \{0\}\},$$

resulting in the equal-weighted aggregation of all judges who scored $\rho > 0$, and the rejection of the contributions of all judges who scored $\rho < 0$. This function has proven useful in several performance-weighted aggregation studies (Afflerbach et al., 2021; Mannes et al., 2014).

Importantly, we can use the step function to provide a formal definition for the *CTE* strategy (Larrick & Soll, 2006). To the extent that an aggregator believes that higher values of ρ predict better judgment quality, the set

$$\{A \mid A = \max(\rho)\}$$

will assign a weight of 1 to the individual—or individuals in the case of a tie—with the highest ρ and a weight of 0 to all others. Aggregators can use the same principle to select any arbitrary value and the function will deterministically select the best judge *as predicted by the performance measure ρ*. In the extreme case where the correlation between ρ and judgment quality is 1, this is equivalent to deterministically selecting the best judge. In practice, however, there will be some deviation between the empirical best judge in the opinion pool and the best judge as predicted by ρ due to sampling error.

In principle, it should always be possible to select a subset of top judges who, when weighted equally, will match the performance of the full crowd with differential performance weights. Himmelstein et al. (2022) illustrate this principle in the context of time-sensitive weighting.

2.3.3 Hybrid Weighting Functions

Although the *weight all* and *select crowd* strategies may appear mutually exclusive, several functions combine elements of both strategies. We refer to these as hybrid weighting functions. For instance, if an aggregator hypothesizes there is some critical value of the performance metric but wants to preserve some discriminability in weighting between both experts and non-experts, they may employ a sigmoid function, defined as

$$\omega_i = \frac{e^{\rho+c}}{e^{\rho+c}+1},$$

where c is the critical value (or inflection point) at which differential weighting is steepest, with the gradient of differential weighting decreasing with increasing distance between ρ and c.

A final function that an aggregator may want to consider is the rectified linear unit (ReLU) or ramp function. This function is defined as

$$\omega_i = \rho^+ = \max(0, \rho),$$

resulting in a weight of 0 for judges with $\rho \leq 0$, increasing weight linearly where $\rho > 0$. The aggregator can trivially modify the formula to accommodate different scales or transformations of ρ as well as different critical values other than zero. Although more commonly associated with the fields of artificial intelligence programming and deep learning, the ReLU function may be useful for aggregators that wish to reject designated non-experts while differentially weighting judges who exhibit at least some level of expertise according to ρ.

2.4 Choosing a Weighting Function

The list of weighting functions we reviewed is not exhaustive. Each is potentially suitable for different conditions and task demands. Regardless, the weighting function chosen by the aggregator represents some hypothesis about the shape of the relationship between a performance measure and some measure of judgment quality (i.e., *BS* or *MAE*). Ideally, the aggregator should target indicators of *information* rather than accuracy itself. If an aggregator believes the relationship is linear, a linear function is logical. If an aggregator believes the relationship is exponential, an exponential function is justified. More general and flexible functions may be employed where the nature of the relationship is uncertain. Arbitrarily complex and specific functions may be employed where the relationship is known and esoteric.

Furthermore, the gradient of the weighting function—the rate at which weight diminishes to zero with increasingly deficient performance—should be determined by evidence that the performance measure is a reliable, sensitive, or specific indicator of expertise. If there is limited evidence, an aggregator should lean towards unweighted aggregation (Armstrong, 2001). If the performance measure is a reliable indicator of expertise, an aggregator may opt for a *weight all* strategy. If the performance measure is a sensitive and specialized indicator of expertise, *select some* or *chase the expert* strategies may be preferable.

In the remainder of the chapter, we discuss in greater detail various indicators of expertise (i.e., variants of ρ) that researchers have investigated or proposed. We discuss evidence in favor of, and against, these indicators as well as the implementation strategies that have worked best in each case.

3 Indicators of Expertise

A key component of performance-weighted aggregation is, of course, the performance measure. As we have discussed, an ideal performance measure is one that is a sensitive, specific, and reliable indicator of expertise. Unsurprisingly, the study of *what* to weight has received a great deal of attention in this context. After all, the best aggregation and weighting strategy will not improve accuracy if there is no underlying relationship between the indicator of expertise and judgment quality. Conversely, if the indicator is sensitive, specific, and reliable, the aggregator will benefit from virtually any of the performance-weighted aggregation strategies we have outlined.

Efforts to research indicators of expertise typically fall into one of three categories. The first is *history-based methods*, in which judges are weighted according to their historical performance on similar judgment tasks or seed events (Budescu & Chen, 2015; Cooke et al., 1988). The second is *disposition-based methods*, in which judges are weighted according to performance on some psychometric inventory or test (Himmelstein et al., 2021; Mellers et al., 2015). The third is *coherence-based methods*, in which judges are weighted according to their internal, logical consistency (Ho, 2020; Osherson & Vardi, 2006; Predd et al., 2008). The choice of an indicator will be determined by the nature of the judgment task itself, as well as pragmatic considerations about what additional data is available or could reasonably be collected. Each of the types of measures demands different degrees of rigor in record keeping, completion of an additional task, or computational power.

3.1 *History-Based Methods*

The most intuitive model of performance-weighted aggregation is to weight the judges according to their past performance. The logic is compelling: If a judge was accurate in past judgment tasks, they are likely to be accurate on similar judgment tasks in the future. Afflerbach et al. (2021) refers to these as history-based models because they rely on past information (i.e., 'seed variables'; Cooke & Goossens, 2008) to predict a judge's expertise and derive aggregation weights. These seed events can be based on the accuracy of earlier judgments on similar tasks (Aspinall, 2010; Budescu & Chen, 2015) or performance on a set of specifically designed calibration judgments (Cooke, 2014).

3.1.1 Cooke's Classical Method

A popular history-based model is Cooke's Classical Model (Colson & Cooke, 2018; Cooke et al., 1988; Cooke, 1991; Cooke & Goossens, 2008). The classical model uses LinOP aggregation that differentially weights judges according to their calibration and information. Calibration, in this context, refers to the correspondence between individual judgment and empirical data. Ideally, the calibration questions should be as closely related to the target judgment as possible and should not be simple general knowledge questions (Armstrong, 2001; Colson & Cooke, 2018). Information, in the context of the model, is the density of the expert's assessment compared to the background distribution. If an expert supplies narrower confidence intervals than their peers, the model assumes they are better informed.

In typical applications, judges answer a series of *target* questions including both the judgment of interest and calibration questions. The calibration questions are either items from the experts' field or related to the target questions. For both target and calibration judgments, judges estimate the fifth, fiftieth, and ninety-fifth percentiles of the relevant distribution (otherwise known as a credible interval). For example, an aggregator interested in the future rise in sea levels due to climate change might query a judge's predictions of sea-level rise in 2050 as the target event and use a nearer-term prediction of sea-level rise in 2025 as the seed event. The calibration score will be based on the actual sea level rise in 2025, while the information score is determined by the width of the credible interval. A judge who supplies a narrow and well-calibrated credible interval will receive a higher weight than one who provides a wider, poorly-calibrated credible interval.

Importantly, the true value of the calibration question(s) is unknown to the judge at the time of elicitation, but it will be known to the aggregator before the resolution of the target judgment of interest. Thus, the aggregator can confirm the calibration and quality of judges in the aggregation pool before the resolution of the judgment of interest, and can weight the individual judges accordingly.

Researchers have successfully applied the Classical Model in several domains such as investment banking, volcanology, public health, and even climate change

(Bamber et al., 2019). However, finding potential calibration questions—particularly those that are closely related to the target judgment—can often be difficult (Colson & Cooke, 2018). The proper application of Cooke's Classical Method also restricts the format of the elicitation method. Consequently, it is unsuited to certain judgment tasks, such as those that do not involve percentile estimates.

3.1.2 Contribution Weighted Model

A more recently developed history-based model is the Contribution Weighted Model (CWM; Budescu & Chen, 2015). The CWM weights forecasts based on each judge's relative performance and contribution to the crowd's accuracy in the past. The CWM weights judges according to their accuracy compared to the group as opposed to an absolute measure of accuracy. A judge *contributes* to the accuracy of the aggregate pool if the accuracy of the aggregate decreases when they are removed. Conversely, if the accuracy of the aggregate pool increases following the removal of that judge, they are down-weighted accordingly. Formally,

$$C_j = \sum_{i=1}^{N_j} \frac{S_i - S_i^{-J}}{N_j},$$

where C_j is the contribution of a given judge (where $j = 1, \ldots, J$), S_i represents the crowd's score or accuracy for a given judgment,[6] and S_i^{-J} represents the score of the crowd without the *j*th judge. The formula calculates the average difference between the crowd's score with and without the *j*th forecaster, across all N_j events. The contribution score can be positive, suggesting the judge improved the crowd's accuracy, or negative, suggesting the judge reduced the average accuracy of the crowd. Importantly, the formula can estimate the contribution of individual judges even if they abstain from some target judgments. The fact that contribution scores are a measure of relative performance insulates the weighting from correlated shifts in judgment accuracy, such as when a judgment task is easy, and performance is good for all judges. The aggregators can use the normalized contribution scores C_J to generate a weighted mean of the crowd's probabilistic judgment, including only those judges with positive contribution scores. This is conceptually similar to the previously described hybrid aggregation functions.

The CWM outperforms the unweighted linear mean of forecasts. It also outperformed a model based on the absolute performance of judges who participated in the Aggregative Contingent Estimation (ACE) project that covered a large variety of geopolitical and economic forecasting tasks (see https://www.iarpa.gov/research-programs/ace) and European Central Bank survey of professional forecasters

[6]Budescu and Chen (2015) used a quadratic scoring rule (de Finetti, 1962) to characterize the crowds accuracy. However, this method is applicable to any proper scoring rule (Bickel, 2007).

(Budescu & Chen, 2015). More recently, Chen et al. (2016) demonstrated that the contributions weights can be interpreted as measures of expertise by documenting that they vary as a function of the training, the teaming of the judges, and the timing of the forecasts. The CWM has also proven useful using diverse weighting strategies. In one study, Mannes et al. (2014) proposed a *select-crowd* strategy in combination with the CWM. Mannes et al. (2014) selected the top five most knowledgeable judges (according to past accuracy), finding that the simple average of this *select crowd* produced robust and accurate judgments across a wide range of settings and tasks.

3.1.3 Discussion

A typical finding for most history-weighted models is that the effectiveness improves as the number of seed events increases (Budescu & Chen, 2015; Chen et al., 2016; Eggstaff et al., 2014). This is intuitive, as a greater number of seed events will produce a more stable and accurate assessment of judges' performance. The effectiveness also depends, unsurprisingly, on the similarity between the seed event and the target judgment of interest (Armstrong, 2001; Colson & Cooke, 2018). As data about past performance becomes particularly rich, it may even be possible to 'weight the weights', placing differential emphasis on topics based on how closely related they are to the target judgment. To the extent that high-quality information about past performance on closely related tasks is available, history-based methods are ideal. This is because they weight judges according to an *objective* measure of true expertise, their past performance, which judges cannot easily fake.

Of course, the need for high-quality seed events to derive performance weights is a major drawback of history-based models. One can only employ these methods if: (1) aggregators track and record the judge's past performance, or (2) aggregators extend the elicitation process with seed tasks. These requirements may make history-based methods difficult, expensive, or outright impossible to apply in many judgment contexts. However, its intuitiveness and the wealth of supporting evidence in favor of these performance-weighting methods nevertheless make them an attractive possibility for aggregators and decision-makers concerned with quality decisions.

3.2 Disposition-Based Methods

The primary weakness of history-based methods is the necessity of either intensive record-keeping or the completion of judgment tasks for seed events upon which to base the weights. In the absence of record-keeping, an alternative solution is to base performance weights on some tests or inventories that judges complete either before or during judgment elicitation (Himmelstein et al., 2021). Mellers et al. (2015) referred to these types of methods as *dispositional methods*, as they utilize information about a judge's disposition rather than the judge's past performance on the judgment task itself. Some potential measures that have been investigated are domain-specific knowledge and psychometric indicators of expertise such as cognitive ability or an actively open-minded thinking style (Himmelstein et al., 2021; Mellers et al., 2014).

3.2.1 Domain Expertise

Weighting individuals according to indicators of domain-specific expertise is an intuitive choice. As Armstrong (2001) showed, even low levels of domain-specific knowledge can provide an effective means of improving aggregated judgment quality with differential weighting, assuming that the measure of domain-specific knowledge was itself valid and reliable. Their 'Rule-Based Forecasting' produced more accurate forecasts than equal-weighted combinations. More recently, Mellers et al. (2015) investigated whether domain-specific political knowledge was indicative of judgment quality in a political forecasting tournament and whether its use as an aggregation weight served to improve the accuracy of combined judgments. Mellers et al. found that percent scores on a series of true-or-false questions were correlated with forecasting accuracy.

3.2.2 Psychometric Indicators of Individual Differences

Another approach to dispositional weighting is to weight judges according to a psychometric measure such as general cognitive ability. It is natural to expect individuals of high cognitive ability to also perform better on certain judgment tasks. Indeed, the studies by Mellers et al. (2015) and Himmelstein et al. (2021) showed that aggregators can harness cognitive ability measures to improve aggregated judgment quality. This is consistent with research showing cognitive ability was one of the few consistent predictors of judgment quality across multiple tasks (Tetlock & Gardner, 2015). Mellers et al. (2015) also found that thinking style, specifically open-mindedness, predicted the quality of judgments and improved the quality of judgments through aggregation. Likewise, this is consistent with research showing that actively open-minded thinking predicted aspects of judgment quality (Baron, 1985; Haran et al., 2010; Mandel & Kapler, 2018), including vulnerability to biases, framing effects, and base-rate neglect. Finally, Himmelstein et al. (2021) showed a negative correlation between overconfidence and accuracy. This relatively new line of research shows that individual differences in thinking style may indeed be an effective basis upon which to base performance-weighted aggregation.

3.2.3 Discussion

The primary advantage of disposition-based methods, compared to history-based methods, is that aggregators do not need to engage in historic record-keeping of judgment quality in related domains to implement the strategy. Rather, aggregators can acquire the data to weight judges during elicitation and apply it immediately. Furthermore, aggregators can apply the method to any judgment task, as the tests and inventories are distinct from, and independent of, the judgment task itself. These results are unsurprising, particularly in the domain-specific case. As we alluded to in the introduction, it is far more likely that an oncologist will correctly diagnose cancer than a layperson, and evidence supports the fact that domain-specific knowledge enhances judgment quality in some cases (Armstrong, 2001; Mellers et al., 2015).

Notably, Tetlock (2005) demonstrated that political science experts on China did not produce better quality judgments about Chinese geopolitics than political scientists with expertise in other regions. And it is also the case that domain-specific experts are not free of noise and bias themselves (Bolger & Wright, 1994). For instance, clinical investigators are overly optimistic: they demonstrate poor accuracy when forecasting scientific and operational outcomes of their clinical trials (Benjamin et al., 2022) and when forecasting the replicability of trial results (Benjamin et al., 2017). As noted earlier, oncologists display virtually no discrimination skill when predicting the efficacy of clinical treatments in randomized controlled trials (Benjamin et al., 2021). Thus, domain expertise does not guarantee judgment accuracy. By contrast, psychometric indicators predict performance on a variety of tasks (Satopää et al., 2021; Tetlock & Gardner, 2015).

Of course, this strength also represents a potential weakness: psychometric dispositional measures are employed specifically when expertise is hard to define or measure. Thus, the dispositional method is, by definition, a less reliable indicator of task-specific judgment quality than a task-specific performance measure. Furthermore, a true domain-specific expert will not necessarily score highest on a measure of general cognitive ability. Thus, dispositional measures and domain-specific expertise will not always align. To the extent that domain-specific measures of expertise are available and are reliable performance indicators, they are likely to be preferable to dispositional measures (Armstrong, 2001). Indeed, Himmelstein et al. (2021) showed that dispositional measures can predict forecasting accuracy in the absence of performance information but, as more historical performance data accumulates, the relative importance of the dispositional measures decreases significantly.

3.3 Coherence-Based Methods

We have shown that history-based and disposition-based methods hold promise for improving the accuracy of aggregated judgments and forecasts. However, both approaches complicate the elicitation and aggregation process. They also may be infeasible in some cases: history-based models require meticulous record-keeping of past performance, and disposition models require the completion of added tests and inventories. In contrast, coherence-weighted models derive aggregation weights from indirect information obtained during the elicitation process itself. Specifically, the method weights judges according to how internally consistent and coherent their judgments are. Simply put, more coherent judges receive greater weight than their less coherent counterparts. Importantly, because coherence depends on indirect information about the judgments themselves, there may be little or no additional elicitation burden or record-keeping needed to implement the strategy.

3.3.1 Coherence Approximation Principle

For any problem where the probabilistic or logical constraints can be mathematically defined, aggregators can use the *coherence approximation principle* (*CAP*; Osherson & Vardi, 2006) to quantify judgment coherence; or more specifically their incoherence. Originally developed to recalibrate incoherent responses, the *CAP* takes a set of elicited, incoherent judgments and finds the nearest equivalent set of judgments that satisfy the relevant logical and probabilistic constraints. The *CAP* returns these coherent probabilities as a recalibrated judgment, as well as the measure of the distance between the elicited and recalibrated judgments. We refer to the resulting distance as the *incoherence metric* (*IM*). Calculation of the *IM* is a constrained optimization problem (Martins & Ning, 2021; Rossi et al., 2006), in which the algorithm identifies the nearest coherent judgment according to an objective function; typically, the Euclidean distance. However, in practice, any distance function could be used (e.g., absolute difference).

Consider Kolmogorov's three axioms of probability (Kolmogorov, 1956; De Finetti, 1937): non-negativity, unitarity, and additivity. Non-negativity states that probabilities cannot be negative. Unitarity (or complementarity in the binary case) states that the summed probability of all elementary events in the sample space must equal 1. Additivity states that a countable sequence of mutually exclusive and exhaustive events E_1, E_2, . . ., En must satisfy the condition

$$\boldsymbol{P}\left(\bigcup_{i=1}^{n}\boldsymbol{E}_i\right)=\sum_{i=1}^{n}\boldsymbol{P}(\boldsymbol{E}_i).$$

To illustrate how the *CAP* applies to judgment tasks, consider the study by Karvetski et al. (2013). In the study, participants judged the probability that each of four different statements about a topic was true. For example, one of the tasks required participants to judge the probability that the following statements about Neil Armstrong and Buzz Aldrin's 1969 moon landing were true:

$P(A)$: Neil Armstrong was the first man to step on the moon.
$P(A^c)$: Neil Armstrong was not the first man to step on the moon.
$P(B)$: Buzz Aldrin was the first man to step on the moon.
$P(A \cup B)$: Either Neil Armstrong or Buzz Aldrin was the first man to step on the moon.

For any arbitrary set of judgments on this topic, we can find the nearest equivalent coherent estimates—and the associated *IM*—by minimizing the following Euclidean distance objective formula

$$IM = \sqrt{\sum_{i=1}^{4} (y_i - c_i)^2},$$

where y is the original elicited judgment and c is its coherentized equivalent, subject to three constraints. First, each probability is subject to the non-negativity constraint. Second, $P(A)$ and $P(A^c)$ are subject to the complementarity constraint

$$P(A) + P\left(A^C\right) = 1.$$

Third, $P(A)$, $P(B)$, and $P(A \cup B)$ are subject to the unitarity constraint:

$$P(A) + P(B) = P(A \cup \mathrm{B}).$$

A coherent judge will not require recalibration and receive an *IM* score of 0, while increasingly incoherent judges will require greater recalibration and have an increasingly large *IM*.

Figure 7.1 displays the *CAP* applied to a subset of the judgments, $P(A)$, $P(\mathrm{B})$, $P(A \cup B)$ that violate the unitarity constraint, represented by the filled in orange dot. For true values, we used the moon landing example above, $x = \{1, 0, 1\}$ for $P(A)$, $P(\mathrm{B})$, and $P(A \cup B)$, respectively, represented by the green circle. The flat grey polygon with vertices {0, 0, 0}, {1, 0 1}, and {0, 1, 1} defines the plane of coherent judgment sets that satisfy the unitarity constraint. Euclidean optimization identifies the shortest distance between the elicited judgments $y = \{1, 1, 1\}$ and some point on this plane, returning $c = \{0.5, 0.5, 1\}$ at $IM = 0.707$, represented by the filled in blue circle. Importantly, $BS(c)$ is strictly superior to $BS(y)$ in all cases using Euclidean optimization (Karvetski et al. 2013; Osherson & Vardi, 2006).

Although *CAP* was developed for judgment recalibration, researchers quickly identified *IM* as a metric suitable for performance-weighted aggregation. In its earliest application for probabilistic forecasts, Predd et al. (2008) showed significant improvements in aggregated group accuracy on tasks ranging from sports predictions to economic forecasts. Similarly, coherence-weighted aggregation improved group accuracy on presidential election forecasts (Wang et al., 2011a, b). Later studies showed its utility when applied to general-knowledge tasks (Fan et al., 2019; Karvetski et al., 2013). This is noteworthy because pooled judgments cannot bracket a probability that is strictly *True* ($P = 1$) or *False* ($P = 0$) and thus the unweighted aggregation is completely ineffective. Moreover, Karvetski et al. (2013) demonstrated that coherence weighting produced substantially larger improvements in accuracy than could be achieved using a dispositional metric that assessed numeracy skill. Finally, group accuracy in complex Bayesian judgment tasks benefitted from

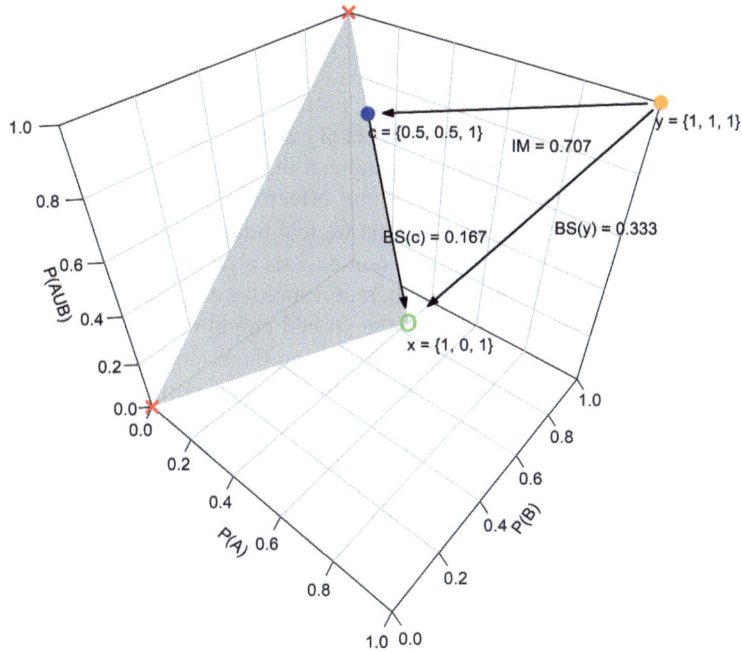

Fig. 7.1 CAP Applied to Set of Moon Landing Judgments Violating the Unitarity Constraint

coherence-weighted aggregation (Collins et al., in press; Karvetski et al., 2020; Mandel et al., 2018). Critically, the benefits of coherence-weighted aggregation accrue in addition to the recalibration process, with each process providing unique improvements. Although these studies demonstrate the efficacy of coherence-weighted aggregation, the relationship between coherence and correspondence is still a matter of debate among decision theorists and researchers (Budescu et al., 2021; Mellers et al., 2017; Weaver & Stewart, 2012; Weiss et al., 2009; Wright & Ayton, 1987).

One reason why coherence-weighting may be effective is that *IM* represents a judge's knowledge of—and willingness to apply—the axioms of probability theory. In this respect, *IM* may act as a dispositional measure of probabilistic numeracy. In other cases, quality judgment may depend on the ability to coherently synthesize multiple pieces of information. For instance, researchers showed that *IM* predicted accuracy on complex Bayesian probability estimation tasks that require the synthesis of multiple pieces of information (Collins et al., in press; Karvetski et al., 2020; Mandel et al., 2018). In these cases, *IM* may act similarly to a dispositional measure of *domain-specific expertise*. Another reason coherence-weighting may be effective is that coherence is a necessary, but insufficient, criterion for perfect judgment (Dunwoody, 2009; Hammond, 2000). Consequently, *IM* is sensitive: if the crowd includes true experts, they will certainly receive the maximal weight. Conversely, however, the necessary but insufficient criterion also reveals a potential drawback of coherence-weighting, in that the *IM* exhibits poor specificity: a coherent judge will not necessarily be correct. Counterintuitively, then, coherence-weighted aggregation often works best when the elicitation process makes coherence harder to achieve (Collins et al., in press; Karvetski et al., 2013).

Karvetski et al. (2013) paradoxically showed that grouping sets of related judgment improved coherence but reduced the effectiveness of coherence-weighted aggregation. In the grouped condition, the logical relationships between the judgments are equally salient to experts and non-experts alike. As a result, both experts and non-experts are likely to provide coherent responses on principle rather than as a consequence of expertise. By contrast, the spaced condition obfuscates the logical relations between related judgments by displacing them temporally. In this condition, an expert who is certain of the correct answer—for example, Neil Armstrong was the first man to land on the moon, $P(A) = 1$—is equally certain about the related queries $P(B)$, $P(A^c)$, and $P(A \cup B)$. They will be coherent, even if they are not cognizant or even aware of the logical relationship between the judgments. By contrast, judges uncertain about $P(A)$ are likely to be uncertain about the related queries $P(B)$, $P(A^c)$, and $P(A \cup B)$. To produce a coherent set of judgments, they must: (1) recognize the underlying coherence relationship between the queries, and (2) correctly recall their responses to previous, related queries. Karvetski et al. (2013) also noted that not all coherence constraints worked equally well. Assessing *IM* according to the complementarity constraint diluted the effectiveness of coherence weighting. This is because individuals will often express epistemic uncertainty by responding 0.5 to both $P(A)$ and $P(A^c)$. This response is incidentally coherent but maximally uncertain. Both examples described above illustrate the potential problems of specificity associated with the use of *IM* in performance-weighted aggregation.

The reliance on sensitivity to improve performance on some judgment tasks suggests that, to the extent that elicitation strategies can minimize the specificity problem, aggregators who employ coherence-weighted aggregation may benefit from a *select crowd* strategy. Specifically, aggregators may wish to consider taking the average of only those judges in the pool who supplied coherent judgments. Thus, future research might investigate whether a *select crowd* of only *coherent* judges might be a useful basis for aggregation.

3.3.2 Probabilistic Coherence Scale

In most applications of the coherence-weighting approach, the degree of (in)-coherence for judges is derived from their answers to a set of items that are part of the main judgment task. Ho (2020) proposed a variation on this theme, by developing and validating psychometrically a measure of probabilistic coherence using items that measure five coherence features:

- Binary additivity (How close is P(Event) + P(not Event) to 1?),
- Trinary consistency (How close is P(A U B) to (P(A) + P(B) when A and B are disjoint?),
- Monotonicity with respect to time (Is P(E) over long time) ≥ P(E) over shorter time?),
- Monotonicity with respect to space (Is P(E) over large area ≥ P(E) over smaller area?),
- Monotonicity with respect to precision and information (Is P(E) imprecise description ≥ P(E) more precise description?).

Ho (2020) administered the coherence scale to two groups of volunteers—judges who took part in past forecasting experiments and tournaments—and found positive correlations between their coherence scores and their accuracy. Consequently, Ho (2020) was able to derive coherence weighting schemes that outperformed an unweighted mean. Most recently, Budescu et al. (2021) administered this scale to several hundred participants in a multi-stage longitudinal forecasting study and found that the coherence score was the best predictor of forecasting accuracy at five distinct forecasting horizons ranging from three months to two weeks.

3.3.3 Discussion

Research on coherence-weighted aggregation is promising. Aggregators may apply it at the level of individual judges and judgments with minimal added burden of record-keeping or elicitation. Applications of coherence-weighted aggregation are, of course, limited to those judgment tasks in which probabilistic coherence constraints are present and relevant, i.e., probability judgments of at least two logically related events, and ideally for constraints that are complex enough or elicited in such a way that they minimize the specificity problem.

Another advantage of coherence-weighted aggregation is that it combines recalibration and performance-weighted aggregation methods. Recall that the *CAP* both generates *IM* and returns a recalibrated set of coherent judgments. Because this recalibration process strictly improves judgment accuracy (Karvetski et al. 2013; Osherson & Vardi, 2006), at least for those problems constrained by axioms Kolmogorov (1956) describes, coherence-weighted aggregation is typically applied to these recalibrated judgments. Thus, coherence-weighted aggregation works in tandem with recalibration in the research we have described. Importantly, this demonstrates that the *CAP* efficiently provides two methods for enhancing judgment quality, each contributing unique improvements.

Although the research provides a good accounting for the mechanism of action underlying coherence-weighted aggregation in several classes of judgment tasks, there is not yet a comprehensive explanation. Future research will need to investigate the precise mechanism of action that underlies the effectiveness of coherence-weighted aggregation in judgments or forecasts of future events. Regardless, aggregators who wish to employ coherence-weighted aggregation should consider how to maximize the sensitivity and specificity of the *IM* during the elicitation process.

4 General Discussion

In this chapter, we have described strategies for combining judgments in an opinion pool. We discussed the merits of *WOC, CTE*, and various performance-weighted alternatives that fall in between these extremes. We have argued that decision-makers have much to gain from the synthesis of the two approaches through performance-weighted aggregation: exploiting the noise-canceling benefits of the *WOC* effects and exploiting the signal-amplifying benefits of detecting *experts* within the crowd. The nature of the judgment task will determine which of the methods—history-based, disposition-based, or coherence-based—as well as the degree of weighting—unweighted, best expert, or varying levels of performance weighting—will produce optimal judgments. Indeed, decision-makers can model their degree of confidence in the validity and reliability of the performance measure using different weighting functions. Performance-weighted aggregation is thus a useful tool for the savvy aggregator to consider.

4.1 Ensemble Methods

Of course, performance-weighted aggregation is just one tool in a decision-maker's box. Importantly, performance-weighted aggregation can be used in combination with other judgment improvement strategies to maximize accuracy. As we demonstrated with the *CAP*, recalibration methods were combined with coherence-weighted aggregation to pronounced effect (Karvetski et al., 2013; Mandel et al., 2018; Osherson & Vardi, 2006). Other recalibration methods such as extremization (Baron et al., 2014; Mandel & Barnes, 2014; Turner et al., 2014) can easily be used in tandem with performance weighting to further enhance judgment quality. Such recalibration methods can be applied to judgments before aggregation or to the aggregated judgments themselves.

Aggregators may also benefit from combining performance-weighted aggregation with different elicitation procedures. We have discussed opinion pools primarily in terms of a simplified, isolated, non-interacting crowd. Indeed, diverse, independent crowds are considered ideal conditions for the formation of wise crowds (Surowiecki, 2004). Independence and diversity avoid the pitfalls of information cascades and groupthink that increase the chance of correlated error. However, structured and sequential sharing of information during elicitation may improve judgment quality (Tump et al., 2020). If knowledgeable individuals respond first, it can lead to a *positive* information cascade that enhances the group's judgment accuracy. Aggregators may also harness the wisdom of the crowd in tandem with the wisdom of the *inner* crowds (Herzog & Hertwig, 2014). Judges can be made to forget their original judgments, or they can be encouraged to consider counterfactuals and alternate realities. In either case, several non-redundant judgments from a single individual can be aggregated for improved judgment. Such judgments may be used in performance-weighted aggregation. Finally, Karvetski et al. (2020) showed that structured elicitation procedures could be exploited using performance-weighted aggregation to further improve judgment quality.

We would caution decision-makers against employing all the tools at their disposal at once. Research has demonstrated that simply combining all the different weighting methods at one's disposal may be ineffective. A 'kitchen sink' approach that weighted several performance metrics at once (e.g., interval width, interval asymmetry, variability in interval width) did not optimize judgment quality relative to simpler methods (Hanea et al., 2021). Furthermore, even less research has examined (or identified) the optimal combinations or *ensembles* combining various elicitation, recalibration, and aggregation (Karvetski et al., 2020). The order they are applied may have important effects, such as whether recalibration should be applied before or after aggregation. Future research that explored the effectiveness of these ensemble methods is needed.

4.2 Conclusion

At the outset, we suggested the default approach of the savvy aggregator should be to rely on *WOC* and utilize unweighted aggregation. However, to the extent that valid and reliable indicators of expertise can be identified, aggregators should increasingly favor a *CTE* approach. We propose that performance-weighted aggregation can serve as a method for reconciling these two extremes. Valid, reliable, and sophisticated indicators of expertise do exist, and decision-makers would be wise to capitalize on them. As larger sets of judgment data become easier and less expensive to collect, and as the computational power at the disposal of aggregators continues to grow, so too will the tilt toward performance-weighted aggregation. With these methods, decision-makers can better ferret out wisdom within the crowd.

References

Afflerbach, P., van Dun, C., Gimpel, H., Parak, D., & Seyfried, J. (2021). A simulation-based approach to understanding the wisdom of crowds phenomenon in aggregating expert judgment. *Business & Information Systems Engineering, 63*(4), 329–348. https://doi.org/10.1007/s12599-020-00664-x

Armstrong, J. S. (2001). Combining forecasts. In *Principles of forecasting: A handbook for researchers and practitioners* (1st ed., p. 21). Kluwer Academic Publishers.

Aspinall, W. (2010). A route to more tractable expert advice. *Nature, 463*(7279), 294–295. https://doi.org/10.1038/463294a

Atanasov, P., Rescober, P., Stone, E., Swift, S. A., Servan-Schreiber, E., Tetlock, P., Ungar, L., & Mellers, B. (2017). Distilling the wisdom of crowds: Prediction markets vs. prediction polls. *Management Science, 63*(3), 691–706. https://doi.org/10.1287/mnsc.2015.2374

Bamber, J. L., Oppenheimer, M., Kopp, R. E., Aspinall, W. P., & Cooke, R. M. (2019). Ice sheet contributions to future sea-level rise from structured expert judgment. *Proceedings of the National Academy of Sciences, 116*(23), 11195–11200. https://doi.org/10.1073/pnas.1817205116

Baron, J. (1985). Rationality and intelligence. *Cambridge University Press.* https://doi.org/10.1017/CBO9780511571275

Baron, J., Mellers, B. A., Tetlock, P. E., Stone, E., & Ungar, L. H. (2014). Two reasons to make aggregated probability forecasts more extreme. *Decision Analysis, 11*(2), 133–145. https://doi.org/10.1287/deca.2014.0293

Benjamin, D., Mandel, D. R., & Kimmelman, J. (2017). Can cancer researchers accurately judge whether preclinical reports will reproduce? *PLoS Biology, 15*(6), 1–17. https://doi.org/10.1371/journal.pbio.2002212

Benjamin, D., Mandel, D. R., Barnes, T., Krzyzanowska, M. K., Leighl, N. B., Tannock, I. F., & Kimmelman, J. (2021). Can oncologists predict the efficacy of treatment in randomized trials? *The Oncologist, 26*, 56–62. https://doi.org/10.1634/theoncologist.2020-0054

Benjamin, D. M., Hey, S. P., MacPherson, A., Hachem, Y., Smith, K. S., Zhang, S. X., Wong, S., Dolter, S., Mandel, D. R., & Kimmelman, J. (2022). Principal investigators over-optimistically forecast scientific and operational outcomes for clinical trials. *PLoS One, 17*(2), e0262862. https://doi.org/10.1371/journal.pone.0262862

Bickel, J. E. (2007). Some comparisons among quadratic, spherical, and logarithmic scoring rules. *Decision Analysis, 4*(2), 49–65. https://doi.org/10.1287/deca.1070.0089

Bolger, F., & Wright, G. (1994). Assessing the quality of expert judgment. *Decision Support Systems, 11*(1), 1–24. https://doi.org/10.1016/0167-9236(94)90061-2

Brier, G. W. (1950). Verification of forecasts expressed in terms of probability. *Monthly Weather Review, 78*, 1–3. https://doi.org/10.1175/1520-0493(1950)078<0001:VOFEIT>2.0.CO;2

Broomell, S., & Budescu, D. V. (2009). Why are experts correlated? Decomposing correlations between judges. *Psychometrika, 74*(3), 531–553. https://doi.org/10.1007/s11336-009-9118-z

Budescu, D. V., & Chen, E. (2015). Identifying expertise to extract the wisdom of crowds. *Management Science, 61*(2), 267–280. https://doi.org/10.1287/mnsc.2014.1909

Budescu, D. V., Himmelstein, M., & Ho, E. (2021, October). Boosting the wisdom of crowds with social forecasts and coherence measures. In *Presented at annual meeting of Society of Multivariate Experimental Psychology (SMEP)*, online.

Chen, E., Budescu, D. V., Lakshmikanth, S. K., Mellers, B. A., & Tetlock, P. E. (2016). Validating the contribution-weighted model: Robustness and cost-benefit analyses. *Decision Analysis, 13* (2), 128–152. https://doi.org/10.1287/deca.2016.0329

Clemen, R. T. (1989). Combining forecasts: A review and annotated bibliography. *International Journal of Forecasting, 5*(4), 559–583.

Clemen, R. T., & Winkler, R. L. (1999). Combining probability distributions from experts in risk analysis. *Risk Analysis, 19*(2), 187–203. https://doi.org/10.1111/j.1539-6924.1999.tb00399.x

Collins, R. N., Mandel, D. R., Karvetski, C. W., Wu, C. M., & Nelson, J. D. (in press). *The wisdom of the coherent: Improving correspondence with coherence-weighted aggregation*. Decision.

Colson, A. R., & Cooke, R. M. (2018). Expert elicitation: Using the classical model to validate experts' judgments. *Review of Environmental Economics and Policy, 12*(1), 113–132. https://doi.org/10.1093/reep/rex022

Cooke, R. M. (1991). *Experts in uncertainty: Opinion and subjective probability in science*. Oxford University Press.

Cooke, R. M. (2014). Validating expert judgment with the classical model. In C. Martini & M. Boumans (Eds.), *Experts and consensus in social science* (Vol. 50, pp. 191–212). Springer. https://doi.org/10.1007/978-3-319-08551-7_10

Cooke, R. M., & Goossens, L. L. H. J. (2008). TU Delft expert judgment data base. *Reliability Engineering & System Safety, 93*(5), 657–674. https://doi.org/10.1016/j.ress.2007.03.005
Cooke, R., Mendel, M., & Thijs, W. (1988). Calibration and information in expert resolution; a classical approach. *Automatica, 24*(1), 87–93. https://doi.org/10.1016/0005-1098(88)90011-8
Davis-Stober, C. P., Budescu, D. V., Dana, J., & Broomell, S. B. (2014). When is a crowd wise? *Decision, 1*(2), 79–101. https://doi.org/10.1037/dec0000004
de Finetti, B. (1937). La prévision: Ses lois logiques, ses sources subjectives. *Annales de l'Institut Henri Poincaré, 7*, 1–68.
de Finetti, B. (1962). Does it make sense to speak of "good probability appraisers"? In I. J. Good (Ed.), *The scientist speculates: An anthology of partly-baked ideas* (pp. 357–363). Wiley.
Dietrich, F., & List, C. (2017). *Probabilistic opinion pooling* (Vol. 1). Oxford University Press. https://doi.org/10.1093/oxfordhb/9780199607617.013.37
Dunwoody, P. T. (2009). Theories of truth as assessment criteria in judgment and decision making. *Judgment and Decision Making, 4*(2), 116–125. https://doi.org/10.1017/S1930297500002540
Eggstaff, J. W., Mazzuchi, T. A., & Sarkani, S. (2014). The effect of the number of seed variables on the performance of Cooke's classical model. *Reliability Engineering & System Safety, 121*, 72–82. https://doi.org/10.1016/j.ress.2013.07.015
Fan, Y., Budescu, D. V., Mandel, D., & Himmelstein, M. (2019). Improving accuracy by coherence weighting of direct and ratio probability judgments. *Decision Analysis, 16*(3), 197–217. https://doi.org/10.1287/deca.2018.0388
Galton, F. (1907). Vox Populi. *Nature, 75*(1949), 450–451. https://doi.org/10.1038/075450a0
Gneiting, T., & Raftery, A. E. (2007). Strictly proper scoring rules, prediction, and estimation. *Journal of the American Statistical Association, 102*(477), 359–378. https://doi.org/10.1198/016214506000001437
Goldstein, R., Almenberg, J., Dreber, A., Emerson, J. W., Herschkowitsch, A., & Katz, J. (2008). Do more expensive wines taste better? Evidence from a large sample of blind tastings. *Journal of Wine Economics, 3*(1), 1–9. https://doi.org/10.22004/ag.econ.37328
Hammond, K. R. (2000). Coherence and correspondence theories in judgment and decision making. In T. Connolly, K. Hammond, & H. Arkes (Eds.), *Judgment and decision making: An interdisciplinary reader* (2nd ed., pp. 53–65). Cambridge University Press.
Han, Y., & Budescu, D. (2019). A universal method for evaluating the quality of aggregators. *Judgment and Decision Making, 14*(4), 395–411. https://doi.org/10.1017/S1930297500006094
Hanea, A. M., McBride, M. F., Burgman, M. A., & Wintle, B. C. (2018). The value of performance weights and discussion in aggregated expert judgments. *Risk Analysis, 38*(9), 1781–1794. https://doi.org/10.1111/risa.12992
Hanea, A. M., Wilkinson, D. P., McBride, M., Lyon, A., van Ravenzwaaij, D., Singleton Thorn, F., Gray, C., Mandel, D. R., Willcox, A., Gould, E., Smith, E. T., Mody, F., Bush, M., Fidler, F., Fraser, H., & Wintle, B. C. (2021). Mathematically aggregating experts' predictions of possible futures. *PLoS One, 16*(9), e0256919. https://doi.org/10.1371/journal.pone.0256919
Haran, U., Moore, D. A., & Morewedge, C. K. (2010). A simple remedy for overprecision in judgment. *Judgment and Decision Making, 5*, 467–476. https://doi.org/10.1017/S1930297500001637
Hastie, R., & Kameda, T. (2005). The robust beauty of majority rules in group decisions. *Psychological Review, 112*(2), 494–508. https://doi.org/10.1037/0033-295X.112.2.494
Hemming, V., Hanea, A. M., Walshe, T., & Burgman, M. A. (2020). Weighting and aggregating expert ecological judgments. *Ecological Applications, 30*(4), e02075. https://doi.org/10.1002/eap.2075
Herzog, S. M., & Hertwig, R. (2014). Harnessing the wisdom of the inner crowd. *Trends in Cognitive Sciences, 18*(10), 504–506. https://doi.org/10.1016/j.tics.2014.06.009
Himmelstein, M., Atanasov, P., & Budescu, D. V. (2021). Forecasting forecaster accuracy: Contributions of past performance and individual differences. *Judgment and Decision Making, 16*(2), 323–362. https://doi.org/10.1017/S1930297500008597

Himmelstein, M., Budescu, D. V., & Han, Y. (2022). The wisdom of timely crowds. In M. Seiffert (Ed.), *Judgment and predictive analytics* (1st ed.). Springer Nature.

Ho, E. H. (2020, June). *Developing and validating a method of coherence-based judgment aggregation*. Unpublished PhD Sissertation. Fordham University.

Jaspersen, J. G. (2021). Convex combinations in judgment aggregation. *European Journal of Operational Research, 299*, 780–794. https://doi.org/10.1016/j.ejor.2021.09.050

Jose, V. R. R., Grushka-Cocayne, Y., & Lichtendahl, K. C., Jr. (2013). Trimmed opinion pools and the crowd's calibration problem. *Management Science, 60*(20), 463–475. https://doi.org/10.1287/mnsc.2013.1781

Kahneman, D., Rosenfield, A. M., Gandhi, L., & Blaser, T. (2016). How to overcome the high, hidden cost of inconsistent decision making. *Harvard Business Review, 94*, 36–43. Retrieved January 28, 2022, from https://hbr.org/2016/10/noise

Kahneman, D., Sibony, O., & Sunstein, C. R. (2021). *Noise: A flaw in human judgment*. Little, Brown Spark.

Karvetski, C. W., Olson, K. C., Mandel, D. R., & Twardy, C. R. (2013). Probabilistic coherence weighting for optimizing expert forecasts. *Decision Analysis, 10*(4), 305–326. https://doi.org/10.1287/deca.2013.0279

Karvetski, C. W., Mandel, D. R., & Irwin, D. (2020). Improving probability judgment in intelligence analysis: From structured analysis to statistical aggregation. *Risk Analysis, 40*(5), 1040–1057. https://doi.org/10.1111/risa.13443

Kolmogorov, A. N. (1956). *Foundations of the theory of probability*. (N. Morrison, Trans.; 2nd English Edition). Chelsea Publishing Company.

Larrick, R. P., & Soll, J. B. (2006). Intuitions about combining opinions: Misappreciation of the averaging principle. *Management Science, 52*(1), 111–127. https://doi.org/10.1287/mnsc.1050.0459

Larrick, R. P., Mannes, A. E., & Soll, J. B. (2011). The social psychology of the wisdom of crowds. In J. I. Krueger (Ed.), *Social judgment and decision making* (pp. 227–242). Psychology Press.

Lorenz, J., Rauhut, H., Schweitzer, F., & Helbing, D. (2011). How social influence can undermine the wisdom of crowd effect. *Proceedings of the National Academy of Sciences, 108*(22), 9020–9025. https://doi.org/10.1073/pnas.1008636108

Makridakis, S., & Winkler, R. L. (1983). Averages of forecasts: Some empirical results. *Management Science, 29*(9), 987–996. https://doi.org/10.1287/mnsc.29.9.987

Mandel, D. R., & Barnes, A. (2014). Accuracy of forecasts in strategic intelligence. *Proceedings of the National Academy of Sciences, 111*(30), 10984–10989. https://doi.org/10.1073/pnas.1406138111

Mandel, D. R., & Barnes, A. (2018). Geopolitical forecasting skill in strategic intelligence: Geopolitical forecasting skill. *Journal of Behavioral Decision Making, 31*(1), 127–137. https://doi.org/10.1002/bdm.2055

Mandel, D. R., & Kapler, I. V. (2018). Cognitive style and frame susceptibility in decision-making. *Frontiers in Psychology, 9*, 1461. https://doi.org/10.3389/fpsyg.2018.01461

Mandel, D. R., Karvetski, C. W., & Dhami, M. K. (2018). Boosting intelligence analysts' judgment accuracy: What works, what fails? *Judgment and Decision Making, 13*(6), 607–621. https://doi.org/10.1017/S1930297500006628

Mannes, A. E., Soll, J. B., & Larrick, R. P. (2014). The wisdom of select crowds. *Journal of Personality and Social Psychology, 107*(2), 276–299. https://doi.org/10.1037/a0036677

Martins, J. R. R. A., & Ning, A. (2021). *Engineering design optimization* (1st ed.). Cambridge University Press. https://doi.org/10.1017/9781108980647

Mellers, B., Ungar, L., Baron, J., Ramos, J., Gurcay, B., Fincher, K., Scott, S. E., Moore, D., Atanasov, P., Swift, S. A., Murray, T., Stone, E., & Tetlock, P. E. (2014). Psychological strategies for winning a geopolitical forecasting tournament. *Psychological Science, 25*(5), 1106–1115. https://doi.org/10.1177/0956797614524255

Mellers, B., Stone, E., Atanasov, P., Rohrbaugh, N., Metz, S. E., Ungar, L., Bishop, M. M., Horowitz, M., Merkle, E., & Tetlock, P. (2015). The psychology of intelligence analysis:

Drivers of prediction accuracy in world politics. *Journal of Experimental Psychology: Applied, 21*(1), 1–14. https://doi.org/10.1037/xap0000040

Mellers, B. A., Baker, J. D., Chen, E., Mandel, D. R., & Tetlock, P. E. (2017). How generalizable is good judgment? A multi-task, multi-benchmark study. *Judgment and Decision Making, 12*(4), 369–381. https://doi.org/10.1017/S1930297500006240

Osherson, D., & Vardi, M. Y. (2006). Aggregating disparate estimates of chance. *Games and Economic Behavior, 56*(1), 148–173. https://doi.org/10.1016/j.geb.2006.04.001

Park, S., & Budescu, D. V. (2015). Aggregating multiple probability intervals to improve calibration. *Judgment and Decision Making, 10*(2), 130–143. https://doi.org/10.1017/S1930297500003910

Peterson, W., Birdsall, T., & Fox, W. (1954). The theory of signal detectability. *Transactions of the IRE Professional Group on Information Theory, 4*(4), 171–212. https://doi.org/10.1109/TIT.1954.1057460

Predd, J. B., Osherson, D. N., Kulkarni, S. R., & Poor, H. V. (2008). Aggregating probabilistic forecasts from incoherent and abstaining experts. *Decision Analysis, 5*(4), 177–189. https://doi.org/10.1287/deca.1080.0119

Predd, J. B., Seiringer, R., Lieb, E. H., Osherson, D. N., Poor, H. V., & Kulkarni, S. R. (2009). Probabilistic coherence and proper scoring rules. *IEEE Transactions on Information Theory, 55*(10), 4786–4792. https://doi.org/10.1109/TIT.2009.2027573

Rossi, F., van Beek, P., & Walsh, T. (2006). Chapter 1—Introduction. In F. Rossi, P. van Beek, & T. Walsh (Eds.), *Foundations of artificial intelligence* (Vol. 2, pp. 3–12). Elsevier. https://doi.org/10.1016/S1574-6526(06)80005-2

Satopää, V. A., Salikhov, M., Tetlock, P. E., & Mellers, B. (2021). Bias, information, noise: The BIN model of forecasting. *Management Science, 67*(12), 7599–7618. https://doi.org/10.1287/mnsc.2020.3882

Silver, N. (2012). *The signal and the noise: Why so many predictions fail—But some don't.* Penguin.

Surowiecki, J. (2004). *The wisdom of crowds: Why the many are smarter than the few and how collective wisdom shapes business, economies, societies, and nations.* Doubleday & Co..

Tetlock, P. E. (2005). *Expert political judgement: How good is it? How can we know?* Princeton University Press.

Tetlock, P. E., & Gardner, D. (2015). *Superforecasting: The art and science of prediction.* Crown Publishers/Random House.

Tump, A. N., Pleskac, T. J., & Kurvers, R. H. J. M. (2020). Wise or mad crowds? The cognitive mechanisms underlying information cascades. *Science Advances, 6*(29), 1–11. https://doi.org/10.1126/sciadv.abb0266

Turner, B. M., Steyvers, M., Merkle, E. C., Budescu, D. V., & Wallsten, T. S. (2014). Forecast aggregation via recalibration. *Machine Learning, 95*(3), 261–289. https://doi.org/10.1007/s10994-013-5401-4

Wallsten, T. S., & Budescu, D. V. (1983). State of the art—Encoding subjective probabilities: A psychological and psychometric review. *Management Science, 29*(2), 151–173. https://doi.org/10.1287/mnsc.29.2.151

Wallsten, T. S., & Diederich, A. (2001). Understanding pooled subjective probability estimates. *Mathematical Social Sciences, 41*(1), 1–18. https://doi.org/10.1016/S0165-4896(00)00053-6

Wang, G., Kulkarni, S. R., Poor, H. V., & Osherson, D. N. (2011a). Improving aggregated forecasts of probability. In *2011 45th annual conference on information sciences and systems* (pp. 1–5). https://doi.org/10.1109/CISS.2011.5766208

Wang, G., Kulkarni, S. R., Poor, H. V., & Osherson, D. N. (2011b). Aggregating large sets of probabilistic forecasts by weighted coherent adjustment. *Decision Analysis, 8*(2), 128–144. https://doi.org/10.1287/deca.1110.0206

Weaver, E. A., & Stewart, T. R. (2012). Dimensions of judgment: Factor analysis of individual differences: Dimensions of judgment. *Journal of Behavioral Decision Making, 25*(4), 402–413. https://doi.org/10.1002/bdm.748

Weiss, D. J., Brennan, K., Thomas, R., Kirlik, A., & Miller, S. M. (2009). Criteria for performance evaluation. *Judgment and Decision Making, 4*(2), 164–174. https://doi.org/10.1017/S1930297500002606

Willmott, C., & Matsuura, K. (2005). Advantages of the mean absolute error (MAE) over the root mean square error (RMSE) in assessing average model performance. *Climate Research, 30*, 79–82. https://doi.org/10.3354/cr030079

Wright, G., & Ayton, P. (1987). Task influences on judgemental forecasting. *Scandinavian Journal of Psychology, 28*(2), 115–127. https://doi.org/10.1111/j.1467-9450.1987.tb00746.x

Yerushalmy, J. (1947). Statistical problems in assessing methods of medical diagnosis, with special reference to X-ray techniques. *Public Health Reports, 62*(40), 1432–1449. https://doi.org/10.2307/4586294

Chapter 8
The Wisdom of Timely Crowds

Mark Himmelstein, David V. Budescu, and Ying Han

Keywords Forecast timing · Aggregation · Item response theory · Crowd selection · Differential weighting

1 Introduction

Who will be the next president of the United States? How will the Euro perform in the next quarter? How many hurricanes will be recorded this season? Who will win the next World Cup? All these questions involve predictions about *future* events. This temporal component of forecasting—that it targets a future point in time—is rather trivial. However, there is a second, more complex, temporal component to any forecast: the present. Unlike the future, which typically refers to a stationary point in time, the present is a moving target. One's belief about electoral outcomes might change when a new poll is released or when new information about the candidates is revealed. The odds of different teams winning the World Cup change as news about key injuries are revealed and potential contenders are eliminated. In other words, the time when one chooses to forecast and the information available at that time can have dramatic implications for the accuracy and usefulness of that forecast.

In this chapter we address several key issues related to changes in the accuracy, variability, and quality of forecasts at different points in time. Many of the topics discussed have been described by different researchers with different concerns, and some have received more direct attention than others. In many cases, what follows is designed to be illustrative rather than definitive, providing outlines of different approaches researchers have taken to solving problems related to the timing of forecasts, recommendations for how this insight can be applied and what topics may warrant further investigation.

Let's define the *resolution date* of a forecast as the fixed point in time at which the forecast is aimed. It is the time by which the outcome of the event is revealed and

M. Himmelstein (✉) · D. V. Budescu · Y. Han
Department of Psychology, Fordham University, Bronx, NY, USA
e-mail: mhimmelstein@fordham.edu

M. Seifert (ed.), *Judgment in Predictive Analytics*, International Series in Operations Research & Management Science 343,
https://doi.org/10.1007/978-3-031-30085-1_8

becomes known.[1] The *time horizon* of a forecast refers to the amount of time remaining between the forecast is made and the resolution date. The resolution date for a question about an election is the date on which the votes are counted, and a winner is announced, while the time horizon is the amount of time remaining between the moment a forecaster makes their prediction and the vote count. The resolution date for the winner of the World Cup is the date on which the champion is determined, while the time horizon is the number of days until the championship match. Because resolution dates are fixed time points, forecasts that are provided later than others (but before the resolution date) are always *closer* to the resolution date than those made earlier, and the time horizon inexorably shrinks until the resolution date is reached. It is generally assumed that as the time horizon becomes shorter, forecasts will become more accurate, on average, than those made earlier, under longer time horizons (Moore et al., 2017; Schnaars, 1984; Ungar et al., 2012), and in fact, such a property can be considered a normative expectation of any forecasting system (Regnier, 2018).

Figure 8.1 illustrates this pattern with data from a recent longitudinal forecasting study: The Hybrid Forecasting Competition (Specifically, Season 2 of the HFC, data collected by the SAGE team, Benjamin et al., 2023; Morstatter et al., 2019). The forecasters[2] in this study were 547 Amazon Mechanical Turk recruits who were incentivized to produce the most accurate forecasts they were capable of. This figure includes the 116 (of 398 total) questions for which there were at least 16 weeks between the first date the question was posed and its resolution date (its maximal time horizon). The average accuracy, measured by the Brier Score, decreased monotonically over time. Figure 8.2 shows a similar effect for the experts who participated in the European Central Bank (ECB) quarterly survey (Garcia, 2003). These analysts were asked to forecast key economic indicators (inflation, unemployment, gross domestic product) over much longer periods at quarterly intervals. This figure summarizes forecasts by 81 (of the 100 total) different analysts who forecasted the inflation during the period of 2010 to 2017.[3] This effect can also be seen in Figure 6 of Regnier (2018), which shows changes to the accuracy of wind-speed probability forecasts over time.

Throughout this chapter, illustrations will focus primarily on these HFC and ECB datasets. They are not intended to be comprehensive, but representative of different types of concerns commonly found in the literature. The second season of the HFC

[1]There are some exceptions to this definition. Most frequently, the question may involve an outcome that either will or will not be revealed before the target date. For example, if we ask judges to forecast the chances that some market index will exceed a certain threshold (say Dow-Jones >36,000) by a certain date, there is a chance that the index will exceed the threshold before that target date. Less frequently, unforeseen circumstances may delay the revelation of the ground truth. For example, in the 2020 US election, control of the Senate was determined two months after the election date because of an unanticipated runoff election.

[2]Throughout this chapter we use the terms *judge* and *forecaster* interchangeably.

[3]This period was selected due to the homogeneity of the elicitation mechanism. All these forecasts used 12 bins spanning the range (−1, 4).

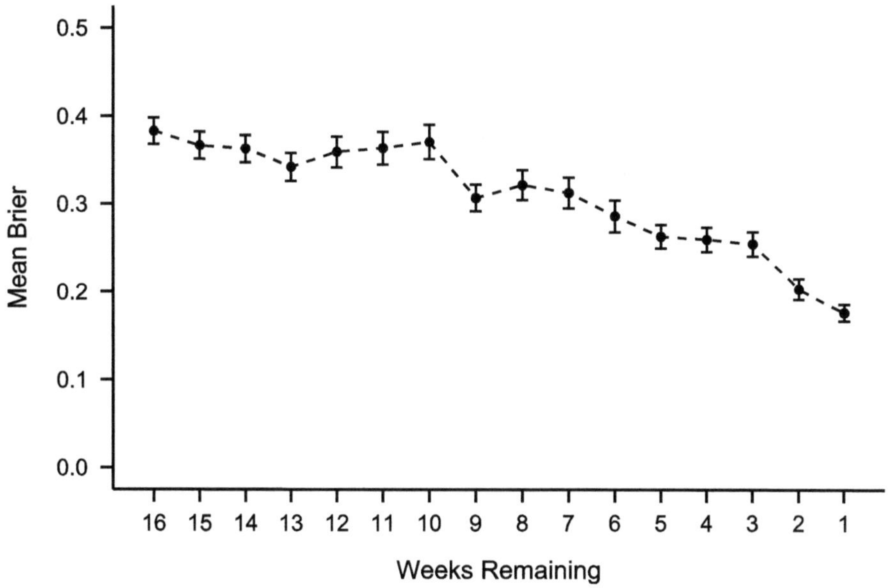

Fig. 8.1 Average accuracy (Brier Score) of N = 547 judges' forecasts as a function of time on 116 questions with at least 16 weeks in length from the HFC Season 2 data

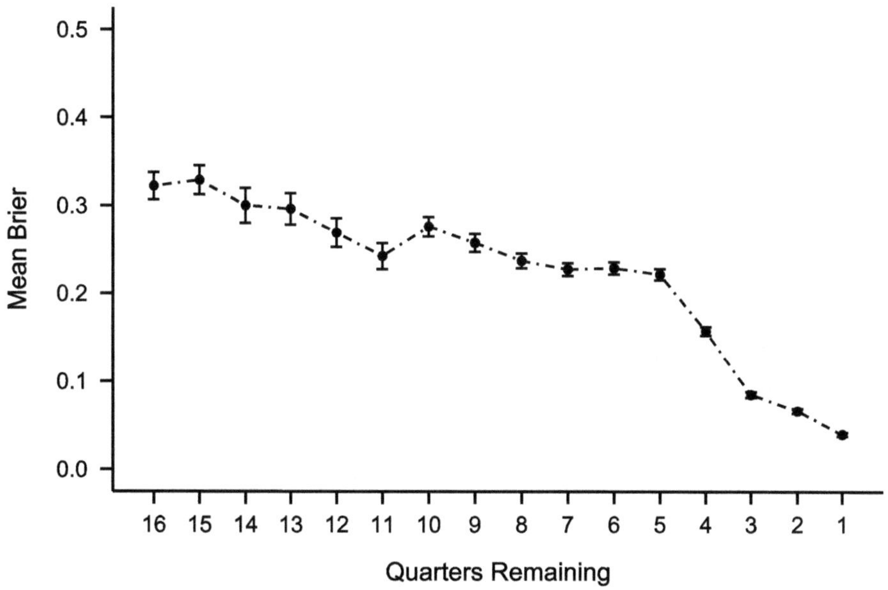

Fig. 8.2 Average accuracy (Brier Score) of N = 81 experts' forecasts as a function of time for inflation in the European Union made between 2010 and 2017 in the ECB quarterly survey of experts

was one of the largest judgmental forecasting tournaments ever conducted, with more forecasting questions than its first season, or the different years of the Aggregative Contingent Estimation program (Bo et al., 2017; Mellers et al., 2015a; Tetlock et al., 2014). The forecasters were not subject-matter experts, maximal time horizons varied from several weeks to several months and each judge forecasted only a subset of all events, at their own convenience, rather than at predetermined and prescribed time points. The ECB forecasting survey, on the other hand, represents a case of experts making forecasts about a single type of event at regular, structured intervals. For each quarter, about half of all eligible analysts made a new forecast, and the time horizons could extend for as long as five years into the future. Among other things, we will use these two sources of information to point out similarities and differences in how the timing of forecasts affects decisions that must be made by forecasters, managers, and aggregators for these different forecasting environments.

1.1 Forecaster Evaluation

Identifying skilled analysts has important practical benefits. The most obvious is resource allocation. Picking out the most adept analysts and matching them to problems they are uniquely suited for, rather than allowing forecasters to toil at problems less aligned to their unique skills, can maximize a team's productivity. A less obvious benefit is the implication for the wisdom of crowds (WoC). It is well understood that aggregating the judgments of many judges can often be a superior alternative to relying on the judgment of a single judge, and forecasting is no exception (Atanasov et al., 2017; Collins et al., 2022; Surowiecki, 2005). While simple unweighted averages of many forecasts perform well, weighted averages, in which the most skilled analysts (and there are many definitions of skill as Collins et al., 2022) are overweighted, improve the accuracy and potentially reduce the cost of these methods even further (Budescu & Chen, 2014; Chen et al., 2016; Mellers et al., 2015a).

General issues related to forecaster evaluation are explored in greater detail by Atanasov and Himmelstein (2022). However, an important practical hurdle in evaluating a group of forecasters is that they typically do not all forecast events at the same time (Garcia, 2003). Laboratory forecasting experiments represent an idealized case where all the judges forecast all the events at the same time. It is unrealistic to expect this for most, if not all, real-world situations. For example, sport prognosticators who predict the outcomes of all the games to be played on a given weekend (NFL, Premier League, etc.) issue their forecasts at various points during the week. Even controlled forecasting studies often adopt a framework in which participants self-select the questions they wish to forecast and do so at different time points. The most famous example was the Good Judgment Project (GJP), the research team who won the IARPA's ACE research competition (https://www.iarpa.gov/index.php/research-programs/ace) and coined the term *Superforecaster* to label particularly skilled forecasters (Mellers et al., 2015b; Tetlock & Gardner,

2016). How does one compare the performance of different judges who forecast the same event at different times and, presumably, with various levels of information? A judge should not be rewarded simply for waiting as long as possible to issue a forecast, even if doing so maximizes their individual accuracy. Nor should one be penalized for providing early forecasts, when they have the largest chance of making major errors. We will discuss and illustrate psychometric and statistical methods that can address the challenge of evaluating and comparing forecasts with different time horizons.

1.2 Time Decay

Many firms, such as financial institutions, rely on an ensemble of forecasts that are aggregated for maximal accuracy (Budescu & Chen, 2014; Chalmers et al., 2013; Ray, 2006). Imagine the year is 2005, and Judy, a financial analyst working for a firm that makes regular predictions about economic outcomes, makes a prediction in first quarter of the year about what she believes the exchange rate between the British Pound and the U.S. dollar will be in the first quarter of 2006. Though she can, she declines to make a new prediction in the second or third quarters of 2005. Perhaps she did not encounter any new information that would cause her to revise her initial belief, or perhaps she simply took a long vacation. However, several other analysts do make predictions in the second and third quarters. Should Judy's first quarter forecast be ignored? Clearly, we would expect it to be less accurate than a third quarter forecast from an equally skilled analyst. But, dropping it would reduce the size of the crowd that contributes to the ensemble model. Larger crowds tend to produce better forecasts, even if some of the individual forecasts are less accurate than others. They also provide a more realistic picture of the degree of disagreement of opinions and protect against spurious consensus. What is an aggregator to do?

In forecasting tournaments, such as the HFC and ACE, forecasts come in at a steady stream over time. Gina, a Superforecaster, might make two predictions about the same election exactly one month apart. During that month, dozens of others will forecast the same question. It could be unwise to ignore a Superforecaster's initial forecast when aggregating the crowd's opinions, simply because it is somewhat older.[4]

This raises the issue of the tradeoff between maintaining a large enough crowd to make a collective prediction system work, and the decay in the accuracy of individual forecasts over time. Aggregation systems require crowds that are large enough to

---

[4]Throughout the chapter we assume that all the forecasters are motivated only by a desire to achieve the highest possible—individual and collective—accuracy, and we ignore alternative motives. Obviously, in competitive environments (e.g., the common knowledge that one forecaster will be promoted next year) knowing that only the most recent predictions count, will induce a variety of strategic considerations regarding the optimal timing of one's predictions that may bias and distort the collective effort.

reliably produce accurate forecasts, but there are very few practical forecasting scenarios in which a sufficiently large number of forecasts are issued simultaneously. Figuring out how to make the best use of all the most up-to-date forecasts available at any given time point is critical. We will address several approaches related to this problem.

1.3 Time and Crowd Size

WoC principles suggest that the crowd's diversity can improve the accuracy of its forecasts (Davis-Stober et al., 2014; Lamberson & Page, 2012). One relatively simple way of doing this is to increase the number of judges. Recruiting a large crowd, while targeting the best forecasters and the most recent forecasts, are well-established methods for improving aggregate forecast quality (Atanasov et al., 2017; Budescu & Chen, 2014; Himmelstein et al., 2021). This raises an additional question: If larger crowds produce more accurate forecasts, can one just use large crowds to offset the difficulty in forecasting questions that are very far from their resolution date? This question is surprisingly under-studied. We will take a deeper dive into the tradeoff between timing and crowd size seeking to determine whether large crowds are a viable solution to long time horizons. The answers come with some surprising results, which can help explain the mechanisms that underly in exactly what ways a crowd is wise, and in what ways it turns out it is not.

2 Evaluating Forecasters Over Time

The notion of forecasting skill has received a lot of attention ever since the Good Judgment Project found they could reliably identify Superforecasters (Mellers et al., 2015a; Tetlock & Gardner, 2016). Over the course of several years, and working with many volunteer forecasters, the researchers were able to reliably identify the best of the best. However, in most applied settings, managers won't have several hundred candidate analysts to choose from, nor several years to evaluate them, so they need to rely on more granular discrimination methods.

Several studies have more broadly evaluated the correlates of forecasting skill, which are covered in more detail by Atanasov and Himmelstein (2022). They identify several key methods for identifying high performing forecasters, the gold standard of which are based evaluating the accuracy of past forecasts for which the ground-truth is known.

2.1 Forecast Timing

An important problem in evaluating forecasters based on past forecasting performance involves the timing of their forecasts. The simplest approach to measuring a forecaster's skill is to calculate the average accuracy of their forecasts, regardless of their timing. But what if a forecaster tends not to forecast questions immediately, but wait until their time horizons shrink; while others tend to make forecasts earlier on, when there's less information available?

Figure 8.3 illustrates this point with the HFC data, a case in which forecasters made their forecasts at different time points throughout the lifespan of the forecasting questions. Each point represents an individual forecaster. The y-axis is their mean accuracy, as measured by their mean Brier score (where lower scores represent better accuracy; Brier, 1950), and the x-axis shows the mean number of days remaining in the questions they forecasted. The correlation between average time remaining and average accuracy was $r = .35$, demonstrating that the timing of the forecasts is an important confounding variable when assessing average accuracy.

Himmelstein et al. (2021) proposed two model-based methods to account for the timing of forecasts, both rooted in psychometric theory. An important consideration in model-based approaches, in which the accuracy of forecasts is the response variable, is that many scoring rules, such as Brier scores, are on scales that may violate fundamental modeling assumptions. For example, because Brier scores are bounded between 0 and 2, linear models that use them as response variables cannot have normally distributed residuals. Therefore, Himmelstein et al. (2021) transformed Brier Scores onto a scale that is more in line with this assumption.

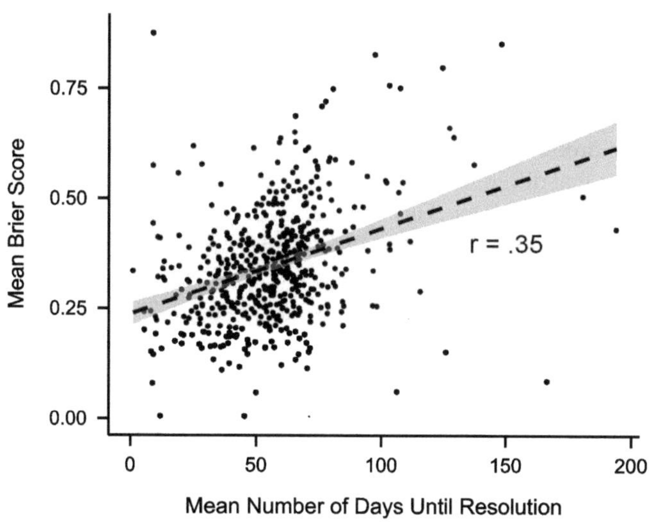

Fig. 8.3 Joint distribution of average accuracy (Mean Brier score) of N = 547 forecasters from Season 2 of the HFC and the average time remaining before resolution in the forecasts they made

First, Brier Scores were transformed to a unit interval, into a metric that could be thought of as simply *accuracy*, as

$$\boldsymbol{Accuracy} = \mathbf{1} - \sqrt{\frac{\boldsymbol{Brier\ Score}}{\mathbf{2}}} \tag{8.1}$$

Because accuracy is on a unit interval, with higher values representing more accurate forecasts, it can be transformed using traditional linking methods, such as the probit or logistic functions. Both Himmelstein et al. (2021) and Merkle et al. (2016) opted for the latter, as it transforms the accuracy measure onto a standard normal distribution, facilitating interpretability.

The simpler of the two models described by Himmelstein et al. (2021) employs *Hierarchical Linear Modeling* (HLM). The HLM approach is, in essence, a regression model predicting the accuracy of each forecast made by a given forecaster (f), for a given question (q) at a given time point (t). Each forecaster gets their own intercept, μ_f, representing their average skill (accuracy), conditional on the other model parameters. To account for the varying difficulties of different events, each question is also given its own conditional intercept μ_q. The key predictor of accuracy is the natural log of time (to resolution). Formally, the model can be written as

$$Probit\left(Accuracy_{f,q,t}\right) = \mu_0 + \beta \log(time) + \mu_f + \mu_q,$$

and can be rearranged to

$$Probit\left(Accuracy_{f,q,t}\right) - \mu_0 - \beta \log(time) - \mu_q = \mu_f.$$

This formula implies that the skill of a given forecaster can be understood by combining information about the accuracy of their forecasts, the timing of the forecasts they made, and the difficulty of the questions they forecasted.

The second, more complex, model is based on an *Item Response Theory* (IRT) approach to the problem, originally developed by (Merkle et al., 2016). The insight of IRT models is that different questions can be uniquely informative about the latent ability of different performers. This specific form also models how the difficulty of an item changes over time. Formally, the model is:

$$\boldsymbol{Probit}\left(\boldsymbol{Accuracy}_{f,q,t}\right) = \boldsymbol{b}_{0,q} + \left(\boldsymbol{b}_{1,q} - \boldsymbol{b}_{0,q}\right)\boldsymbol{e}^{\boldsymbol{b}_2 \boldsymbol{t}_{f,q}} + \boldsymbol{\lambda}_q \boldsymbol{\Theta}_f.$$

The different b parameters represent each question's difficulty (event's unpredictability) over time. The question's maximal difficulty, when the time horizon is extremely long, is represented by $b_{0,\ q}$. The question's minimal difficulty, just before the resolution date, is $b_{1,\ q}$. The rate of change in difficulty between those two anchor values is represented by b_2. The λ_q parameter represents how well a given question discriminates between different forecasters, and Θ_f represents the latent skill level of a given forecaster.

Himmelstein et al. (2021) pitted these two model-based techniques against the simpler approach of taking the average accuracy of all the forecasts made by each judge,[5] as well as how well accuracy could be predicted based on dispositional information alone. The hierarchical and IRT based methods showed very similar correlations with the accuracy of future forecasts (r = .54 in each case), and significantly better than the simple method (r = .48) or dispositional information (e.g., measures of fluid reasoning ability and political knowledge) by itself (r = .21).

Although accounting for the timing of forecasts can help discriminate the skill of individual forecasters, Himmelstein et al. (2021) found it was not necessary for optimally weighting different forecasters. Using simple standardized accuracy measures as performance weights, which did not account for forecast timing, produced aggregate forecasts that were very similar in terms of accuracy to the more sophisticated model-based methods described here. This result is consistent with the notion that, when devising performance weights, it is much more important to identify the key predictors than to find the optimal estimation method for those weights (Budescu & Chen, 2014; Dawes, 1979).

2.2 *Information Accrual*

Many performance evaluations take place at one time. A student takes the SAT in a single seating, for example. But the temporal component of prediction means the amount of information about the people making the predictions is constantly changing as more and more questions pass their resolution date. Himmelstein et al. (2021) found that the predictive utility of a forecaster's past accuracy changes dramatically as more questions resolve. For example, in the HFC, as the first forecasting questions were resolved, and ground-truth based performance information began to accrue, it could explain around 10% of the variability in future accuracy—roughly the same as dispositional trait information about the forecasters (see also Atanasov & Himmelstein, 2022). However, as more questions were resolved, past performance information became an increasingly reliable signal of forecasting ability, eventually explaining as much as 50% of the variability in future accuracy.

In the interest of aggregation, dynamic models can automatically account for changes to the information environment as time passes. For example, it has been shown that forecasters can be evaluated in terms of their *contribution* to the aggregate wisdom of the crowd (Budescu & Chen, 2014; Chen et al., 2016). The so-called contribution-weighted model for aggregating forecasters allows forecasters to be either weighted or selected based on how much they contribute to the total

[5] In this simple approach, the difficulty of each question was controlled by standardizing the accuracy values within each question, so they always had a Mean = 0 and SD = 1, to account for the possibility that some forecasters tended to forecast more difficult questions than others. However, differences in accuracy related to time were not accounted for.

crowd wisdom, on average. This average, however, is a moving target, that is constantly changing as more ground truth information is obtained. By allowing aggregation models the flexibility to account for changing information environments, aggregators can make the best use of all information available at any given time point.

A relatively recent innovation are proposals for new *intersubjective* scoring rules, such as *proper proxy* scoring rules (Witkowski et al., 2017) or *surrogate* scoring rules (Liu et al., 2020), to evaluate the quality of forecasts in the absence of ground truth resolution (see Atanasov & Himmelstein, 2022). The idea behind such scoring rules is to extract some information about the quality of a forecast by comparing it to the aggregate crowd forecast. For example, Witkowski et al. (2017) propose a variant of the Brier Score which references the squared difference between the c probabilities (p) provided by a forecaster's (f) forecast and the probabilities associated with the aggregate (a)

$$\boldsymbol{Proxy_f} = \sum_{i=1}^{c} \left(\boldsymbol{p_{i,f}} - \boldsymbol{p_{i,a}}\right)^2.$$

Intersubjective scores are especially appealing in situations where forecasts are made at points that are extremely distant to their resolution, where it is not feasible to evaluate the quality of individual forecasters in a timely manner. The intersubjective approach to forecaster evaluation is appealing, and addresses some important temporal challenges, but research on the subject is in early stages.

2.3 *Reliability of Forecaster Assessment*

The notion of reliability is a pillar of psychometric theory (Cronbach et al., 1963). Measures that are reliable are consistent in the sense that measurement on one occasion, or by one instrument, will generalize to other occasions or other instruments designed to measure the same construct. Estimation of reliability in forecaster assessment has received mostly implicit attention. It is important to use reliable scales to measure dispositional information (Himmelstein et al., 2021; Mellers et al., 2015a), but what about the reliability of ground-truth based performance information? The temporal component is a substantial challenge in assessing the reliability of forecasters' accuracy. Unlike a standardized test, like the SAT, no two sets of forecasting problems administered at different points in time can ever be identical.

Himmelstein et al. (2021) touch on this problem by showing how the different accuracy metrics they assessed in the HFC results correlated over time. Table 8.1 shows the test-retest correlations of the various metrics, obtained by chronologically splitting the data in half (the first half represents the first 50% of questions to resolve). This can be viewed as a type of *temporal reliability*. Atanasov and Himmelstein (2022) also show how forecasters' accuracy is correlated across randomly split samples, a form of *split-half* reliability.

Table 8.1 Temporal accuracy of forecasting metrics obtained by correlating the first 50% of questions and second 50% of questions in HFC Season 2

Method	Test-retest correlation
Simple	.54
Hierarchical	.60
IRT	.66

N = 398 questions, 547 forecasters

Future researchers may wish to further investigate the feasibility of establishing reliable methods of assessing the accuracy of forecasters. Such methods should account for the fact that, because of the inherent temporal invariance, no single standardized assessment battery can ever exist. This might involve standardizing item structures, or wording stems; as well as eliciting forecasts about similar types of events, or types of events for which expected accuracy is to be evaluated. For example, do items that have different numbers of response categories tend to produce more reliable estimates of individual accuracy? Would particular descriptions of how the ground truth will be determined provide more consistent and clear information? Are there forecasting domains which tend to produce more reliable personal accuracy estimates than others? These are all important questions to explore in establishing the reliability of ground-truth based forecaster assessments.

2.4 *Recommendations*

Probabilistic forecasting has been well established as a unique skill, and this point is further explored by Atanasov and Himmelstein (2022). However, there are different reasons one might wish to evaluate the skill of forecasters, and they may favor different approaches. If the goal is to simply identify the best individual performers, methods which account for the timing of the forecasts produce more stable estimates of how they are likely to perform in the future. However, this level of sophistication may not be necessary if the goal is to simply generate optimal performance weights. For aggregators concerned with performance weighting, the added benefit of accounting for forecast timing may be marginal. So, for managers deciding which analysts to assign to certain roles, accounting for forecast timing may be important, but for aggregators deciding how much weight to assign to each of their analysts from a constant (especially a large) pool, this concern may not be quite so important. Instead, aggregators may want to employ dynamic models, which can keep up with changing information environments. Future research should also direct attention towards intersubjective methods of accuracy assessment, as well as methods of establishing the reliability of various approaches to forecaster evaluation.

3 The Timeliness of Crowds

Regnier (2018) defines several key normative properties of multi-period forecasting systems, one of which is *efficiency*. Any forecast at time t should contain all relevant information from time points prior to t. Put another way: one should not be able to improve the accuracy of a given forecast based on prior forecasts of the same event made by the same system. Second, it should be *martingale*. That is, future forecasts should have an expected revision of 0. Put another way: one should not be able to predict how the system's future forecasts will differ from present forecasts. If one could, it would indicate present forecasts are suboptimal. And finally, given these properties, a forecasting system will have the intuitive property of *strict improvement*—forecasts made at later time points would be expected to be more accurate than forecasts at earlier time points.

When many analysts forecast the same events, they often will not do so at the same time, which can lead to violations of these properties in an aggregation system. In forecasting tournaments, such as the HFC, participants had the freedom to forecast any unresolved questions whenever, and as often as, they liked. In some cases, fewer than 1% of the 547 judges forecasted a given question in a week. Even in the best-case scenario, only 48% of the judges forecasted an event during a week-long period. The professional forecasters surveyed by the ECB made their forecasts at more regular intervals, upon receiving periodic invitations from the organizers. Still, of the 100 professional forecasters surveyed, between 39 and 69 failed to make new forecasts in each quarter.

This situation presents a dilemma for information aggregators. Any time a member of the crowd revises their forecast, the aggregate can be revised. Older forecasts will generally be less accurate than more recent ones, but different forecasters may have different information, and tend to contribute more positively or negatively to the aggregate than others (Budescu & Chen, 2014; Chen et al., 2016). How does one choose criteria for retention? Where is the temporal inflection point at which a forecast is expected to harm rather than benefit the aggregate? Removing forecasts expected to benefit to the aggregate would compromise efficiency, but so would retaining forecasts expected to harm the aggregate. Furthermore, when older forecasts are discarded, the revision to the aggregate cannot be assured to be martingale—it is possible that the discarded forecasts contained information that would make the direction of the revision predictable, depending on the criteria by which they are removed.

How should forecasts that are out of date and have not been revised be treated? What tools are available for aggregating forecasts based on their timing? The literature proposes two classes of methods for accounting for forecast recency when aggregating judgments: a *selection* method (Atanasov et al., 2017) and a *weighting* method (Baron et al., 2014; Ungar et al., 2012).

3.1 Selection Methods

Atanasov et al. (2017) analyzed results from the Good Judgment Project and found that selecting 20% of all the forecasts, based on their recency, optimized the accuracy of the aggregate forecast. They identified this value by performing a sensitivity analysis that compared how accuracy was affected by different retention thresholds and found that 20% produced the most accurate aggregation.

This method is appealing both in its simplicity and its cost effectiveness. It doesn't require any special calculations, and it implies that maximization of the aggregate accuracy of many forecasts does not necessarily require the largest sample possible. It was, in fact, preferable to *reduce* the total crowd size in cases where the forecasts were made at different time points, rather than retain unrevised forecasts that were not as recent as others.

The sensitivity analysis performed was, of course, specific to the forecasting tournament setting, in which there were hundreds of available forecasts made over a long time, with a high degree of variability in terms of their timing (and recency). What about a setting that has a smaller crowd to begin with? Or one in which the forecasters' timing varies in a different way? It is difficult to determine if the 20% selection threshold would be optimal under these different conditions. The most general form of this approach can be specified as a two-parameter rule:

$$N_{Select} = Max(N_{min}, T(N_{total})).$$

Here N_{min}, is the minimal number of judges to retain and T is the chosen proportion threshold (between 0 and 1) of how much of the crowd to retain when the number of available forecasts exceeds N_{min}. These two values can be chosen based on the specific circumstances of each application.

In principle, this approach allows the aggregate to retain some efficiency—it is much less likely information about past revisions can improve upon current revisions when only a sufficient number of recent forecasts are retained. However, this approach can lead to violations of the martingale property. Any time a new forecast is made, older forecasts are removed from the aggregate. If the new forecast is from a forecaster who wasn't already included in the aggregate, it may be possible to predict the directional impact of their inclusion based on their past tendencies that had been selected out of the current aggregate.

3.2 Weighting Methods

Another approach does not completely discard less recent forecasts but underweights their contribution to the aggregate (Baron et al., 2014; Ungar et al., 2012). This is a slightly more complex approach, but it has some potential advantages. By down-weighting less recent forecasts, the information they contain is not totally lost. So, if

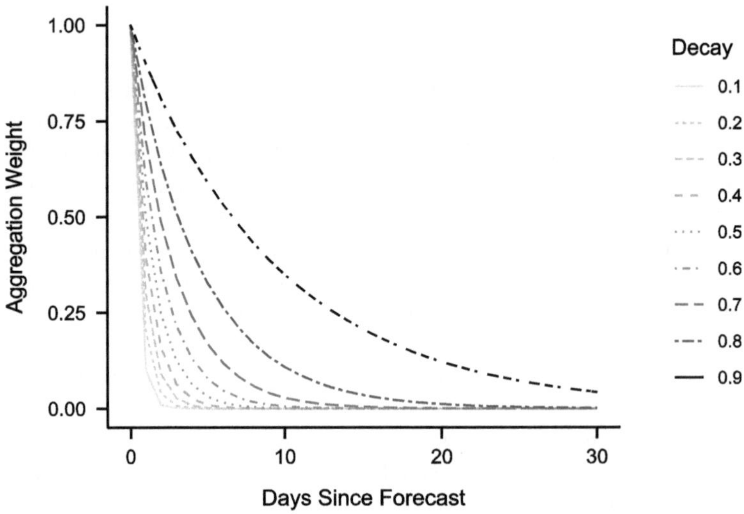

Fig. 8.4 Weight assigned to forecasts based on different values of the decay parameter as a function of time

a forecaster had unique information about an event, and never felt the need to revise their original forecast, it can still contribute to the crowd aggregate, but because it cannot account for any new information that accumulates over time, its influence is reduced.

A popular method for recency weighting is based on *exponential decay*, where a judge's most recent forecast is given a weight, w, based on a decay rate, $0 \leq d \leq 1$, raised to a power based on the time that has passed since the forecast was made, t, so the weight is $w = d^t$.

For example, Baron et al. (2014) used a value of $d = 0.6$. Let t be the number of days[6] that had passed since the forecast was made. On the day the forecast is made, $t = 0$, and that forecast receives a weight of $w = d^t = 0.6^0 = 1$. The next day, $t = 1$, and that forecast receives a weight of $w = d^t = 0.6^1 = 0.6$. The following day, $t = 2$, and that forecast receives a weight of $w = d^t = 0.6^2 = 0.36$. Each day that passes, the forecast's weight is reduced, meaning it contributes less to the global aggregate of the crowd forecast. Figure 8.4 illustrates how the choice of the decay parameter influences the weight forecasts receive as time passes.

Baron et al. (2014) chose the value of $d = 0.6$ to be illustrative, rather than definitive. As with the selection parameter described by Atanasov et al. (2017), the optimality of their choice may be context-specific. One principled heuristic for defining the decay parameter is to select a lower boundary for weights at given time point, w_t, defined as $w_t = d^t$, implying that $\log(d) = \frac{\log(w_t)}{t}$. For example, if we

[6]Clearly, the use of days is arbitrary and is made for illustration purposes. The approach applies to any relevant time unit (week, month, quarter, etc.) that makes sense in given context.

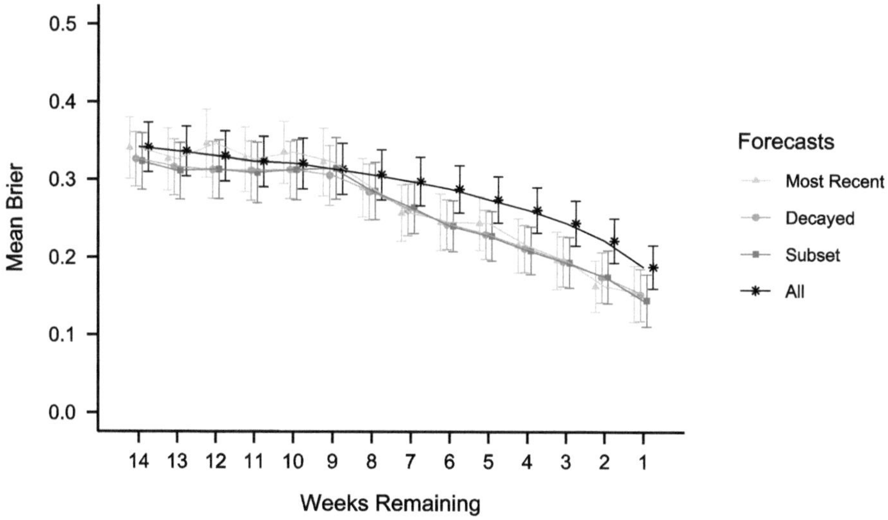

Fig. 8.5 Comparison of four aggregation methods for 116 HFC Season 2 questions at least 16 weeks in length with standard errors: (**a**) using only forecasts made during a particular week, (**b**) using a decay parameter of 0.6 to down-weight less recent forecasts, (**c**) selecting the 20% of most recent forecasts, (**d**) use all forecasts regardless of when they were made

wish to ensure that all $w_t \geq 0.05$ when $t \leq 10$, we must make $d \geq \exp\left(\frac{log(0.05)}{10}\right) = 0.741$. If, instead, we wish to ensure $w_t \geq 0.05$ when $t \leq 20$, we must make $d \geq \exp\left(\frac{log(0.05)}{20}\right) = 0.861$. This consideration allows the aggregator to define the duration they consider relevant, reflecting the fact that various decay rates will converge very quickly to 0, as shown in Fig. 8.4.

Like the selection method, this approach preserves a degree of efficiency that would be lost by discarding all dated forecasts. And, like the selection method, it can lead to violations of the martingale property, for much the same reasons. It may become possible to make predictions about future aggregate revisions based on information about the tendencies of the forecasters who trigger them.

3.3 Comparing Methods

Figure 8.5 compares the selection and decay approaches based on all 116 events from the HFC Season 2 data that had time horizons of at least 16 weeks when they were first posed. The y-axis shows the mean aggregate Brier scores across all of these events, and the x-axis their consecutive time horizons at one-week intervals. The various curves describe the performance of (a) only the week's most recent forecasts (any forecasts made more than a week prior were removed), (b) all unrevised forecasts, weighted with a decay parameter of $d = 0.6$, (c) the most recent

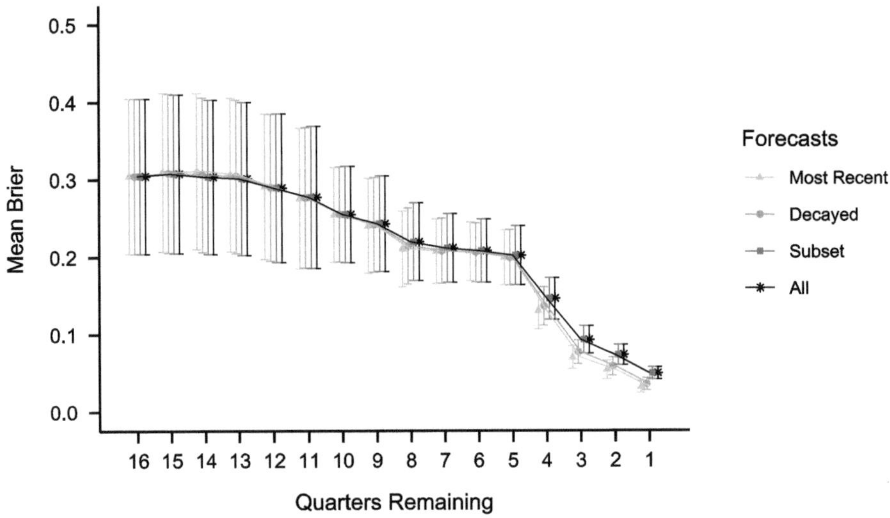

Fig. 8.6 Comparison of four aggregation methods for ECB inflation based on forecasts made between 2010 and 2017 with standard errors: (**a**) using only forecasts made during a particular quarter, (**b**) using a decay parameter of 0.6 to down-weight less recent forecasts, (**c**) selecting the 20% of most recent forecasts, (**d**) use all forecasts regardless of when they were made

20% of unrevised forecasts, and (d) all unrevised forecasts, regardless of recency. The key takeaway is that at each time point, accounting for recency is critical, but that over-sensitivity to recency can also lead to violations of strict improvement. Both the decay and selection methods, which retain some information from previous weeks, appear to preserve the strict improvement property and are almost indistinguishable from one another.[7] It is also apparent that retaining older forecasts becomes less important as the time horizon approaches 0.

Figure 8.6 applies the same method (employing the same parameters) to the inflation forecasts from the European Central Bank data. In this case we used all forecasting horizons from 2010 to 2017. The y-axis again represents the average aggregate Brier Scores, and the x-axis the shrinking time horizons for each consecutive quarter. This data is different in several important ways: it includes forecasts with much longer time horizons, more analysts made forecasts at each opportunity, and forecasts were updated four times a year, rather than at weekly intervals. These results also highlight the benefit of accounting for recency, but primarily as the time horizons become relatively small. Differences in the four approaches don't even appear until forecasts are about a year from their resolution date, and inefficiency is much more difficult to detect, perhaps due to the larger proportion of the total pool of judges who revise their beliefs at each point in time.

[7] It is possible to show, under reasonable assumptions about the rate of information accrual over time, that one can always find a decay rate that will match the accuracy of any selection procedure.

3.4 A Probabilistic Hybrid Method

It is possible to envision a hybrid method that combines some of the advantages of the selection and weighting approaches. The simple idea is to avoid the differential weighting which can be computationally complex in some cases but use the same principle to select only a subset of the older forecasts. In essence the decay rate determines what fraction of older judgments will be retained. For example, using the same parameter ($d = 0.6$), one would retain all the forecast from the most recent week, 60% of the forecasts from the previous week, 36% of the forecasts from two weeks ago, etc. The simplest way of selecting a subset of older forecasts is random sampling. This approach opens the door for selection of multiple samples which can be used to derive confidence intervals for the aggregate. However, one can envision more complex schemes where the selected subset is determined by other considerations. For example, if there is a high degree of agreement among the crowd members, one could treat this as a sign of consensus, and could perhaps develop a scheme that retains forecasts based on their similarity to one another. On the other hand, if there is a large amount of disagreement, the opposite might be desirable, and one might prefer to maximize diversity of opinion.

3.5 Martingale Violations

Methods that account for forecast recency, such as selection and decay, preserve a degree of efficiency in WoC forecasting systems while excluding the most dated forecasts. However, they cannot guarantee that the martingale property of forecasting systems is satisfied. This is because revisions to the aggregate are not conditionally independent of prior revisions when triggered by new forecasts from forecasters whose old forecasts were previously discarded. Knowing what they believed in the past can tell you about what they might believe in the future, and when their new forecasts enter an aggregate that has discarded their past forecasts, those prior beliefs can inform how the aggregate will change.

A possible solution could involve peer imputation methods. Feng and Budescu (2021, 2022) have shown that it is possible to impute the beliefs of a given judge who fails to forecast a given problem based on their similarity to other judges from past problems. Perhaps such an approach can be incorporated into selection or decay methods, such that as time passes, a judge's imputed forecast is selected, or receives more weight, than their original forecast. If the bias of these estimates can be shown to be negligible, it could be a method for preserving the martingale property of WoC forecasting systems while accounting for recency. This method has not been studied in the context of changes to forecast accuracy over time, but may be a worthwhile direction for future research.

3.6 Recommendations

The results presented in this section were designed to be more illustrative than definitive, and the differences between the needs of the HFC analysts and the ECB analysts are a good example of why. Unless all analysts regularly update their beliefs in time series forecasting, team leaders must make judgment calls about forecasts that are less recent than others. Clearly, there are benefits to emphasizing forecast recency, and very dated forecasts should rarely, if ever, be included when pooling crowd wisdom. But there can also be consequences for discarding useful information. Balancing these concerns is not trivial, and depends on several issues, such as the size of the analyst pool, the frequency of forecast revision, and distance between forecasts and the resolution date. There also remains work to be done in developing optimal strategies under commonly anticipated conditions, and developing methods that preserve desirable and normative properties of effective forecasting systems.

4 Crowd Size and Timing

Next, we examine how the size of a crowd interacts with the timeliness of forecasts. The literature on forecasting aggregation has established two principles: sufficiently large crowds will tend to be more accurate than their constituent individuals (Davis-Stober et al., 2014; Wagner & Vinaimont, 2010) and forecasts that are issued closer to resolution will tend to be more accurate than those provided earlier (Himmelstein et al., 2021; Moore et al., 2017; Regnier, 2018; Schnaars, 1984; Ungar et al., 2012). This begs the question: can long term forecasts be made more accurate by increasing the size of the crowd? To our knowledge, this question has not been systematically studied in the literature. In this section, we illustrate how crowd size and timing trade off in terms of accuracy and demonstrate that these factors contribute to accuracy in quite different ways.

4.1 Resampling the Crowd

To study the relationship between crowd size and timing on accuracy, we require a special set of conditions that are not very common in forecast settings. We need a series of forecasting questions in which forecast timing varies with respect to question resolution, sufficient time between forecast timing and resolution, and sufficiently large crowd such that a diverse set of subsamples of crowds can be compared. The HFC and ECB datasets are, again, excellent examples. In the HFC data, we selected the 122 forecasting questions which had time horizons of at least 12 weeks when first posed and had at least 40 total forecasts made during the first week of their lifespan. During each consecutive week, we randomly sampled 2, 4,

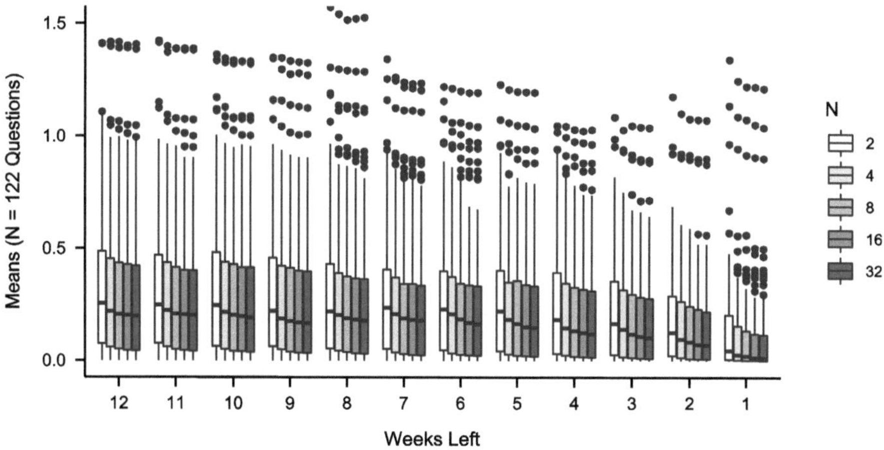

Fig. 8.7 Box plots for mean accuracy of aggregate forecasts based on 500 resamples of different sizes (2, 4, 8, 16, 32) at different time points in the HFC Season 2 data (N = 511 forecasters, 122 questions at least 12 weeks in length)

8, 16 or 32 of the most recent forecasts made on each question and computed the aggregate accuracy of that subsample. We repeated this procedure 500 times in each case.

Figure 8.7 shows the distribution of aggregate *mean* accuracies at each time point. Clearly, accuracy improves with time, for all sample sizes. While the mean accuracy improves as a function of sample size, the effect is small compared to the effect of time and mostly reflects the difference between small ($n = 2, 4, 8$) and larger ($n = 16, 32$) crowds. Figure 8.8 shows the distribution of the *standard deviations* of accuracies at each time point. Here we observe strong effects of sample size and very small effects of time (mostly for $n = 2, 4$). The clear picture emerging here is that while time is a major driver of the accuracy across questions, crowd size is driving the variability in accuracy. Importantly, the two factors don't have similar effects and crowd size can't be used to substitute for new forecasts close to the resolution.

Figure 8.9 reflects a very similar result based on the ECB inflation data. Using the target resolution date of 2016, this figure shows the distributions of accuracy based on the same procedure (because there is only one target horizon, there are not multiple different questions to average over). The same pattern is observed: as time passes, the resampled crowds tend to become more accurate, but at each time point, there is much less variability in the accuracy of larger crowds than of smaller crowds.

To formalize this result, we fit a beta regression model (with random intercepts for person and item) on the HFC data, where log(time), $\log_2$(sample size), and their interaction predicted both the expectation and variability of accuracy based on the same resamples. For the purposes of this model, Brier scores were transformed onto the accuracy scale (Eq. 8.1). Figure 8.10 shows the expected distributions of

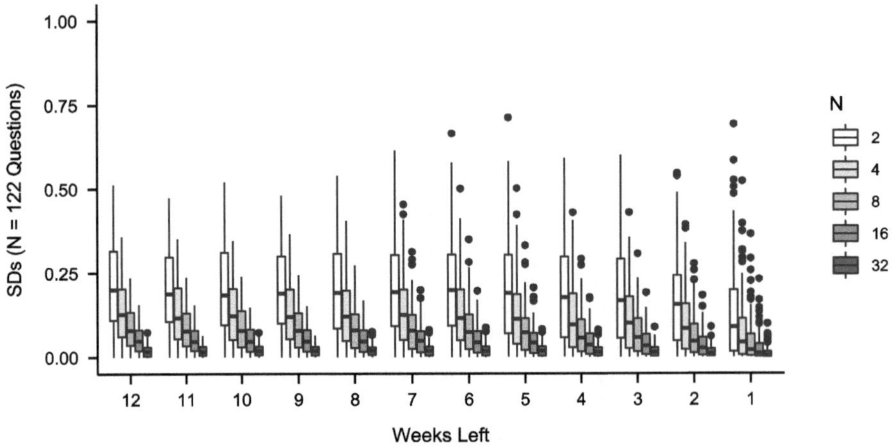

Fig. 8.8 Box plots for standard deviations of the accuracy of aggregate forecasts based on 500 resamples of different sizes (2, 4, 8, 16, 32) at different time points in the HFC Season 2 data (N = 511 forecasters, 122 questions at least 12 weeks in length)

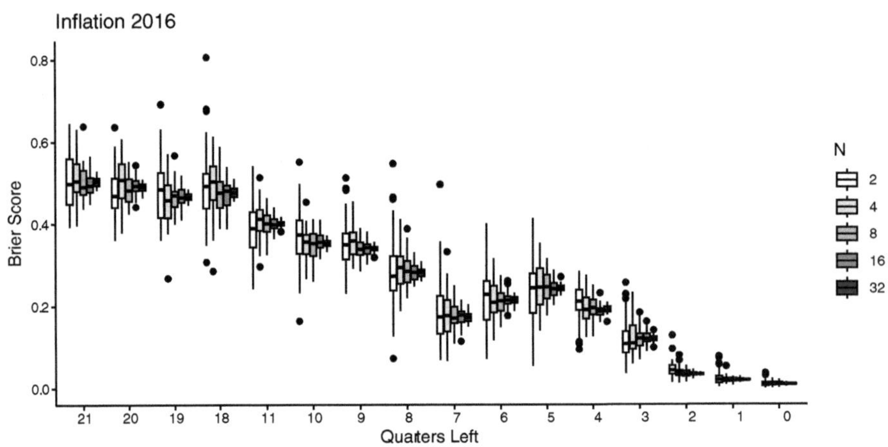

Fig. 8.9 Box plots of the accuracy of aggregate forecasts based on 500 resamples of different sizes (2, 4, 8, 16, 32) at different time points for the ECB inflation indicator with target forecast horizon of 2016

accuracy as a function of time and sample size, and Fig. 8.11 re-transforms the results to the Brier scale.

These figures summarizing the model clarify the relationship between crowd size and accuracy: larger crowds improve accuracy by reducing the variability and the likelihood of large errors, but do little to improve the mean (expected) accuracy. The centers of the distributions in these figures largely overlap but are considerably wider

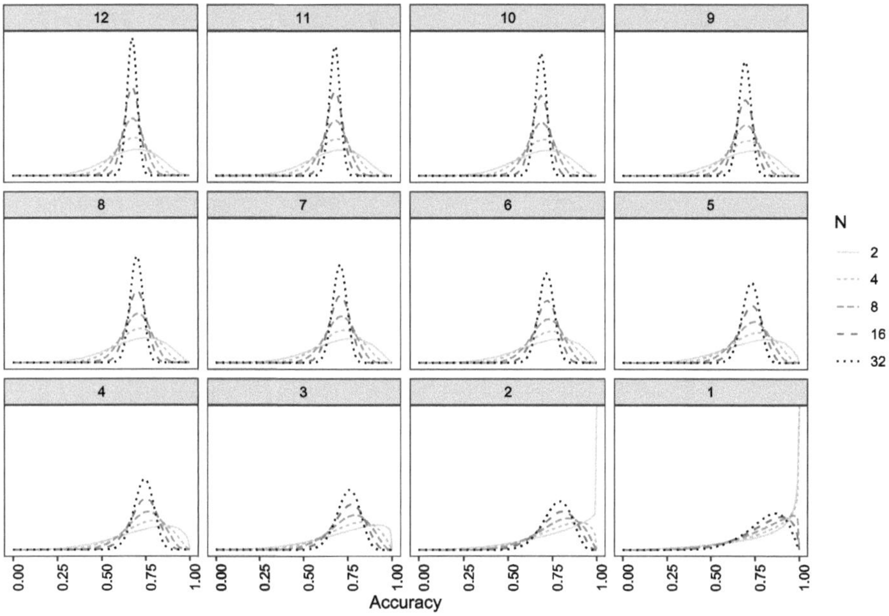

Fig. 8.10 Predicted distributions of aggregate *Accuracy* based on the 500 resampled sample sizes (2, 4, 8, 16, 32) at different time points from beta regression with log(time) and log2(N) as predictors

for smaller crowds. An important, and often desirable, property of Brier scores is that they are sensitive to large errors. So, on the Brier score scale (Fig. 8.11), more variable distributions include many more potentially punishing scores when the crowd is small. Larger crowds have much narrower distributions, leading to Brier score distributions containing very few extremely poor scores.

This result is shown from a slightly different perspective in Fig. 8.12, which contains the expectation and standard deviation of the beta distributions estimated by the model for all combinations of sample size and time to resolution. This display highlights that time is a major driver of the expectation parameter, while crowd size is the major driver of the standard deviation parameter.

5 General Discussion

The fact that information bases change with the passage of time is a unique feature of judgmental forecasting. The added uncertainty associated with longer time horizons as compared to shorter time horizons will often lack an epistemic remedy. Finding more informed and more experienced forecasters or, simply, more forecasters in such situations, can only go so far. This is a challenge not just for the forecasters, but

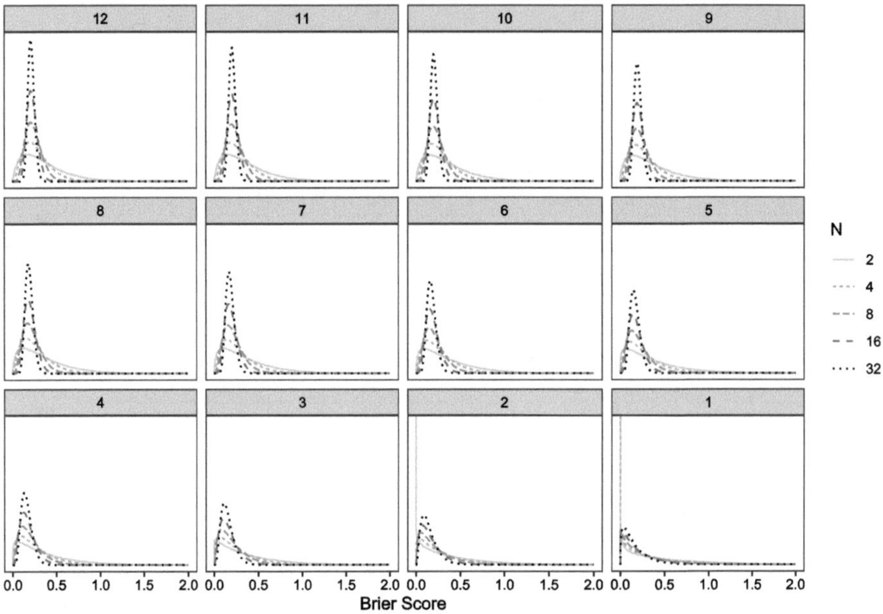

Fig. 8.11 Predicted distributions of aggregate Brier scores based on the 500 resampled sample sizes (2, 4, 8, 16, 32) at different time points from beta regression with log(time remaining) and log2 (N) as predictors

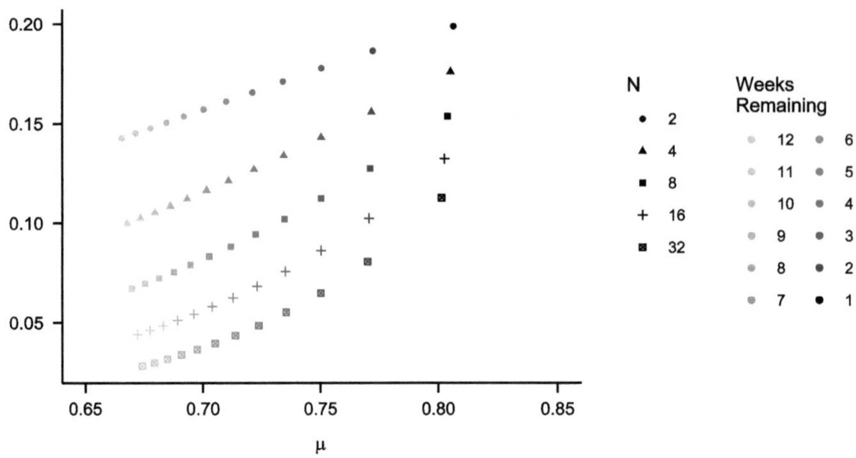

Fig. 8.12 Joint distributions of expectation (μ) and standard deviation (σ) parameters from beta regression conditioned on log (weeks remaining) and log2(N)

also forecast users and evaluators as well. How can judges who make forecasts at different times, about different problems, be compared to one another? How should one decide which forecasts to retain if they were made at different times? Can one use larger crowds to offset the difficulty of forecasting questions that are far from their resolution?

5.1 Signal Sources

What are the features that make the uncertainty associated with forecasting so unique? Steyvers et al. (2014) frame the evaluation of forecasters as a signal detection problem: better forecasters better distill signal from noise. Their results also indicate that as time passes, it becomes easier for the average forecaster to separate signal from noise, suggesting that the temporal distance associated with time horizons entails a unique type of noise. However, variability between forecasters in signal detection also implies that this variability can be further decomposed, and some forecasters are consistently better at signal distillation than others.

Taken together, these can be conceived of as two distinct features of signal detection, both of which change as a function of time: signal *availability* and signal *accessibility*. Signal availability refers to the non-epistemic aspect of forecasting: the maximal degree to which uncertainty could be reduced at a given time point. The amount of available signal for a forecasting problem immediately prior to resolution is the inverse of the notion of *irreducible uncertainty* (Fox & Ülkümen, 2011; Hammond, 1996; Stewart, 2000). Signal accessibility refers to the epistemic aspect of forecasting: the ease with which individual agents can discern the available signal at a given moment. It is a slight reframing of the perspective on why some forecasters consistently outperform others.

The wisdom of crowds is a method for distilling available signals that has been accessed by many agents. If all agents accessed signals with equivalent precision and randomly distributed errors, the errors would cancel out, leaving only available signal. However, because some forecasters access and process signals better than others, and because forecasters access signal at different time points (when the amount of available signal changes), accounting for these confounds is a key method for improving the aggregate wisdom over a simple average.

5.2 Bias

One dimension of crowd wisdom unexplored by this chapter is bias. We implicitly assumed that signal accessibility errors are conditionally independent. However, there is evidence that this may not always be the case. Several biases involving forecasting of time series events have been identified by the literature, which tend to be exacerbated in high noise (low accessibility) environments. For example, people will often add noise to an implied trend in an attempt to represent uncertainty (Harvey, 1995; Harvey et al., 1997), demonstrate over-sensitivity to recent events (Reimers & Harvey, 2011), and reduce the extremity of empirical trends (Harvey & Reimers, 2013). Such biases could create non-independent accessibility errors.

Importantly, de Baets and Vanderheyden (2021) found low correlations between forecasting biases and dispositional information commonly associated with forecasting skill. These results suggest that, particularly in high noise (low availability) environments, such as when forecasts are made far from their resolution, if there is little information available about past individual performance, assessing bias susceptibility may be an important component of identifying forecasters who are likely to produce more accurate forecasts.

5.3 Beyond Judgmental Forecasting

This chapter has focused on judgmental forecasting processes. However, this is not the only mechanism by which the wisdom of crowds can be elicited. Another increasingly popular method is prediction markets (Atanasov et al., 2017; Maciejovsky & Budescu, 2020) in which traders bet on future event outcomes by buying and selling contracts commensurate with their expectations about the likelihood of those events. Contract prices provide a representation of the market, or crowd expectations about the events being traded, and because markets are inherently dynamic, changes in price over time will inherently represent changes in crowd belief about an event over time.

A few recent studies have compared the efficacy of judgmental forecasting to prediction markets in extracting the wisdom of crowds. For example, Atanasov et al. (2017) found that prediction markets performed better than simple averages of judgmental forecasts. However, several statistical techniques, such as performance weighting, recency selection, and extremization, improved judgmental forecast averages to the point where they surpassed the accuracy of prediction markets. Interestingly, both Atanasov et al. (2017) and Sethi et al. (2021) found that markets tend to produce more accurate forecasts for longer time horizons, while judgmental forecasting aggregates produced more accurate forecasts for shorter time horizons. Maciejovsky and Budescu (2020) also show that people trust markets less than human judges, possibly because of lack of familiarity with markets and their perceived opacity.

5.4 *Summary of Recommendations*

5.4.1 Evaluating Forecasters

Accounting for forecast timing when comparing forecasters is important for accurately discriminating the skill levels of different forecasters. Model-based techniques, such as the HLM and IRT approach, can de-confound forecast timing in the comparison of forecasters to one another (Himmelstein et al., 2021). Application of these methods may be more important for selecting the most highly skilled analysts than for developing optimal performance weights in a WoC aggregation.

5.4.2 Information Accrual

When evaluating forecasters, the amount of information one has about their skill will change over time. When relatively little information is available, dispositional information can be at least as important as information about past accuracy. However, as information about past accuracy accrues, it becomes a far superior predictor of future accuracy (Himmelstein et al., 2021). Dynamic models, such as the contribution weighted model, can continually take advantage of the information available at a given time point in forming WoC aggregations (Budescu & Chen, 2014; Chen et al., 2016). Intersubjective scoring techniques also represent an appealing innovation for evaluating forecasting skill in the absence of ground truth resolutions (Himmelstein et al., 2023; Liu et al., 2020; Witkowski et al., 2017).

5.4.3 Forecast Recency and Aggregation

In situations where judges make forecasts at different times, aggregators must decide how to handle forecasts that are older than others. Researchers have proposed methods for selecting only a subset of the most recent forecasts (Atanasov et al., 2017), or for down-weighting forecasts that are less recent than others (Baron et al., 2014; Ungar et al., 2012). Both methods discount out-of-date forecasts while maintaining the size of the crowd at or above a given information threshold. They also maintain the efficiency (Regnier, 2018) of the WoC forecasting system. It is possible to show that, in some cases, these methods can be made identical, and methods that combine the two approaches are conceptually appealing, if untested, as well.

5.4.4 Time and Crowd Size

It may be tempting to use larger crowds to offset expected accuracy losses associated with forecasts that have very long time horizons, but this approach is unlikely to be helpful. The benefits of increasing crowd size involve reducing the variability of the aggregate and decreasing the chances a very large error will be made due to sampling error. Perhaps methods more designed to tap into hidden information (e.g., Lichtendahl et al., 2013; Palley & Soll, 2019; Prelec et al., 2017) can be of use in de-confounding the accuracy of aggregate forecasts from their timing. However, it remains highly likely that some information can only be revealed with the passage of time.

Acknowledgments We wish to thank Drs. Pavel Atanasov, Daniel Benjamin, Nigel Harvey, Jack Soll, Mark Steyvers, and Thomas S. Wallsten for useful comments on an early version of this chapter.

References

Atanasov, P., & Himmelstein, M. (2022). Talent spotting in predictive analytics. In M. Seifert (Ed.), *Judgment in predictive analytics*. Springer.

Atanasov, P., Rescober, P., Stone, E. R., Swift, S. A., Servan-Schreiber, E., Tetlock, P., Ungar, L., & Mellers, B. A. (2017). Distilling the wisdom of crowds: Prediction markets vs. prediction polls. *Management Science, 63*(3), 691–706.

Baron, J., Mellers, B. A., Tetlock, P. E., Stone, E., & Ungar, L. H. (2014). Two reasons to make aggregated probability forecasts more extreme. *Decision Analysis, 11*(2), 133–145.

Benjamin, D. M., Morstatter, F., Abbas, A. E., Abeliuk, A., Atanasov, P., Bennett, S., Beger, A., Birari, S., Budescu, D. V., Catasta, M., Ferrara, E., Haravitch, L., Himmelstein, M., Hossain, K. T., Huang, Y., Jin, W., Joseph, R., Leskovec, J., Matsui, A., et al. (2023). Hybrid forecasting of geopolitical events. *AI Magazine, 44*(1), 112–128. https://doi.org/10.1002/aaai.12085

Bo, Y. E., Budescu, D. V., Lewis, C., Tetlock, P. E., & Mellers, B. A. (2017). An IRT forecasting model: Linking proper scoring rules to item response theory. *Judgment and Decision making, 12*(2), 90–104.

Brier, G. W. (1950). Verification of forecasts expressed in terms of probability. *Monthly Weather Review, 78*(1), 1–3.

Budescu, D. V., & Chen, E. (2014). Identifying expertise to extract the wisdom of crowds. *Management Science, 61*(2), 267–280.

Chalmers, J., Kaul, A., & Phillips, B. (2013). The wisdom of crowds: Mutual fund investors' aggregate asset allocation decisions. *Journal of Banking & Finance, 37*(9), 3318–3333.

Chen, E., Budescu, D. V., Lakshmikanth, S. K., Mellers, B. A., & Tetlock, P. E. (2016). Validating the contribution-weighted model: Robustness and cost-benefit analyses. *Decision Analysis, 13*(2), 128–152.

Collins, R. N., Mandel, D. R., & Budescu, D. V. (2022). Performance-weighted aggregation: Ferreting out wisdom within the crowd. In M. Seifert (Ed.), *Judgment in predictive analytics*. Springer.

Cronbach, L. J., Rajaratnam, N., & Gleser, G. C. (1963). Theory of generalizability: A liberalization of reliability theory. *British Journal of Statistical Psychology, 16*(2), 137–163.

Davis-Stober, C., Budescu, D. V., Dana, J., & Broomell, S. (2014). When is a crowd wise? *Decision, 1*(2), 79–101.

Dawes, R. M. (1979). The robust beauty of improper linear models in decision making. *American Psychologist, 34*(7), 571–582.

de Baets, S., & Vanderheyden, K. (2021). Individual differences in the susceptibility to forecasting biases. *Applied Cognitive Psychology, 35*(4), 1106–1114.

Feng, Y., & Budescu, D. V. (2021). Using selected peers to improve the accuracy of crowd sourced forecasts (abstract). *Multivariate Behavioral Research, 56*(1), 155–156.

Feng, Y., & Budescu, D. V. (2022). *Using selected peers to improve the accuracy of crowd sourced forecasts*. Manuscript submitted for publication. Department of Psychology, Fordham University.

Fox, C. R., & Ülkümen, G. (2011). Distinguishing two dimensions of uncertainty. In W. Brun, G. Keren, G. Kirkebøen, & H. Montgomery (Eds.), *Perspectives on thinking, judging and decision making: A Tribute to Karl Halvor Teigen*. Universitetsforlaget.

Garcia, J. A. (2003). An introduction to the ECB's survey of professional forecasters. *ECB Occasional Paper, 8*, 3–36.

Hammond, K. R. (1996). *Human judgment and social policy: Irreducible uncertainty, inevitable error, unavoidable injustice*. Oxford University Press on Demand.

Harvey, N. (1995). Why are judgments less consistent in less predictable task situations? *Organizational Behavior and Human Decision Processes, 63*(3), 247–263.

Harvey, N., & Reimers, S. (2013). Trend damping: Under-adjustment, experimental artifact, or adaptation to features of the natural environment? *Journal of Experimental Psychology: Learning, Memory, and Cognition, 39*(2), 589–607.

Harvey, N., Ewart, T., & West, R. (1997). Effects of data noise on statistical judgement. *Thinking & Reasoning, 3*(2), 111–132.

Himmelstein, M., Atanasov, P., & Budescu, D. V. (2021). Forecasting forecaster accuracy: Contributions of past performance and individual differences. *Judgment & Decision Making, 16*(2), 323–362.

Himmelstein, M., Budescu, D. V., & Ho, E. H. (2023). The wisdom of many in few: Finding individuals who are as wise as the crowd. *Journal of Experimental Psychology: General*. Advance online publication. https://doi.org/10.1037/xge0001340

Lamberson, P. J., & Page, S. E. (2012). Optimal forecasting groups. *Management Science, 58*(4), 805–810.

Lichtendahl, K. C., Grushka-Cockayne, Y., & Pfeifer, P. E. (2013). The wisdom of competitive crowds. *Operations Research, 61*(6), 1383–1398.

Liu, Y., Wang, J., & Chen, Y. (2020). Surrogate scoring rules. In *Proceedings of the 21st ACM Conference on Economics and Computation*, pp. 853–871.

Maciejovsky, B., & Budescu, D. V. (2020). Too much trust in group decisions: Uncovering hidden profiles by groups and markets. *Organization Science, 31*(6), 1497–1514.

Mellers, B. A., Stone, E. R., Atanasov, P., Rohrbaugh, N., Emlen Metz, S., Ungar, L., Bishop, M. M., Horowitz, M., Merkle, E., & Tetlock, P. E. (2015a). The psychology of intelligence analysis: Drivers of prediction accuracy in world politics. *Journal of Experimental Psychology: Applied, 21*(1), 1–14. https://doi.org/10.1037/xap0000040

Mellers, B. A., Stone, E. R., Murray, T., Minster, A., Rohrbaugh, N., Bishop, M. M., Chen, E., Baker, J., Hou, Y., Horowitz, M., & Others. (2015b). Identifying and cultivating superforecasters as a method of improving probabilistic predictions. *Perspectives on Psychological Science, 10*(3), 267–281.

Merkle, E. C., Steyvers, M., Mellers, B. A., & Tetlock, P. E. (2016). Item response models of probability judgments: Application to a geopolitical forecasting tournament. *Decision, 3*(1), 1–19.

Moore, D. A., Swift, S. A., Minster, A., Mellers, B., Ungar, L., Tetlock, P., Yang, H. H. J., & Tenney, E. R. (2017). Confidence calibration in a multiyear geopolitical forecasting competition. *Management Science, 63*(11), 3552–3565.

Morstatter, F., Galstyan, A., Satyukov, G., Benjamin, D., Abeliuk, A., Mirtaheri, M., Hossain, K. S. M. T., Szekely, P., Ferrara, E., Matsui, A., Steyvers, M., Bennet, S., Budescu, D., Himmelstein, M., Ward, M., Beger, A., Catasta, M., Sosic, R., Leskovec, J., et al (2019). SAGE: A hybrid geopolitical event forecasting system. In *Proceedings of the Twenty-Eighth International Joint Conference on Artificial Intelligence, IJCAI-19*, pp. 6557–6559. https://doi.org/10.24963/ijcai.2019/955

Palley, A. B., & Soll, J. B. (2019). Extracting the wisdom of crowds when information is shared. *Management Science, 65*(5), 2291–2309.
Prelec, D., Seung, H. S., & McCoy, J. (2017). A solution to the single-question crowd wisdom problem. *Nature, 541*(7638), 532–535.
Ray, R. (2006). Finance, the wisdom of crowds, and uncannily accurate predictions. *Investment Management and Financial Innovations, 3*(1), 35–41.
Regnier, E. (2018). Probability forecasts made at multiple lead times. *Management Science, 64*(5), 2407–2426.
Reimers, S., & Harvey, N. (2011). Sensitivity to autocorrelation in judgmental time series forecasting. *International Journal of Forecasting, 27*(4), 1196–1214.
Schnaars, S. P. (1984). Situational factors affecting forecast accuracy. *Journal of Marketing Research, 21*(3), 290–297.
Sethi, R., Seager, J., Cai, E., Benjamin, D. M., & Morstatter, F. (2021). Models, markets, and the forecasting of elections. *SSRN*. https://doi.org/10.2139/ssrn.3767544
Stewart, T. R. (2000). Uncertainty, judgment, and error in prediction. In D. Sarewitz, R. A. Pielke, & R. Byerly (Eds.), *Prediction: Science, decision making, and the future of nature* (1st ed., pp. 41–57). Island Press.
Steyvers, M., Wallsten, T. S., Merkle, E. C., & Turner, B. M. (2014). Evaluating probabilistic forecasts with Bayesian signal detection models. *Risk Analysis, 34*(3), 435–452.
Surowiecki, J. (2005). The wisdom of crowds. *Anchor*.
Tetlock, P. E., & Gardner, D. (2016). *Superforecasting: The art and science of prediction*. Random House.
Tetlock, P. E., Mellers, B. A., Rohrbaugh, N., & Chen, E. (2014). Forecasting tournaments: Tools for increasing transparency and improving the quality of debate. *Current Directions in Psychological Science, 23*(4), 290–295.
Ungar, L., Mellers, B. A., Satopää, V., Tetlock, P. E., & Baron, J. (2012). The good judgment project: A large scale test of different methods of combining expert predictions. In *2012 AAAI Fall Symposium Series*.
Wagner, C., & Vinaimont, T. (2010). Evaluating the wisdom of crowds. *Proceedings of Issues in Information Systems, 11*(1), 724–732.
Witkowski, J., Atanasov, P., Ungar, L. H., & Krause, A. (2017). Proper proxy scoring rules. In *Proceedings of the Thirty-First AAAI Conference on Artificial Intelligence*, pp. 743–749. AAAI.

Part III
Contextual Factors and Judgmental Performance

Chapter 9
Supporting Judgment in Predictive Analytics: Scenarios and Judgmental Forecasts

Dilek Önkal, M. Sinan Gönül, and Paul Goodwin

Keywords Scenario · Judgment · Forecast · Uncertainty

1 Introduction

Despite advances in predictive analytics there is widespread evidence that algorithm-based forecasts are often judgmentally adjusted (Fildes & Goodwin, 2007; Fildes & Petropoulos, 2015; Fildes et al., 2009). Ostensibly, these adjustments are made to take into account factors not included in the algorithmic forecasts. For example, in demand forecasting the judgments may reflect the estimated effects of promotion campaigns, price changes, the activities of competitors or government policies (Fildes & Goodwin, 2007). However, other latent factors can also lead to adjustments such as a false perception of systematic variation in time series noise that has been filtered out by the algorithm (Lawrence et al., 2006). Moreover, adjustments can be made on the basis of rumors, unsubstantiated narratives or the experience of a single similar past event that has no or little diagnostic value (Fildes et al., 2019).

Against this background, scenarios can potentially play a crucial role in supporting the role of judgment-where predictive analytics are involved. Scenarios are narrative-based tools that provide structured alternative accounts of how the events might unfold. They enable people to think about the future and envisage the uncertainties associated with it. As such, they can provide useful guidance when making judgmental forecasts, or judgmental adjustments to algorithmic forecasts, and when assessing the bounds of uncertainty (Önkal et al., 2013).

In this chapter, we review the literature which has examined the role of scenarios where short to medium term forecasts have to be made, with a focus on situations

D. Önkal (✉) · M. S. Gönül
Newcastle Business School, Northumbria University, Newcastle upon Tyne, UK
e-mail: dilek.onkal@northumbria.ac.uk

P. Goodwin
School of Management, University of Bath, Bath, UK

M. Seifert (ed.), *Judgment in Predictive Analytics*, International Series in Operations Research & Management Science 343,
https://doi.org/10.1007/978-3-031-30085-1_9

where forecasts based on predictive analytics are available. We then present a behavioral experiment that aims to extend our knowledge of this area by exploring the effects of scenario tone and extremity on judgmental forecasts. We also examine the effects of sharing scenarios with team members on individual and team forecasts. The chapter concludes by discussing practical implications for incorporating scenario approaches to support judgment in predictive analytics.

2 Literature Review

In long-range strategic planning the exercise of formulating scenarios is intended to alert managers to uncertainties they may be unaware of, enabling them to make agile plans for alternative futures (Wright & Goodwin, 2009). Scenarios come in many guises ranging from quantitatively-orientated combinations of probabilistic outcomes to narratives expressed only in qualitative terms that depict alternative plausible futures (van der Heijden, 2005). It is this latter form that concerns us here, but ironically people who pioneered these types of scenarios intended them to be used in situations where forecasts were likely to be unreliable because of extreme uncertainties and the unlikelihood that patterns of the past would continue into the future. Such scenarios are usually produced in groups of two to four enabling them to depict the estimated bounds of uncertainty. For example, in the simplest so-called extreme case approach (Goodwin & Wright, 2014), we may simply have a worst-case (pessimistic) and best-case (optimistic) scenario.

Despite the origins of scenario planning, researchers have recently explored the effectiveness of using scenarios to support judgmental forecasts that are produced for relatively short-term futures. Scenarios can add information to forecasts, particular where forecasts would otherwise be based only on time-series information and depend on the assumption that prevailing trends will continue even though there may be no underlying rationale to justify such a belief. It is also known that narratives tend to engage people more than statistical information (Schnaars & Topol, 1987; Taylor & Thompson, 1982; Anderson, 1983) so the information they contain is more likely to be absorbed by people when forming judgments. For example, Taleb (2005) has argued that narratives are salient while statistics are invisible.

In forecasting, prediction intervals are the most common way of representing the range of uncertainty associated with a forecast, but intervals estimated either by judgmental forecasters or analytical methods tend to be too narrow so that the extent of uncertainty is underestimated. Where judgmental forecasters are concerned this phenomenon is known as overconfidence. This raises the question: can the provision of scenarios lead to better calibrated prediction intervals—and more accurate point forecasts -in shorter term forecasting? For example, could judgmental adjustments to forecasts based on predictive algorithms improve the quality of the forecasts when scenarios are available to forecasters, in particular by reducing overconfidence (Zentner, 1982)?

While researchers such as Bunn and Salo (1993) are enthusiastic about the use of scenarios in forecasting the more recent literature suggests that the benefits of scenario provision are not guaranteed. For example, in one experiment forecasters were shown a time series depicting the past sales of a product alongside best-case and worst-case scenarios that provided reasons why the future sales might increase or decrease (Goodwin et al., 2019a). It is known that judgmental forecasters tend to overreact to the most recent movement in a time series even though this may result from randomness (Bolger & Harvey, 1993). The experiment revealed that this was exacerbated when scenarios were provided. For example, a best-case scenario led to even greater weight being attached by judgmental forecasters to a recent rise in the time series, while a worst case scenario led to greater attention being paid to a recent fall. In each case, the scenario suggesting a movement in the direction inconsistent with the latest time series movement was discounted. The result was a reduction in point forecast accuracy. Moreover, there was no evidence that the calibration of the forecasters' prediction interval was improved by the provision of worst and best-case scenarios and they tended to underestimate the level of noise in the series.

The provision of opposing scenarios was also explored by Goodwin et al. (2019b). In this case participants were asked to make point and interval forecasts under the assumption that either a best-case or worst-case scenario would prevail. Some participants were additionally provided with an opposing scenario. For example, they were asked to make forecasts given that a best-case scenario would happen, but they were also provided with a worst-case scenario, which should have been irrelevant to their judgment. However, the effect of having access to the opposite scenario led to a narrowing of prediction intervals (i.e., greater overconfidence), the opposite of what was desirable, with the effect being greater where scenarios were more extreme. This result was consistent with earlier studies finding that confidence in forecasts increases as more information is available even when the information is irrelevant (e.g., Hall et al., 2007; Oskamp, 1965). This study also found that point forecasts made on the assumption that a given scenario would prevail were subject to a contrast effect (Sherif et al., 1958). When an opposing scenario was also provided, forecasts based on the original scenario became more extreme. For example, forecasts based on a best-case scenario were more optimistic when a worst-case scenario was also provided and vice versa. Again, the effect was greater where more extreme scenarios were given. This is consistent with the psychology literature suggesting that contrast effects are more likely to occur where differences between the target stimulus (in this case the scenario assumed to prevail) and the context (the opposing scenario) are greater (Sherif et al., 1958). Where the two stimuli are less distinct, and may even be perceived as overlapping, an assimilation effect is more likely. In the case of scenarios this would imply that a point forecast based on a best-case scenario would be less optimistic if a worst-case scenario was also provided and vice versa. (Chien et al., 2010).

Both of the above studies differed in two ways from routine forecasting practice. First, participants were not provided with an algorithm-based point forecast, they had to rely solely on their own judgment. Secondly, they involved individual forecasters, but in practice forecasts are often agreed and negotiated by groups of people, for example at forecast review meetings (Fildes & Goodwin, 2021). The use of scenarios by both individuals and teams of forecasters who also received algorithm-based forecasts was examined in a study by Önkal et al. (2013). In this study, participants were provided with time series and model-based one-period-ahead point forecasts of the demands for a series of products. Depending on the treatment they were assigned to, they received either no scenario, a worst- or best-case scenario or both scenarios. After being invited as individuals to use the provided information to make their own judgmental point and best- and worst-case forecasts, the participants met in teams of two to agree on consensus forecasts. The provision of opposing scenarios led to the judgmental point and best- and worst-case forecasts being closer to the model-based point forecasts both in case of individuals and teams. In the case of point forecasts this could be advantageous in that research suggests that people tend to overly discount model-based point forecast in favor of their own judgment (Fildes et al., 2019). However, the results also indicate that the provision of either single or opposing scenarios generally led to narrower prediction intervals because people also stayed closer to the model-based point forecasts when assessing the bounds of possible demand. Hence access to scenarios led to an increase in overconfidence. Moreover, when asked to assess the probability that actual demand would lie outside their estimated intervals (in the form of a surprise index) participants who saw scenarios provided lower surprise indices, which again implies that scenarios increase confidence in forecast even when this may be unwarranted.

These findings raise a number of additional questions. To what extent can increasing the strength of a single best- or worst-case scenario encourage people to make a large adjustment to a model-based point forecast when estimating a given bound of a prediction interval and does this differ for worst- and best-case scenarios? Secondly, in cases such as sales series, we would expect (say) a worst-case scenario to lead to a downwards adjustment to a model-based point forecast when estimating the lower bound of an interval. However, what is its effect, if any, on the estimate of the upper bound of possible sales? Thirdly, the provision of worst- and best-case scenarios should not cause people to modify model-based point forecasts as such forecasts are associated with the central tendency of underlying probability distributions, rather than their bounds. Nevertheless, do people make such modifications? To address these questions an experiment was conducted to investigate the effects of scenario tone and extremity on individual and team-based judgmental predictions The details are given next.

3 Methodology

Participants were business students from Bilkent University who were taking a forecasting course. Predictive context involved making demand forecasts for different models of mobile phones and tablets produced by a mobile telecommunications company.

Artificially constructed time-series were used to portray past demand while controlling the levels of trend and uncertainty presented to the participants. These series involved six combinations with three levels of trend (positive, negative, and stable) and two levels of uncertainty (low and high). A total of 18 series (presenting three iterations of six combinations of trend and uncertainty) were constructed, with participants receiving these series in a randomized order.

The procedure for constructing these artificial time series was akin to previous work (e.g. Goodwin et al., 2019b; Gönül et al., 2006; Önkal et al., 2009, 2013), with the following formula used in generating the time series:

$$y(t) = 125 + bt + error(t) \qquad t = 0, 1, .., 20 \tag{9.1}$$

The trend coefficient, *b,* was set at -5 for negative trends, 0 for stable trends and +5 for series those with positive trends. The error was normally distributed with zero mean and a standard deviation of either 10% for low uncertainty or 20% for high uncertainty. Double exponential smoothing method with optimal smoothing parameters was employed to generate the model-based predictions.

3.1 *Experimental Design*

A total of 118 participants completed the experiment, which consisted of an individual forecasting phase (Phase 1), followed by a team forecasting phase (Phase 2) and ending with a final/preferred individual forecasting phase (Phase 3).

3.1.1 Phase 1: Individual Forecasts

First phase of the experiment consisted of participants being presented with 18 time-series plots showing past demand over 20 periods for different products, accompanied by scenarios and model-based point forecasts for period 21.

Participants were randomly assigned to one of four experimental groups that differed on the level of optimism/pessimism in the presented scenarios:

IGroup 1 (IG1)—Weak Optimistic Scenarios: 31 participants received scenarios with weak optimistic tone entitled 'best-case scenario'

IGroup 2 (IG2)—Strong Optimistic Scenarios: 28 participants received scenarios with strong optimistic tone entitled 'best-case scenario'

IGroup 3 (IG3)—Weak Pessimistic Scenarios: 30 participants received scenarios with weak pessimistic tone entitled 'worst-case scenario'

IGroup 4 (IG4)—Strong Pessimistic Scenarios: 29 participants received scenarios with strong pessimistic tone entitled 'worst-case scenario'

The strong/weak levels of optimism/pessimism were achieved through varying the tone of the scenario vignettes by manipulating the wording, while keeping the content and the amount of information the same across the scenarios for each product. The scenarios were validated by practitioners working in the same sector as the task context (i.e., mobile telecommunications).

Having received the relevant material, the participants were asked to make the following forecasts for period 21 for each product: (i) point forecast, (ii) best-case forecast, and (iii) worst-case forecast.

3.1.2 Phase 2: Team Forecasts with Scenario Discussions

To reflect the social context in which forecasts are typically formed in organisations, participants were assigned to two-person teams in the second phase of the experiment. There were four groups of teams based on the level of optimism or pessimism combinations in team members' scenarios:

TGroup 1 (TG1)—Weak Optimistic vs Weak Pessimistic Scenarios: 17 teams with one member from *IG1* (weak optimistic scenario) and another member from *IG3* (weak pessimistic scenario) in Phase 1

TGroup 2 (TG2)—Weak Optimistic vs Strong Pessimistic Scenarios: 14 teams with one member from *IG1* (weak optimistic scenario) and another member from *IG4* (strong pessimistic scenario) in Phase 1

TGroup 3 (TG3)—Strong Optimistic vs Weak Pessimistic Scenarios: 13 teams with one member from *IG2* (strong optimistic scenario) and another member from *IG3* (weak pessimistic scenario) in Phase 1

TGroup 4 (TG4)—Strong Optimistic vs Strong Pessimistic Scenarios: 15 teams with one member from *IG2* (strong optimistic scenario) and another member from *IG4* (strong pessimistic scenario) in Phase 1.

Each participant was given the same 18 time-series plots, the same model-based forecasts and the corresponding scenarios from Phase 1 when they made their individual predictions. Participants were requested to discuss the given model-based forecasts, past demands, and their differential scenarios as a team and arrive at consensus forecasts in the form of point, best-case, and worst-case predictions.

3.1.3 Phase 3: Final/Preferred Individual Forecasts After Scenario Discussions

Upon completing the team consensus forecasts for each product, each participant was requested once again to convey their 'preferred individual predictions'. This was done to explore whether the team predictions reflected their final forecasts or whether their preferred predictions deviated from the predictions made as a team. These forecasts would also show whether/how their final forecasts were different from their initial predictions (prior to the team discussion and knowledge of other member's scenarios), yielding insights into potential effects of sharing other's scenarios. An exit questionnaire was filled out after all phases were complete.

3.2 Results

3.2.1 Assessments of Scenario Tone

The exit questionnaire asked the participants to rate the tone of the scenarios they received on a 5-point scale between *1 = strongly pessimistic* and *5 = strongly optimistic*. Since participants did not see any labels regarding the strength of the scenarios, these ratings serve as a manipulation check for the experimental groups. The results are shown in Table 9.1.

ANOVA analysis reveals that for the assessments of scenario tone, there was a significant main effect of the scenario type (optimistic/pessimistic) ($F_{1,114} = 134.98$, $p < .0001$) and a significant interaction between scenario type (optimistic/pessimistic) and scenario strength (weak/strong) ($F_{1,114} = 9.54$, $p = .003$) as shown in Fig. 9.1.

Table 9.1 Mean ratings for scenario tone

	Weak optimistic (IG1) [n = 31]	Strong optimistic (IG2) [n = 28]	Weak pessimistic (IG3) [n = 30]	Strong pessimistic (IG4) [n = 29]
Mean ratings for scenario tone (1:Strongly pessimistic; 5: Strongly optimistic)	3.90	4.50	2.53	2.14

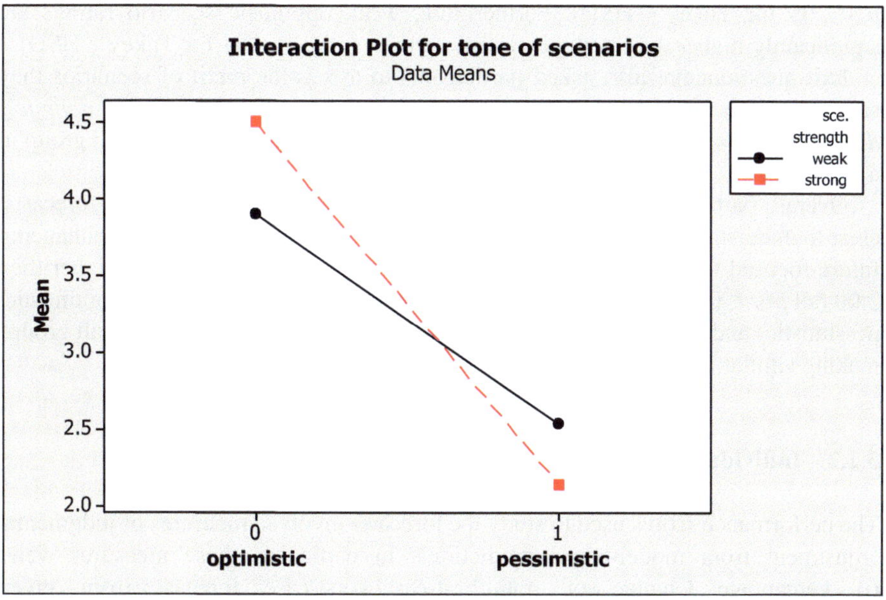

Fig. 9.1 Interaction Plot for Scenario Tone

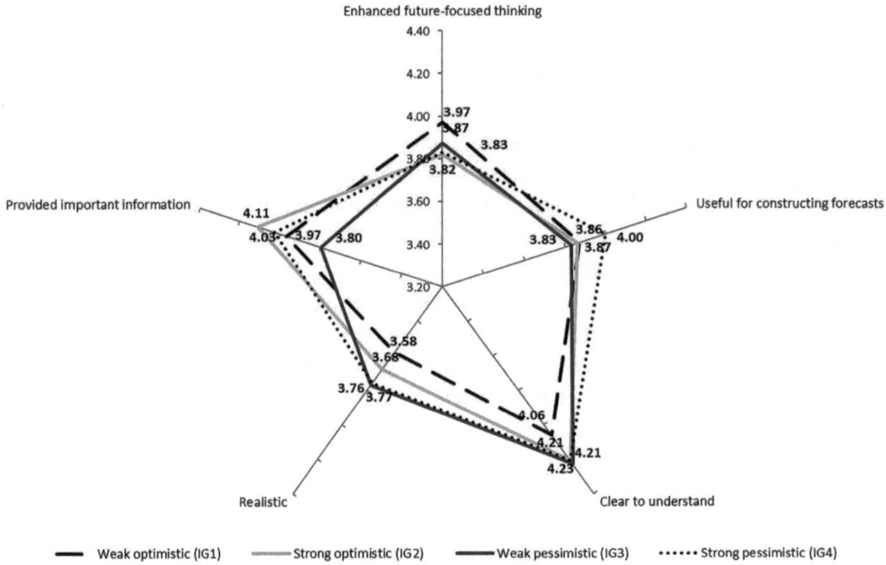

Fig. 9.2 Mean ratings of scenario assessments

Furthermore, there was a significant difference among these groups ($F_{3,114} = 47.48$, $p < .0001$). Tukey's pairwise comparisons indicate that strong optimistic scenarios received significantly higher ratings than the other groups (all $p < .05$ for Tukey's HSD). Furthermore, weak optimistic scenario ratings are significantly higher than both pessimistic groups (all $p < .001$ for Tukey's HSD).

Exit questionnaire also asked participants to assess the merit of scenarios they received. Each question involved a rating on a 5-point scale from *1* = *definitely disagree* to *5* = *definitely agree*. Figure 9.2 gives the mean ratings for each group in a radar chart.

Overall, participants found the given scenarios useful for constructing forecasts, clear to understand, realistic, providing important information, as well as enhancing future-focused thinking (mean ratings for each dimension significantly higher than 3.00 (all p's < 0.01)). There were no significant effects of scenario type (optimistic/pessimistic) and scenario strength (weak/strong), with participants across all groups making similar assessments.

3.2.2 Individual Forecasts

The performance scores used to study the forecasts involved measures of judgmental adjustment from model-based predictions. In particular, these measures were: (i) Percentage Change of point/best-case/worst-case forecast from given model-based forecasts (measuring the overall direction of adjustments from

Table 9.2 Mean performance scores for initial individual forecasts

	Weak optimistic (IG1) [n = 31]	Strong optimistic (IG2) [n = 28]	Weak pessimistic (IG3) [n = 30]	Strong pessimistic (IG4) [n = 29]
% change of point forecasts from model-based point forecasts	2.62%	4.22%	−4.36%	−6.74%
Absolute deviation bet. Point forecasts and model-based point forecasts	6.19	7.74	8.78	10.14
% change of best-case forecasts from model-based point forecasts	20.15%	25.84%	19.00%	17.53%
Absolute deviation bet. Best-case forecasts and model-based point forecasts	13.54	16.35	15.10	16.26
% change of worst-case fore-casts from model-based point forecasts	−16.39%	−15.26%	−25.06%	−25.24%
Absolute deviation bet. Worst-case forecasts and model-based point forecasts	15.34	15.79	22.49	24.53

model-based forecasts; with positive scores showing upward adjustments and negative scores displaying downward adjustments); (ii) Absolute Deviation of point/best-case/worst-case forecast from given model-based forecasts (measuring the average magnitude/size of adjustments from model-based forecasts). Table 9.2 provides a summary of the mean performance scores on these measures.

Point forecasts made individually at Phase 1 showed that scenario type (optimistic/pessimistic) had a significant main effect for the direction of adjustments (measured as percent change of point forecasts from given model forecasts; $F_{1,114} = 73.56$, $p < .0001$) as well as the magnitude of adjustments (measured as absolute deviation of point forecasts from given model forecasts; $F_{1,114} = 11.56$, $p = .001$). Only significant main effect of scenario strength (weak/strong) was found for magnitude of adjustments ($F_{1,114} = 3.94$, $p = .05$) and not for adjustment direction. All interactions were insignificant. These findings imply that the initial point forecasts are adjusted differently (in terms of both size and direction) depending on whether optimistic or pessimistic scenarios are given to participants. As could be expected, strength of optimism/pessimism did not alter the direction of judgmental adjustments but did make a significant difference on adjustment size.

In arriving at their best-case forecasts, participants' changes from model predictions were not affected by the scenario type/strength. Regardless of whether they received optimistic or pessimistic scenarios and regardless of the strength of these scenarios, all individual forecasts indicated a sizable uplift from the given model-based point forecasts.

Interestingly, when arriving at their worst-case forecasts, changes from model predictions differed for optimistic vs pessimistic scenarios on both the direction ($F_{1,114} = 38.76$, $p < .0001$) and magnitude ($F_{1,114} = 37.80$, $p < .0001$) of adjustments. Regardless of scenario strength, individuals receiving optimistic scenarios deviated less negatively from model-based point predictions, while those given pessimistic scenarios gave worst-case forecasts that were more extreme comparatively.

3.2.3 Team Forecasts with Scenario Discussions

Teams made consensus forecasts after discussing the past demand series and model-based forecasts as well as their differential scenarios for each product. In order to understand how team forecasts in Phase 2 differed from team members' initial individual forecasts in Phase 1, difference scores (i.e., [team forecast—individual forecast]) were calculated. Table 9.3 summarizes these analyses.

Difference score analyses in Table 9.3 reveals that when one member receives an optimistic scenario (regardless of whether it is weak/strong), and the second member receives a pessimistic scenario (irrespective of weak/strong), there are pronounced variations in how the initial predictions are adjusted to arrive at team forecasts. For point forecasts, those receiving (strong/weak) pessimistic scenarios increase their initial predictions, while those receiving (strong/weak) optimistic scenarios always lower their initial forecasts, so that the differences between the team members' difference scores are all significant. Similar findings prevail for worst-case forecasts, with the only exception being when a strong optimistic scenario is combined with the other member getting a weak pessimistic scenario, in which case both members increase their initial forecasts to arrive at higher team worst-case predictions. It appears that one member getting a strong optimistic scenario affects the other member with weak pessimistic scenario to share the other's optimism, and increase their initial forecasts in arriving at team worst-case predictions. Interestingly, for best-case predictions, the only significant difference in adjustment behavior happens when a member receives strong optimistic scenario while the other receives strong pessimistic scenario, in which case each member is affected by other's scenario and adjusts accordingly.

3.2.4 Final/Preferred Individual Forecasts After Scenario Discussions

Phase 3 required the participants to give their 'final/preferred individual predictions'. These final forecasts provide essential information on how the preferred predictions deviate from the predictions made in teams and enable us to track participants' forecasting journey from initial predictions to team predictions to final preferred predictions after scenario discussions. Figures 9.3, 9.4, and 9.5 track these adjustments for point, best-case and worst-case forecasts across groups in terms of

Table 9.3 Mean difference scores between team forecasts and initial individual forecasts

Team group	Individual group	Difference between team and individual point forecasts	Difference between team and individual best-case forecasts	Difference between team and individual worst-case forecasts
One member receives weak optimistic scenario; other member receives weak pessimistic scenario (TG1)	Weak opt. (IG1)	−2.57 (17)	−0.73 (17)	−4.06 (17)
	Weak pess. (IG3)	3.22 (17)	0.46 (17)	5.04 (17)
	Diff significant?	$t_{305} = -8.01$, $p < .0001$	n.s.	$t_{305} = -8.60$, $p < .0001$
One member receives weak optimistic scenario; other member receives strong pessimistic scenario (TG2)	Weak opt. (IG1)	−4.70 (14)	−3.66 (14)	−4.89 (14)
	Strong pess. (IG4)	3.29 (14)	−9.00 (14)	6.29 (14)
	Diff significant?	$t_{251} = -8.44$, $p < .0001$	n.s.	$t_{251} = -8.14$, $p < .0001$
One member receives strong optimistic scenario; other member receives weak pessimistic scenario (TG3)	Strong opt. (IG2)	−3.15 (13)	−6.62 (13)	2.43 (13)
	Weak pess. (IG3)	2.92 (13)	−2.67 (13)	4.51 (13)
	Diff significant?	$t_{233} = -7.03$, $p < .0001$	n.s.	n.s.
One member receives strong optimistic scenario; other member receives strong pessimistic scenario (TG4)	Strong opt. (IG2)	−4.02 (15)	−4.21 (15)	−6.80 (15)
	Strong pess. (IG4)	4.91 (15)	2.89 (15)	4.93 (15)
	Diff significant?	$t_{269} = -8.57$, $p < .0001$	$t_{269} = -3.76$, $p < .0001$	$t_{269} = -9.37$, $p < .0001$

percentage change from the model-based predictions, giving a common base across the three experimental phases.

As portrayed in Fig. 9.3, point forecast adjustments across all participants show significant changes between initial individual predictions and team predictions (all $p < .05$), signaling that the team discussions lead the members to move their point forecasts in the direction of the other member, arriving at a team forecast that lies between the two members' initial forecasts. When asked about their final/preferred individual forecasts after the team discussions, there is some deviation from the collaborative predictions; but whether this difference is significant or not appears to depend on the scenario combinations. In particular, when individuals are initially given pessimistic scenarios, their final point forecasts revert back to their initial individual predictions (prior to team discussions). Thus, regardless of the strength of pessimism (weak/strong) in their scenario, sharing the (weak/strong) optimistic scenario and discussions with other team member leads to temporary significant shifts when deciding on joint team forecasts (all $p < .05$), but this effect seems to

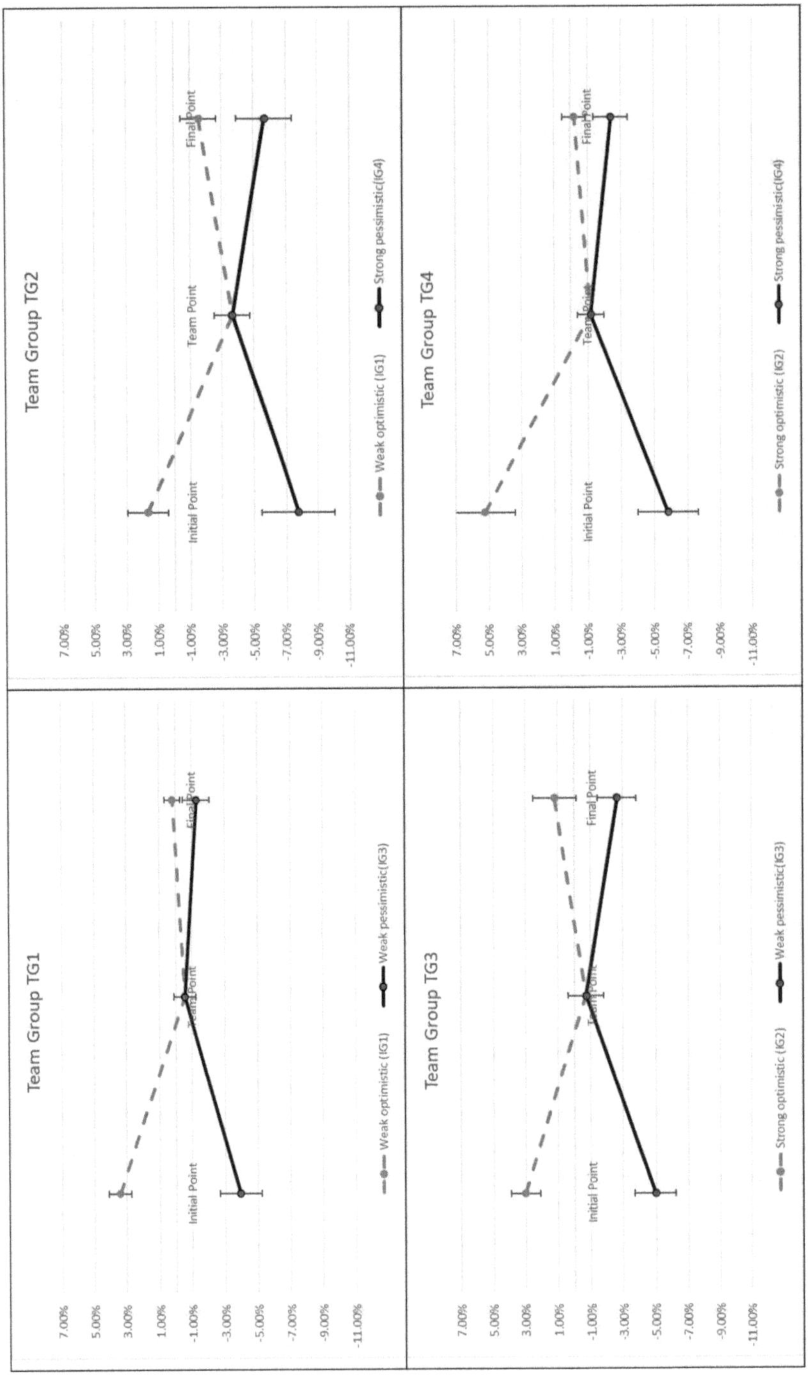

Fig. 9.3 Adjustment track of POINT FORECASTS (measured as % change from model-based predictions; error bars designate standard error)

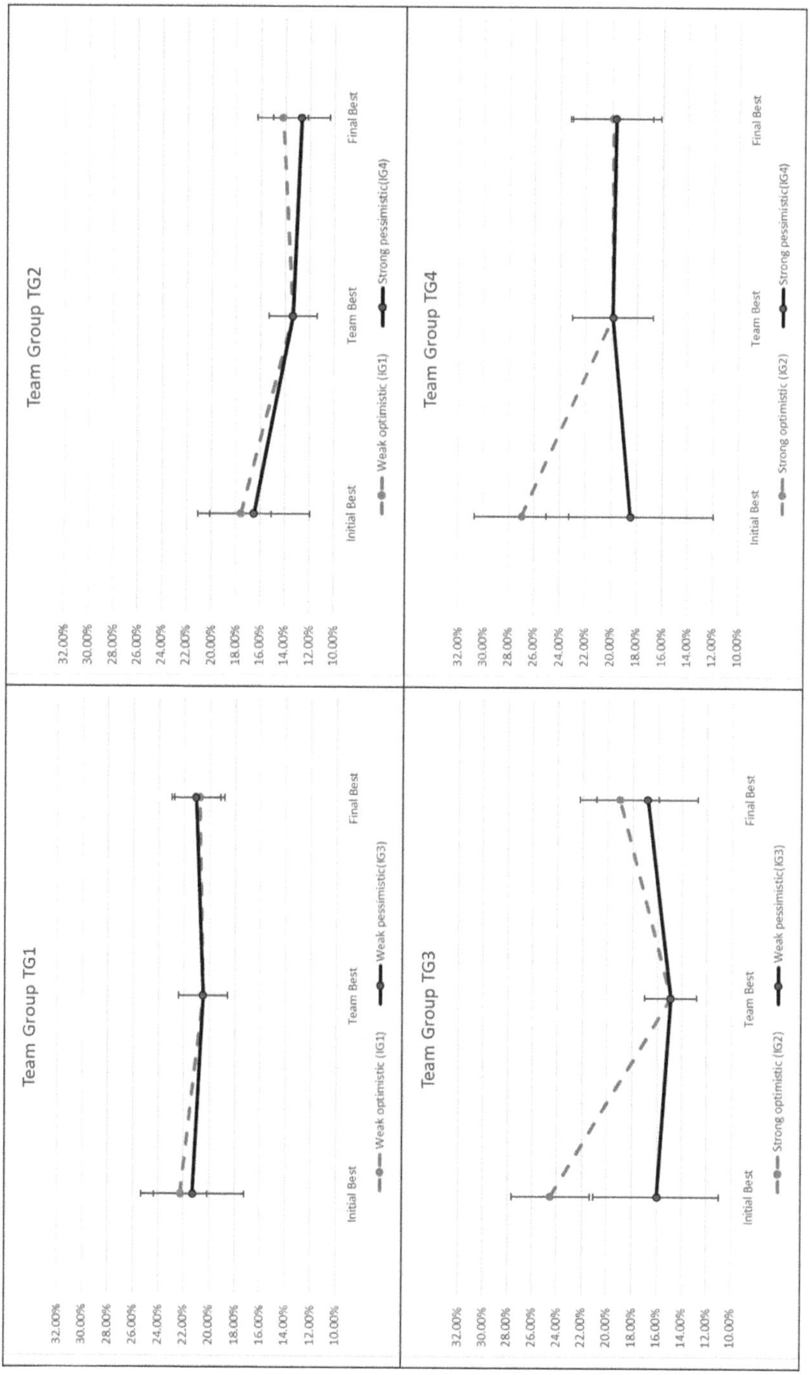

Fig. 9.4 Adjustment track of BEST-CASE FORECASTS (measured as % change from model-based predictions; error bars designate standard error)

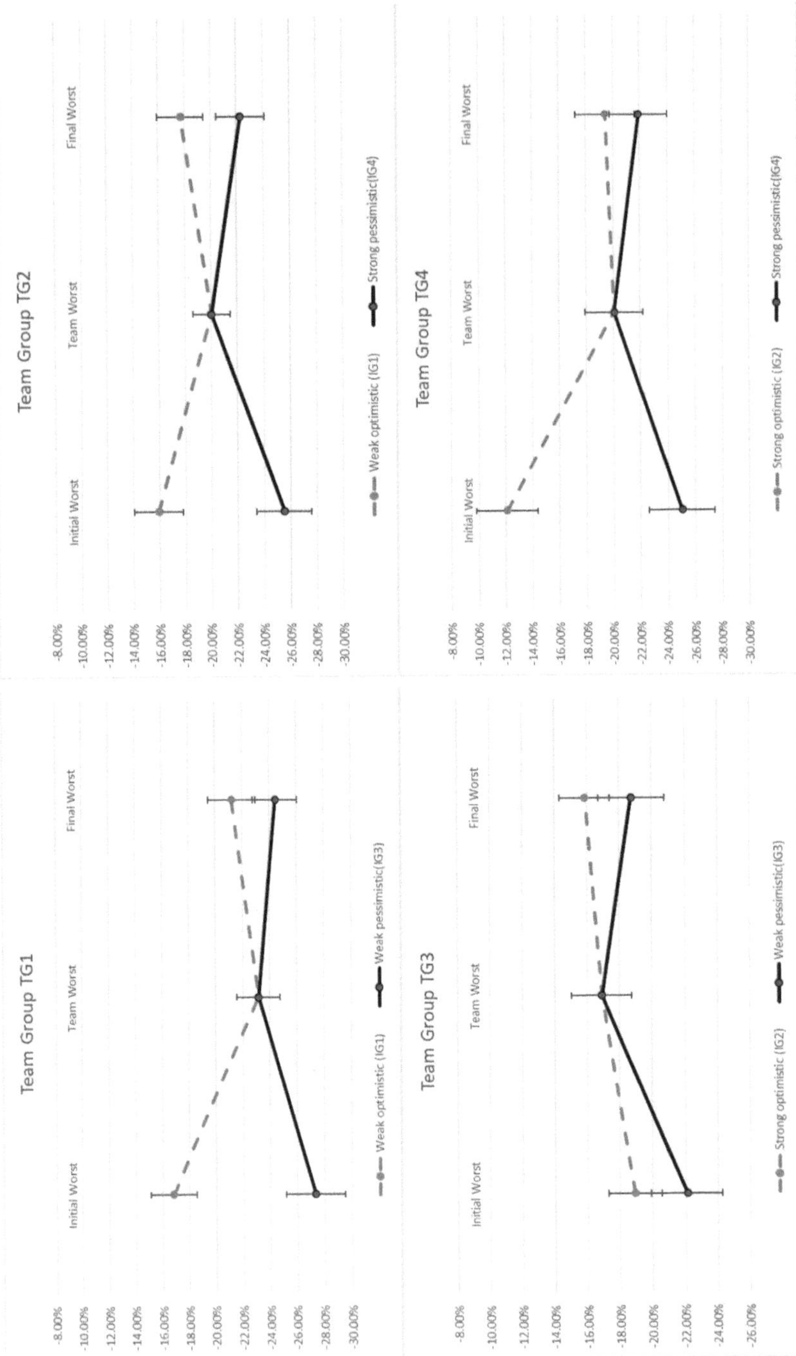

Fig. 9.5 Adjustment track of WORST-CASE FORECASTS (measured as % change from model-based predictions; error bars designate standard error)

disappear when asked about final forecasts (with final preferred predictions showing no significant changes from their original individual forecasts; all $p > .05$). While receiving the other's optimistic scenario appears to inject a bit of positive bias, this does not persist for those individuals initially given pessimistic scenarios.

The situation is reversed for the individuals initially given optimistic scenarios. While these individuals also shift their point forecasts towards the direction of the other member to arrive at team predictions (all $p < .02$), their preferred forecasts stay relatively closer to team predictions. This is reflected in significantly dampened final point forecasts as compared to initial forecasts ($t_{16} = 3.81$, $p = .001$ for participants receiving weak optimistic scenario teaming up with the other receiving weak pessimistic scenario; $t_{13} = 2.34$, $p = .018$ for participants receiving weak optimistic scenario teaming up with the other receiving strong pessimistic scenario; $t_{14} = 3.03$, $p = .005$ for participants receiving strong optimistic scenario teaming up with the other receiving strong pessimistic scenario). The only exception is when an individual given a strong optimistic scenario teams up with the other member receiving weak pessimistic scenario, in which case the final preferred forecasts stay similar to their initial forecasts before team discussions.

Figure 9.4 shows that best-case forecasts signal a similar behavior. Best-case predictions from individuals with (weak/strong) pessimistic scenarios appear to be unchanged throughout the phases of this study. These individuals do not appear to alter their best-case forecasts in reaching team predictions and their final predictions are at similar levels to their initial individual forecasts (all $p > .05$). On the other hand, for individuals receiving optimistic scenarios, best-case forecasts appear to go down to match the pessimistic member's levels in the team forecasts, with their final preferred best-case forecasts being significantly lower than their initial predictions ($t_{13} = 2.04$, $p = .032$ for participants receiving weak optimistic scenario teaming up with the other receiving strong pessimistic scenario; $t_{12} = 2.06$, $p = .031$ for participants receiving strong optimistic scenario teaming up with the other receiving weak pessimistic scenario; $t_{14} = 2.11$, $p = .026$ for participants receiving strong optimistic scenario teaming up with the other receiving strong pessimistic scenario). The only exception is when an individual given a weak optimistic scenario teams up with the other member receiving weak pessimistic scenario, in which case the final preferred best-case forecasts stay similar to the initial forecasts before team discussions for both members.

Worst-case forecasts signal real differences among the groups. Those receiving pessimistic scenarios appear to be affected by the other person's optimistic scenarios for the first time. In general, individuals receiving (weak/strong) pessimistic scenarios appear to relax their worst-case forecasts after team discussions, which then translate to their preferred final forecasts as well ($t_{16} = -2.18$, $p = .022$ for participants receiving weak pessimistic scenario teaming up with the other receiving weak optimistic scenario; $t_{13} = -2.00$, $p = .033$ for participants receiving strong pessimistic scenario teaming up with the other receiving weak optimistic scenario; $t_{12} = -2.08$, $p = .030$ for participants receiving strong weak pessimistic teaming up with the other receiving strong optimistic scenario). The only exception is when an individual with prior strong pessimistic scenario is matched with a team member receiving strong optimistic scenario, in which case the pessimistic scenario appears to overtake the optimistic one, and the initial worst-case predictions persist through both the team and final predictions. Interestingly, this combination (of strong pessimistic with strong optimistic scenario) is the only case when the individual originally given an optimistic scenario shifts their worst-case forecasts so that their final predictions are worse (more extreme) than their initial forecasts ($t_{14} = 3.79$, $p = .001$). This shows that being exposed to other member's strong pessimistic scenario produces a significant effect in making their worst-case forecasts pull toward the other member's predictions, and this effect persists when asked to state their preferred forecasts at the end of the study. For the other team combinations, optimistic scenario member's initial and final worst-case predictions lie at similar levels.

4 Discussion

Findings from the behavioral experiment detailed above were instrumental in understanding the effects of scenarios on individual and group-based judgmental predictions. Scenario manipulations were successful in that participants correctly identified the type (optimistic/pessimistic) and strength (weak/strong) of their scenarios. Also, scenarios were assessed as useful for constructing forecasts, clear to understand, realistic, providing important information, as well as enhancing future-focused thinking.

The effect of providing a single scenario did cause individual participants to change the model-based point forecasts, even though there was no normative case for doing this Nevertheless, as Table 9.2 shows, the percentage changes were relatively small. The results showed that the *initial individual point forecasts* are adjusted differently (in both size and direction) depending on whether participants received optimistic or pessimistic scenarios. Strength of optimism/pessimism made a significant difference on the magnitude of adjustments while not affecting their adjustment direction.

For their *individual best-case forecasts*, participants' changes from model predictions were not affected by the scenario type/strength, with all showing a considerable uplift from the model-based point forecasts suggesting that the scenarios were of some help in assisting participants to take into account uncertainty. However given the noise standard deviations of 10% or 20% shown in Eq. (9.1), best and worst case outcomes would be expected to be at least roughly 30% and 60% higher or lower than the model based forecasts, but none of the typical adjustments in Table 9.2 come near to these values, suggesting overconfidence. Moreover, increasing the strength of the scenario was not conducive to increasing participants' estimates of the level of uncertainty. The most notable difference was in *individual worst-case forecasts,* where the adjustments from model predictions differed for optimistic vs pessimistic scenarios on both the direction and magnitude. Irrespective of scenario strength, participants receiving optimistic scenarios adjusted less negatively from model-based point predictions, while those with pessimistic scenarios gave more extreme worst-case forecasts.

Significant variations were also found in adjustments for team forecasts. For *team point forecasts*, significant changes between initial individual predictions and team predictions revealed that the team discussions led the members to move their point forecasts in the direction of the other member, arriving at a team forecast that was between the two members' initial forecasts. For *final/preferred individual point forecasts* after the team discussions, individuals given pessimistic scenarios revert back to their initial individual predictions (prior to team discussions). Irrespective of the strength of pessimism (weak/strong) in their scenario, sharing the (weak/strong) optimistic scenario and discussions with other team member leads to temporary significant shifts when deciding on joint team forecasts, with this effect disappearing when asked about final forecasts. While receiving the other's optimistic scenario appears to inject a bit of positive bias, this does not persist for those individuals initially given pessimistic scenarios. Participants given optimistic scenarios also shift their point forecasts towards the direction of the other member to arrive at team predictions; furthermore, this team effect persists, with their final forecasts staying close to team predictions. Interestingly, the only case when the final preferred forecasts stay similar to initial forecasts is when an individual given a strong optimistic scenario teams up with the other member receiving weak pessimistic scenario. Overall, these findings show how influential pessimistic scenarios can be (irrespective of strong/weak framing), so that the discussions of an alternative optimistic scenario do not mitigate the effectiveness of the original pessimistic scenario. On the other hand, optimistic scenarios yield similar potent effects only when strong framing is used and only when this strong optimism meets up with a weak pessimistic scenario in the team discussions. The bottom line is pessimism prevails in point forecasts; with participants leaning towards overweighing their pessimistic scenarios, this is true when they are initially given only the pessimistic scenarios to start with, but it is also true when they are given optimistic scenarios but then encounter pessimistic scenarios. The only barrier to this could be when strong optimistic scenarios are initially given and they match up with weak pessimistic scenarios in team discussions.

A similar adjustment behavior is observed for *best-case predictions*. Individuals with (weak/strong) pessimistic scenarios do not appear to change their individual best-case forecasts in reaching team predictions; their final predictions also stay close to their initial forecasts. However, individuals receiving optimistic scenarios appear to lower their best-case forecasts to match the pessimistic member's levels in the team forecasts, with their final preferred best-case forecasts being significantly lower than their initial predictions. The only exception is when an individual given a weak optimistic scenario joins another member receiving weak pessimistic scenario, in which case the final preferred best-case forecasts stay similar to the initial forecasts before team discussions for both members.

Differences among the groups are most striking for *worst-case predictions*. Overall, participants receiving pessimistic scenarios appear to be affected by the other person's optimistic scenarios, while the optimistic scenario member's initial and final worst-case predictions lie at similar levels. The only combination when the dynamics change is when an individual with strong pessimistic scenario joins a team member receiving strong optimistic scenario. In this case, the pessimistic scenario appears to overtake the optimistic one, and the initial worst-case predictions persist through both the team and final predictions. Also, the individual with the optimistic scenario shifts their worst-case forecasts so that their final predictions are worse (more extreme) than their initial forecasts. These findings reveal that being exposed to other member's strong pessimistic scenario produces a significant effect in making their worst-case forecasts pull toward the other member's predictions, and this effect persists when asked to state their preferred forecasts at the end of the study.

The results carry important implications for the theory and practice of incorporating scenarios to support judgment in predictive analytics. It is evident that point forecasts by themselves signal pseudo-certainty and hide critical information about the uncertainty embedded in the prediction. Eliciting best-case and worst-case forecasts (in addition to point predictions) exposes the scope of uncertainty surrounding the forecaster's predictive judgment and may be an 'ecologically valid' alternative (that is used more commonly by practitioners) to asking for prediction intervals that typically provide narrow and overconfident bounds (Önkal & Bolger, 2004).

Overall, capturing predictive judgment via different elicitation formats is essential to expert knowledge elicitation (Alvarado Valencia et al., 2017). Communication of such judgments depicting the spectrum of uncertainty through alternative predictive formats also increases decision-makers' trust in forecasts (Gönül et al., 2012; Önkal et al., 2019). Incorporating scenarios is another decision support tool that enhances users' trust in the given predictions as well as enabling better communication of forecasts (Önkal et al., 2013; Petropoulos et al., 2021). Our findings point to the potential consequences of framing scenarios with varying levels of optimism and pessimism, along with possible biases such portrayals can carry. These yield insights into designing a support toolbox to provide effective feedback and to debias predictive judgment.

5 Conclusion

This chapter focused on examining the effects of scenarios on judgmental predictions given by individuals and teams. The literature review and the results of the experiment show that scenarios can be a mixed blessing in that they can encourage participants to adhere more closely to model-based point forecasts, though they do not eliminate non-normative changes to these forecasts, but in doing so they may also lead to, or fail to prevent, overconfidence when estimating prediction intervals. Our findings suggest more research is needed to find effective ways of incorporating scenarios to support judgmental forecasts. Providing effective uncertainty aids to improve predictive judgment, if they can be designed, would be a highly rewarding approach to enhancing the theory and practice of predictive analytics.

Given the prevalence of group forecasts in practice, current results highlight the need for detailed investigations of forecast sharing and scenario sharing on forecasts given by teams and by individuals following group discussions. Do team discussions accompanied by sharing scenarios have lasting effects on individual members? What role do extreme scenarios play? How can the optimism/pessimism continuum be adapted to benefit the framing of scenarios? How can organizational forecasting processes be re-designed to incorporate use of scenarios as an essential tool? How can scenario sharing be supported with asymmetric informational cues? Which elicitation format combinations and which ordering of these formats would be more effective in reflecting the shared information extracted from scenarios? Answers to these questions will enable better use of judgment in navigating uncertainties and strengthening its role in predictive analytics.

References

Alvarado Valencia, J. A., Barrero, L. H., Önkal, D., & Dennerlein, J. (2017). Expertise, credibility of system forecasts and integration methods in judgmental demand forecasting. *International Journal of Forecasting, 33*, 298–313.

Anderson, C. A. (1983). Abstract and concrete data in the perseverance of social theories: When weak data lead to unshakeable beliefs. *Journal of Experimental Social Psychology, 19*, 93–108.

Bolger, F., & Harvey, N. (1993). Context-sensitive heuristics in statistical reasoning. *The Quarterly Journal of Experimental Psychology Section A, 46*(4), 779–811.

Bunn, D. W., & Salo, A. A. (1993). Forecasting with scenarios. *European Journal of Operational Research, 68*(3), 291–303.

Chien, Y. W., Wegener, D. T., Hsiao, C. C., & Petty, R. E. (2010). Dimensional range overlap and context effects in consumer judgments. *Journal of Consumer Research, 37*(3), 530–542.

Fildes, R., & Goodwin, P. (2007). Against your better judgment? How organizations can improve their use of management judgment in forecasting. *Interfaces, 37*(6), 570–576.

Fildes, R., & Goodwin, P. (2021). Stability in the inefficient use of forecasting systems: A case study in a supply chain company. *International Journal of Forecasting, 37*(2), 1031–1046.

Fildes, R., & Petropoulos, F. (2015). Improving forecast quality in practice. *Foresight: The International Journal of Applied Forecasting, 36*, 5–12.

Fildes, R., Goodwin, P., Lawrence, M., & Nikolopoulos, K. (2009). Effective forecasting and judgmental adjustments: An empirical evaluation and strategies for improvement in supply-chain planning. *International Journal of Forecasting, 25*(1), 3–23.

Fildes, R., Goodwin, P., & Önkal, D. (2019). Use and misuse of information in supply chain forecasting of promotion effects. *International Journal of Forecasting, 35*(1), 144–156.

Gönül, M. S., Önkal, D., & Lawrence, M. (2006). The effects of structural characteristics of explanations on use of a DSS. *Decision Support Systems, 42*, 1481–1493.

Gönül, M. S., Önkal, D., & Goodwin, P. (2012). Why should I trust your forecasts? *Foresight: The International Journal of Applied Forecasting, 27*, 5–9.

Goodwin, P., & Wright, G. (2014). *Decision analysis for management judgment.* Wiley.

Goodwin, P., Gönül, M. S., & Önkal, D. (2019a). When providing optimistic and pessimistic scenarios can be detrimental to judgmental demand forecasts and production decisions. *European Journal of Operational Research, 273*, 992–1004. https://doi.org/10.1016/j.ejor.2018.09.033

Goodwin, P., Gönül, M. S., Önkal, D., Kocabıyıkoğlu, A., & Göğüş, I. (2019b). Contrast effects in judgmental forecasting when assessing the implications of worst- and best-case scenarios. *Journal of Behavioral Decision Making, 32*, 536–549. https://doi.org/10.1002/bdm.2130
Hall, C. C., Ariss, L., & Todorov, A. (2007). The illusion of knowledge: When more information reduces accuracy and increases confidence. *Organizational Behavior and Human Decision Processes, 103*(2), 277–290.
Lawrence, M., Goodwin, P., O'Connor, M., & Önkal, D. (2006). Judgemental forecasting: A review of progress over the last 25 years. *International Journal of Forecasting, 22*, 493–518.
Önkal, D., & Bolger, F. (2004). Provider-user differences in perceived usefulness of forecasting formats. *OMEGA: The International Journal of Management Science, 32*, 31–39.
Önkal, D., Goodwin, P., Thomson, M., Gönül, M. S., & Pollock, A. (2009). The relative influence of advice from human experts and statistical methods on forecast adjustments. *Journal of Behavioral Decision Making, 22*, 390–409.
Önkal, D., Sayım, K. Z., & Gönül, M. S. (2013). Scenarios as channels of forecast advice. *Technological Forecasting and Social Change, 80*, 772–788.
Önkal, D., Gönül, M. S., & DeBaets, S. (2019). Trusting forecasts. *Futures & Foresight Science, 1*, 3–4. https://doi.org/10.1002/ffo2.19
Oskamp, S. (1965). Overconfidence in case-study judgments. *Journal of Consulting Psychology, 29*(3), 261–265. https://doi.org/10.1037/h0022125
Petropoulos, F., Apiletti, D., Assimakopoulos, V., Babai, M. Z., Barrow, D. K., Ben Taieb, S., Bergmeir, C., Bessa, R. J., Bikaj, J., Boylan, J. E., Browell, J., Carnevale, C., Castle, J. L., Cirillo, P., Clements, M. P., Cordeiro, C., Cyrino Oliveira, F. L., De Baets, S., Dokumentov, A., Ellison, J., Fiszeder, P., Franses, P. H., Frazier, D. T., Gilliland, M., Gönül, M. S., Goodwin, P., Grossi, L., Grushka-Cockayne, Y., Guidolin, M., Guidolin, M., Gunter, U., Guo, X., Guseo, R., Harvey, N., Hendry, D. F., Hollyman, R., Januschowski, T., Jeon, J., Jose, V. R. R., Kang, Y., Koehler, A. B., Kolassa, S., Kourentzes, N., Leva, S., Li, F., Litsiou, K., Makridakis, S., Martin, G. M., Martinez, A. B., Meeran, S., Modis, T., Nikolopoulos, K., Önkal, D., Paccagnini, A., Panagiotelis, A., Panapakidis, I., Pavía, J. M., Pedio, M., Pedregal Tercero, D. J., Pinson, P., Ramos, P., Rapach, D., Reade, J. J., Rostami-Tabar, B., Rubaszek, M., Sermpinis, G., Shang, H. L., Spiliotis, E., Syntetos, A. A., Talagala, P. D., Talagala, T. S., Tashman, L., Thomakos, D., Thorarinsdottir, T., Todini, E., Trapero Arenas, J. R., Wang, X., Winkler, R. L., Yusupova, A., & Ziel, F. (2021). Forecasting: Theory and practice. *International Journal of Forecasting*. arxiv.org
Schnaars, S. P., & Topol, M. T. (1987). The use of multiple scenarios in sales forecasting: An empirical test. *International Journal of Forecasting, 3*(3–4), 405–419.
Sherif, M., Taub, D., & Hovland, C. I. (1958). Assimilation and contrast effects of anchoring stimuli on judgments. *Journal of Experimental Psychology, 55*(2), 150–155.
Taleb, N. (2005). *The black swan*. Random House.
Taylor, S. E., & Thompson, S. C. (1982). Stalking the elusive "vividness" effect. *Psychological Review, 89*(2), 155–181.
Van der Heijden, K. (2005). *Scenarios: The art of strategic conversation*. Wiley.
Wright, G., & Goodwin, P. (2009). Decision making and planning under low levels of predictability: Enhancing the scenario method. *International Journal of Forecasting, 25*(4), 813–825.
Zentner, R. (1982). Scenarios, past, present and future. *Long Range Planning, 15*(3), 12–20.

Chapter 10
Incorporating External Factors into Time Series Forecasts

Shari De Baets and Nigel Harvey

Keywords Time series forecasting · Judgment · External events · Model transparency

1 Introduction

People have always been on a search for what the future holds. While the question has remained the same throughout the ages, the methods used to forecast what is going to happen have changed. The Pythia's forecasts (the priestess, or Oracle, of Delphi) were largely based on visions created by the inhalation of hallucinogenic gasses (De Boer & Hale, 2000; Broad, 2006). In modern days, we like to believe this has been replaced by more rational, objective methods. The fumes have been replaced by the digital language of 0 and 1s, the Pythia by the data scientist, and the visions by neat graphs and estimated numbers. The place of worship is no longer the Temple of Apollo in Delphi, but the forecasting software on our computers. Yet, as advanced and rational as we seem to be compared to the priestesses from the ancient Greek world, we have not left the human role in predicting the future behind. While technology races forward with Big Data, Machine learning and AI, the modern-day Pythias continue to use their interpretation, or judgment, when faced with outcomes of these technological innovations.

Over the years, surveys have shown time and again that judgment remains a quintessential part of forecasting in business practice. Nearly three quarters of businesses report the use of judgment in forecasting, by itself (a minority) or in combination with a statistical forecast (Fildes & Petropoulos, 2015). The remaining quarter, indicating the sole use of statistical methods, should be taken with a grain of

S. De Baets (✉)
Open University of the Netherlands, Heerlen, The Netherlands
e-mail: shari.debaets@ou.nl

N. Harvey
University College London, London, UK

M. Seifert (ed.), *Judgment in Predictive Analytics*, International Series in Operations Research & Management Science 343,
https://doi.org/10.1007/978-3-031-30085-1_10

salt. After all, judgment can intervene at many phases of the forecasting process: cleaning the data, defining parameters, selecting criteria and selecting the statistical method, to name but a few. Whether this is a good thing or not, is a long-standing question within the forecasting community. Like a pendulum, the views on the value of judgment in forecasting have swung from pessimism to optimism and back. Those who hold a more pessimistic view often point to the many biases that are associated with judgment. We are overoptimistic in our forecasts, and overvalue positive information. We see patterns where there are none, leading us to replicate noise in our predictions. We place too much value on the last data point in a historic series, staying too close to it even when it is an outlier. And often, we just fiddle around with a forecast and make small unnecessary, but damaging, changes to a forecast. To add insult to injury, we are (over)confident while we are doing so.

Judgment-pessimists tend to favor the statistical approach to forecasting. And can we blame them? Models, unlike humans, are logical and systematic in their processing of information. Moreover, they can handle much larger amounts of data than a human can dream of. They are consistent, less error-prone, and reliable. Yet, however central forecasting software has become in organizational life, systems are not perfect. Statistical models perform well in a stable environment, with plenty of data, and a continuous trend. Unfortunately, that is not how the world works. We live in a world of uncertainty and noise, of missing data and changing trends. The pendulum swings to the judgment-optimists. In contrast to models, judgment is capable of incorporating contextual information and changes to trends: a predicted heatwave will surely alter ice cream sales; a competitor launching a rival product will require some unwelcome downward adjustment of the sales forecasts; the marketing team launching a campaign, on the other hand, will require an upward adjustment of the sales forecasts. And so we have come to the core of this chapter: the external event that disturbs the time series. It is what judgment-pessimists and judgment-optimists agree on: in comparing, and combining, statistical methods with judgment, this is where judgment has its opportunity to shine.

In what follows, we first categorize the important dimensions along which external events vary. We then consider the dimensions along which their effects on time series vary. Next we summarize recent research on the role of judgment in forecasting from series disrupted by external events. Finally, we discuss recent statistical approaches to this type of forecasting task.

2 External Events

What precisely are external events, in the context of forecasting? Perhaps the most basic definition is those events that disturb the historical baseline in a significant manner (i.e., base rate distractors; Fildes et al., 2019). It is an event that is not automatically represented in the data available to a statistical model, or if available, not distinguished by the model as being a separate effect. Such events add complexity to time series, which makes them difficult to handle. In the supply chain world, the outbreak of the pandemic combined with an underestimation of demand from the automobile industry led to a shortage of semiconductors, throwing the entire automotive supply chain off course. A subsequent fire in a leading computer chip manufacturer's factory did not help the situation. But it need not be negative—at least for the company. The pandemic and concurrent lockdown caused a significant size change in Levi's customer base. Together with an increased need for casual wear, demand for their products increased significantly (Reuters, 2021). The spread of the USA's tradition of Black Friday across the globe has caused stores everywhere to see an increased spike in sales in this period. Moving beyond the supply chain, the financial crash caused by Lehman Brother's bankruptcy in 2008 and the housing crisis, sent the whole world reeling. Geopolitical events such as the Brexit vote in 2016, the emergence of China as world's largest economy, and the Greek debt crisis are all external events that had a significant impact. Disastrous environmental events can hit equally hard. Japan's "Triple Disaster" of 2011, a 9.0 magnitude earthquake, a tsunami crashing the North-Eastern shoreline, and the Fukushima nuclear meltdown, wreaked havoc on the Japanese economy. And who saw the COVID-19 pandemic of 2020 coming? Supply and demand were hit hard, travel and entertainment were diminished due to governmental lockdown regulations, and unemployment peaked due to massive lay-offs.

All events are, however, not all the same. While an argument could be made for classifying them according to domain (e.g. geopolitical events, business events, or calendar events), this would not serve our purposes. A useful taxonomy of external events in the context of forecasting requires us to consider the properties of the event, and how those properties impact potential forecasts. To create such a taxonomy, we need to identify the important dimensions along which external events themselves vary and along which impacts of those events on time series vary.

2.1 Event Characteristics

External events that disrupt the forecasting process can differ along various dimensions, particularly *magnitude, duration, regularity, frequency, and predictability.*

2.1.1 Magnitude and Duration

The magnitude or size of an event is one of the most important aspects of it. This is because, generally speaking, larger events have a larger effect on the time series. However, duration of the event is also important. Its effects may be less straightforward. When an event extends over multiple periods, its effects may change direction. For example, a promotion is likely to boost sales initially but later, it may depress them (e.g., Hewage et al., 2022). In the extreme case, duration of an event may be unlimited; in other words, there may be a step change in the context within which forecasts are made. This may or may not produce a permanent change in a time series.

2.1.2 Regularity and Frequency

While *regularity* and *frequency* of events may seem intertwined, this is not necessarily so. To clarify this, we can look at the event-time series in the figure below.

In the upper left quadrant, we have those events that are both regular and frequent. We can think of a demand series that is subject to bi-monthly promotional activities organized by the marketing team.

In the lower left quadrant, where events are irregular yet frequent, we have sufficient data to use, but the intervals are not set. This could be the occasional promotion, that is not tied to a specific week or month. Alternatively, we need only to think of those holidays that are associated with gift-giving, that are frequent, yet spread out irregularly over the year. Toy companies experience a yearly peak on the fixed dates of Valentine's day, the Easter holidays and the end-of-year holidays and the more flexible, but still determined well in advance, dates of Chinese New year, Eid and Diwali. One could also think about the effects of the publication of the quarterly financial results, which often results in game-playing by the companies to represent their financial figures more suited to their needs.

In the upper right quadrant, events are regular but less frequent. The events still occur at more or less set intervals, but do so less often. In practice, this will make a difference because we will have less historical event data from which to extrapolate. An example is the occurrence of elections—when all is well, these happen at regular intervals but they are not frequent.

In the lower right quadrant, we are dealing what is perhaps the most difficult situation, but one that is, unfortunately, also realistic: the irregular, infrequent occurrence of external events. A prime example is the emergence of a new competitor on the market. One should hope this does not happen every day, but it is certainly not unheard of. Perhaps slightly more frequent is the launch of a competing product by existing competitors.

2.1.3 Predictability

Regularity and frequency have a great influence on *predictability*. One can assume that the events that are frequent and regular have higher predictability than those that are infrequent and irregular. But is that so? A straightforward example from the first quadrant of Fig. 10.1, regular and frequent promotions, can be complicated by a number of factors. An increase in promotional activities may lead to an overflow of products on to the market and consumers may then start to pace their demand accordingly, leading to an oscillatory effect on demand and subsequently, production. In addition, the type of promotion may differ. Thus, something that is

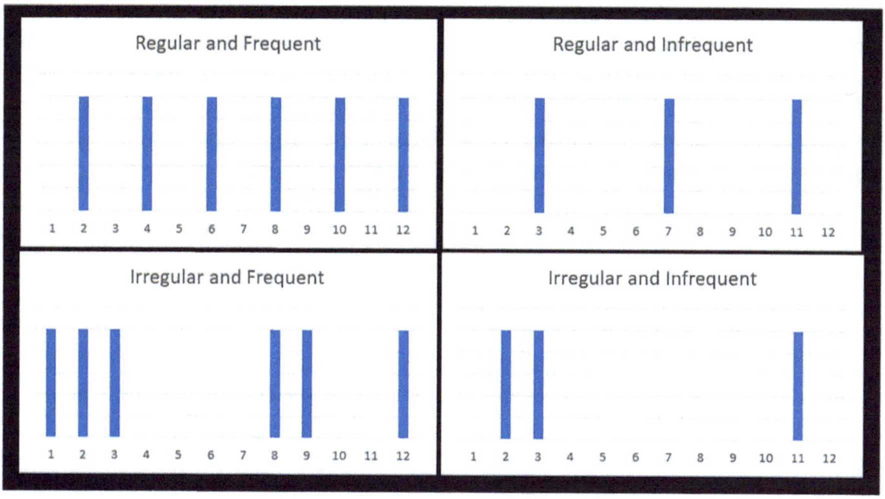

Fig. 10.1 Regularity and frequency of events

seemingly straightforward and predictable, can turn out to be less predictable than originally estimated.

Predictability centers on the question of whether we can see an event coming. In some cases, particularly where we have control over the event's occurrence (e.g., promotions that we or our colleagues organize), we can be certain (or almost certain) that the event will occur. In other cases, there might be indications or early warning systems in place that make an event *somewhat* predictable. We are thus looking at a continuum, ranging from those events that are fully predictable at one end to those that are impossible to predict at the other end. Consider a binary event: we may be able to say that it definitely will occur or definitely will not occur (fully predictable), we may be able to say that one of those outcomes is more likely than the other (somewhat predictable), or we may have no information about the likelihood of the event (unpredictable). To make plans, we need forecasters to be able to assess the likelihood that an event will occur.

When the regularity of events is established or when we have control over the occurrence of events, we can usually assess that likelihood as being close to zero or one. However, things are not always this straightforward. For example, imagine that my elderly father has been going to the cinema every first Friday of the month and has been doing so for years on end. If asked to predict the chance that he would go the next first Friday of next month, I might state that there is a 100% chance that he would indeed do so. However, when making that forecast, what I did not know was that he has just broken his leg, leaving him housebound and thus, unable to go to the cinema. When estimating the likelihood of events occurring, forecasters must factor in "broken-leg" effects of this type (Kleinmuntz, 1990).

Estimating the likelihood of the occurrence of an irregular event over which one has no control is even more difficult. However, it appears that some forecasters are better at doing this than others. Work on geopolitical forecasting has shown that some people are 'superforecasters': they are exceptionally skilled at assigning realistic probabilities to possible events occurring (Tetlock & Gardner, 2015). There are various reasons for this, including a natural preference to think probabilistically rather than deterministically, a propensity to decompose events into components each of which is necessary for the event to occur, and greater cognitive complexity (Karvetski et al., 2021; Mellers et al., 2015).

When events are predictable we can develop plans that take their anticipated occurrence into account. Even when they are somewhat predictable (e.g., 75% chance of occurring), we can still act proactively by developing contingency plans. In other words, we have two scenarios; if the event occurs, we assume that it will perturb the time series in a particular way and make our forecast taking this perturbation into account; if the event does not occur, we assume that the time series will remain unperturbed and make our forecast accordingly. If we need a single forecast that takes account of the possibility of the event occurring, we can weight the forecasts associated with each scenario by the probability of the scenario occurring and then aggregate them. For example, our sales forecast for a product may be 100 units unless our main competitor markets a rival product; there is a 75% chance of this happening and, if it does, we would forecast our sales to be 80 units. Thus, before knowing what our competitor will do, our sales forecast would be $(25\% \times 100) + (75\% \times 80) = 85$. If events are unpredictable, we cannot take this proactive approach by adjusting our forecasts. Instead, when something unexpected disrupts the time series—our factories may be flooded, or a pandemic may severely disrupt supply chains—we must be reactive and generate our forecasts anew.

What types of cues do people use when making probabilistic forecasts for the occurrence of an event? We can make a distinction here between cues or signals that indicate what is happening *now*, and cues or warning signals indicating what might happen in the *future*. Let us first consider cues indicating what is happening at the moment. The expression "where there's smoke, there's fire" provides a good starting point. The fire is already happening on the ground, but you may not know it. A smoke plume in the distance however, is a clear indication that something is not right. In the stock market world, a close eye is kept on the VIX (Chicago Board Options Exchange's CBOE Volatility Index)—a measure of current volatility based on S&P 500 index options (Wikipedia, 2022). It gives an indication of the mood of the market at the present time: a low VIX means all is reasonably quiet on the market front, there is stability and long-term growth; a high VIX indicates investor fear and panic—which earned it its apt nickname: the 'fear gauge' or 'fear index'. Consider next a cue or signal that something might happen in the future. Staying in the realm of banking, press releases on monetary policies are carefully analyzed with regard to word usage, body language and tone, as indicators of imminent movements on the financial markets (Gorodnichenko et al., 2021). On a larger scale, cues are used in conservation efforts. For example, when the size of North Sea cod on the fish market became smaller, this was recognized as an indication that the cod population would collapse if nothing was done. As a result, intervention measures and regulations were put in place and cod stocks are now recovering (Clements et al., 2019).

Judgment analysts (Cooksey, 1996) have examined how and how well people use cues to make predictions about uncertain criterion values. Their approach is broadly based on multiple regression and could be used to study forecasters' use of cues to estimate likelihood of occurrence of uncertain events. Consider a case where a forecaster estimates the likelihood of occurrence of an event that could disrupt a time series on the basis of information about one or more cues. They do this for a number of series. The available cues are treated as the independent variables or predictors in a regression and the judged likelihood of occurrence are regarded as the dependent variable. Regression of the judged likelihood on to the cues gives us information about the cues that forecasters use and the weight that they place on each one. We can then examine how these factors differ between different forecasters and study whether better forecasters place more weight on particular cues.

Certain factors may disrupt forecasters' use of cues to make estimates of the probability that an event will occur. Prominent among these is the hindsight bias or the "I-knew-it-all-along" effect. After a predicted event occurs, people tend to overestimate the likelihood of the event's occurrence that they gave before the event. In contrast, after a predicted event fails to occur, they tend to underestimate the likelihood of the event's occurrence that they gave before the event (Fischhoff, 1975). These effects suggest that we distort our memories in a way that implies that we think that we are better at making predictions than we actually are. For example, on 20 February 2022, a geopolitical forecaster might have estimated that there was a 60% likelihood of Russia undertaking a full invasion of Ukraine. When asked on 25 February to recall the likelihood that they had given for this event occurring, they would tend to give a value higher than 60%, such as 70%.

Consider now what effect this hindsight bias would have on someone's repeated probabilistic forecasting for the occurrence of an unpredictable event that would disrupt a time series. They might initially predict that the event has a 60% likelihood of occurring; it then occurs and, because of the hindsight bias, they recall their estimate of the event occurring as 70%. They then use this 70% value as the probabilistic forecast of the event occurring a second time; it does occur again and, because of the hindsight bias, they recall their estimate of the event occurring as 80%. Thus, over a number of iterations, repeated occurrence of an uncertain event would progressively drive up probabilistic forecasts. In contrast, repeated non-occurrence of the event would progressively drive probabilistic forecasts downwards.

2.2 *Event Impact*

The impact or effect of an external event may itself vary in various ways. Here we focus on dimensions of magnitude, direction, duration, and type of impact.

2.2.1 Magnitude

Generally speaking, larger events are likely to have larger effects—but only up to a point. Let us again turn to one of our simplest of examples, the effect of a predictable promotion on a sales series. It is reasonable to expect that a larger promotion will have a larger effect. Suppose we measure the size of a promotion by the level of funding assigned to it and the size of its effect by the number of sales above a no-promotion baseline occurring in the period associated with the promotion. If we graphed the relationship, we might expect to see a linear increase in the latter as the former increased. However, any such increase is unlikely to continue indefinitely—there is an absolute limit on the number of people who can be incentivized to buy more of a product. So the relationship is likely to be negatively accelerating: at some point, the effect on sales of a given increase in the size of the promotion, though still positive, will start to decrease until finally, it will level off after a certain point.

2.2.2 Direction

What about the *direction* of the effect? It is natural to expect promotions to produce an increase in sales and, in fact, they usually do so in periods immediately after they are implemented. However, in post-promotional periods, they can have the opposite effect (Hewage et al., 2022). This is likely to be because promotions lead some consumers to bring forward purchases that they had intended to make anyway at some future point. As a result, sales drop at that future point. Furthermore, one might expect that the more effective a promotion was in the promotional periods (perhaps because it was larger), the greater the drop in sales in the post-promotional periods.

2.2.3 Duration

Another aspect is the *duration*. The effect of an event may be transient—for instance, a promotion may produce a spike in sales in the month in which it occurs. At the other end of the scale, some interventions may permanently affect the level of the time series—a fire destroying a factory would have this effect. In between a spike and a permanent step change, there will be interventions that are not permanent but have continued effects over a number of periods. The relation between the duration of an event and the duration of its effect is an empirical issue that is likely to vary across domains. For example, if a promotion is maintained over a number of periods, its effect on each successive period is likely to decrease as the number of customers who have taken advantage of it increases and the remaining pool still susceptible to it therefore shrinks. On the other hand, invaders of a country in Eastern Europe are likely to find that laying siege to a city has more permanent effects the longer the siege goes on.

2.2.4 Type

Finally, interventions can have different *types* of effects on time series. Up to now, we have considered only effects of those events on the (e.g., sales) level of a series. But interventions can also change the trend, variance, sequential dependence, or some other feature of the series. While the Russian IT market had been showing a steady growth, even during the pandemic (Gerden, 2021), the war with Ukraine and the subsequent economic sanctions meant a nearly total halt of operations by the major global tech companies in Russia (Rubio-Licht et al., 2022). As an example of a different type of effect, volatility (variance) of financial series is typically increased when world events increase the uncertainty of the trading environment.

When external factors produce a regime change in a time series, previous data useful for forecasting may no longer be useful. Thus, it is important for forecasters to monitor time series and be able to detect such regime changes when they occur. Their ability to do so depends on the features of the time series and the nature of the regime change. For example, people are poor at detecting changes in the level of times series and their ability to do this is even lower when points in the series are sequentially dependent (Matyas & Greenwood, 1990; Speekenbrink et al., 2012). Furthermore, people are much better at detecting increases than decreases in variance of series, a phenomenon that has implications for perception of changes in financial risk (Harvey et al., 2018).

Given the difficulties that people have in using their judgment to detect regime change in time series, algorithmic process control measures are used in some applications (e.g., quality control, nuclear power stations). Time series are monitored to see whether any data points are outside their expected range. A single point may be an outlier just due to noise. However, there are statistical approaches to picking up more systematic deviations that act as early warning systems and indicate that something untoward has occurred. In some of these cases (e.g., quality control), this triggers an automatic correction. In other cases, (nuclear power, life support systems), it triggers a warning or emergency signal that draws in a human judge to deal with the situation. So, in reactive cases, automatic, human or hybrid monitoring systems can provide important solutions to the problem of determining whether an external event has occurred that must be incorporated into forecasts.

3 The Role of Judgment in Dealing with External Events

How do we deal with external events in our time series? We can use our judgment. Though ever increasingly sophisticated formal analysis techniques exist, judgment is usually involved at some stage or another. Whether it is cleaning of the data prior to statistical processing, selecting parameters or analogies, or post-model adjustment, judgment is present. Below, we outline some recent research into the role of judgment has at various stages of the forecasting process. We start with work dealing with unaided and aided forecasting from series that are disrupted by external events; this comprises the majority of research on this topic. We then move on to two other aspects of the forecasting process in which judgment plays a role. One is the use of judgment to clean data of the effects of occasional external events so that baseline forecasts can be produced. The other is the use of judgment to decide which past event provides the best guide for forecasting the effect of the next event. This judgment usually relies on assessments of similarity between past events and the upcoming event and can therefore be regarded as an analogical process. (There are other ways in which judgment is used when forecasting from series disrupted by external events but they remain to be investigated.)

3.1 Judgmental Adjustment of Statistical Forecasts from Series Disrupted by External Events

First and foremost, there is the omnipresent judgmental adjustment of a statistical model. While this theoretically should be the ideal way of working together with statistical forecasting models, judgment is often flawed. Adjustments to the statistical forecasts are not limited to rational adjustments for external events. We tinker with forecasts to show that we pay attention to the task and because we want to have control—resulting in unnecessary small changes that damage accuracy (Önkal & Gönül, 2005). But let us focus on the adjustments that are made for the right reasons: to account for an external event. How well do people do this?

A first study of interest is that of Lim and O'Connor (1996). They set up an experiment where people were asked to make a simple forecast from time series given (a) causal information (the presence of an external event) relevant to the forecasting period, or (b) a statistical forecast that did not take the causal factor into account, or (c) both the causal information and the statistical forecast. Participants then adjusted their original forecast to take this new information into account. It was found that, though forecasts were improved after receiving causal information, this improvement was small and no greater than when a statistical forecast was provided. Failure of forecasters to take sufficient account of causal information can be attributed to three factors. First, an adjustment task was used. Research has shown that, when such tasks are used, participants are conservative (Harvey & Fischer, 1997): in other words, they fail to move away from their initial forecast to take sufficient account of new information. Second, forecasters had to learn about the effect of the causal information over time. For each series, they were presented with causal information relevant to only one forecast period. As a result, effects of that information could be appreciated only by mentally integrating them over different series seen on successive occasions. As no outcome feedback was given, this would not have been easy—indeed, forecasting performance declined rather than improved over time. Third, causal information relevant to the forecast was always present. As Lim and O'Connor (1996) point out, this is not typical of real forecasting tasks where causal information is typically available sporadically. Sporadic availability enables forecasters to compare outcomes when causal information has been available with those when it is not: this is likely to enable them to appreciate its influence more effectively.

Goodwin and Fildes (1999) used an experimental design that was not affected by these problems. They employed a simple extrapolative forecasting task rather than an adjustment task; no learning was necessary because information about the size of causal factors (external factors) affecting past data points was presented as vertical bars on the same graph as the time series that had to be forecast; causal information was sporadic, affecting about half the time periods in the presented series. In their experiment, series were either simple (independent points scattered around a constant mean) or complex (linear trend with a multiplicative seasonal pattern superimposed) and contained either high or low noise. The relation between the size of the promotion and its effect in elevating sales was always linear but was either weak or strong. All participants received outcome feedback because they made repeated forecasts for the same series (i.e., outcomes in the series were updated before each new forecast was made). Some participants received statistical forecasts. When these forecasts were given, they were shown for the whole of the presented series as well as for the upcoming required forecast. They were based on simple exponential smoothing (for series without trends) or on the Holt-Winters method (for series with trends): as a result, they took no account of the effects of the events (in this experiment: promotions).

Goodwin and Fildes (1999) restricted their analyses to absolute error and considered periods with and without promotions separately. They report two main findings. First, forecasts for non-promotional periods were worse when promotions had a stronger effect on sales: this was presumably because effects of promotions appeared to add a higher level of noise to the series when they were stronger and this impaired forecasting for periods without promotions. Second, provision of statistical forecasts had no effect on forecasts for periods with promotions and reduced forecast error for periods without promotions only when series were complex or contained high noise. Thus, despite using a different experimental design from that of Lim and O'Connor (1996), Goodwin and Fildes' (1999, p. 49) conclusions were similar: "The main finding of this study is that, while judgmental forecasters benefited from the availability of statistical forecasts under certain conditions, they almost always made insufficient use of these forecasts. … In Lim and O'Connor's study subjects had already made an initial forecast before they were presented with the statistical forecast. Our study suggests that this underweighting prevails even when the statistical forecast is presented before the judgmental forecast has been formed".

Goodwin et al. (2011) developed Goodwin and Fildes' (1999) research. Series were again either relatively simple (ARIMA (0, 1, 1)) or more complex (linear trend with a multiplicative seasonal pattern superimposed) and contained either high or low noise. Presentation of the time series and the promotional events was the same as before, though it appears that promotions were less frequent. In this study, participants' forecasts made with statistical forecasts were not compared to those made without them. Instead, the aim was to investigate whether better use would be made of statistical forecasts when large changes away from those forecasts were restricted (i.e., forbidden) or when guidance was given about the appropriateness of making changes away from those forecasts. (In the latter condition, participants were told that their intention to make a change on a non-promotion period or of not making one on a promotion period was likely to reduce forecast accuracy.) Thus, participants could use statistical forecasts without restriction or guidance, with restriction but no guidance, or with guidance but no restriction.

Analyses showed that guidance failed to improve performance and that restriction impaired it. Generally, judgmental forecasts for promotion periods were better than raw statistical forecasts whereas those for non-promotion periods were worse. These effects did not interact with series type though it took participants longer to make forecasts from more complex series.

Perhaps the strength of judgment lies in the absence of bias, rather than the general error. Trapero et al. (2013) analyzed data obtained from a manufacturing company. What marked their work out was that they were able to obtain statistical forecasts and final forecasts for both promotional and non-promotional periods: "this is the first case study to employ organizational data for verifying whether judgmental forecasts during promotional periods achieve lower forecasting errors than their statistical counterparts" (Trapero et al., 2013, p. 235). The dataset comprised 18,096 data triplets (i.e., statistical forecast, final forecast, outcome) from 169 SKUs (stock keeping units). Eight percent of the triplets were for promotional periods. Statistical forecasts were based solely on time series information and so took no account of the effect of promotions. Because of this and because promotional periods were relatively rare, accuracy of statistical forecasts was lower for promotional periods than for non-promotional ones. Analyses showed that final forecasts were less accurate than statistical ones, particularly for promotional periods.

Mean percentage error scores showed that statistical forecasts were somewhat too high for non-promotional periods but too low for promotional periods. This was because the way in which those forecasts were produced meant that they did not distinguish between promotional and non-promotional periods. Overall, adjustment produced final forecasts that were overestimates and considerably higher than the statistical forecasts. The authors attribute this to optimism. This is reasonable: a tendency to over-forecast desirable quantities (e.g., sales, profits) is well-established (Eggleton, 1982; Lawrence & Makridakis, 1989; Harvey & Bolger, 1996). More detailed analysis showed that small negative adjustments improved and large positive adjustments impaired accuracy on non-promotional periods whereas small positive adjustments improved and other adjustments impaired accuracy on promotional periods. These patterns are to be expected given the above-mentioned biases in the statistical forecasts,

Trapero et al. (2013, p. 239) asked whether managers could "have analyzed patterns of past promotions and tried to project the results for forecasting similar future product promotions"? For all their advantages, studies that analyze data from organizations cannot answer this question because no records are kept of the information (e.g., length of time series, causal factors) that forecasters used when they made their adjustments. However, Goodwin and Fildes' (2011) survey of company forecasting behavior showed that forecasts are often based on very short data series (e.g., six points). Also, as Lee et al. (2007) point out, forecasters may rely on their memory for information about the size of effects previously produced by different types of promotion. With availability of longer time series of past sales and more explicit information about the effects of past promotions, it is possible that optimism would be moderated—on the other hand, this additional information might lead to the appearance of other types of biases.

De Baets and Harvey (2018) examined biases in the way that forecasters made forecasts from (simulated) un-trended sales series that included promotions. Promotions, and hence their effects varied in size. However, forecasters were provided with historical data series with and without promotions and so they were able to see the size of effect expected from a promotion of a given size. Forecasts were required either for periods on which promotions were planned or for those in which they were not planned. Forecasters systematically under-forecast when promotions were planned but systematically over-forecast when they were not. It appears that they use the overall mean level of past sales as a judgment anchor and mentally adjust upwards from this when a promotion is expected and adjust downwards from it when no promotion is expected. However, when anchor-and-adjust heuristics are used, adjustments are typically insufficient (Tversky & Kahneman, 1974): this explains the biases that were obtained. The relative size of the under-adjustment and over-adjustment biases depended on the proportion of periods that contained promotions in the historical data series. More promotions increased the overall mean of that series and hence the mental anchor used in forecasting: as a result, under-forecasting on promotional periods decreased but over-forecasting on non-promotional periods increased. Providing forecasters with statistical forecasts improved their accuracy but did so not by decreasing these biases but by decreasing the random error or scatter that they contained.

In summary, forecasters need to be given information about the size of effects of promotions in order to include those effects in their forecasts. They can be given this information either by observing the past effects of promotions (guidance) or by being given information about the accuracy of their forecasts (feedback). However, while such information is helpful, errors still remain. Specifically, biases arising from optimism and anchoring effects combine with random error or noise in people's judgments (Kahneman et al., 2021) to reduce forecast accuracy. Provision of statistical forecasts reduces the contribution of the latter component of overall error (i.e., noise).

3.2 *Using Judgment to Select and Clean Data to Produce Baseline Forecasts*

Early in the forecasting process, judgment plays its role in selecting and cleaning data. Before a baseline forecast can be produced using statistical methods, series must be cleaned of disruptive effects produced by external events. Webby et al.'s (2005) experiment examined how good people are at using their judgment to perform this task. Participants were given a time series that was subject to sporadic perturbations arising from external factors. Among these external factors were those that elicited an upward motions of the time series, such as promotions, and downward motions, such as a recessions or strikes. Participants in the study were asked to clean out the effects from the presented series so that it was reduced to its historical baseline. How did people fare? Though adjustments were made in the right direction by moving towards the baseline, movements were insufficient. It is likely that this was because people were again using the anchor-and-adjust heuristic by employing the trend line of the underlying time series as a mental anchor and then making adjustments relative to it. The effects reported by Webby et al. (2005) arose because adjustments showed the typical insufficiency that is observed when this heuristic is used (Tversky & Kahneman, 1974). Furthermore, these anchoring effects were superimposed on the sort of optimism bias that we discussed above when extrapolated from the underlying time series. This optimism bias was greater when more events perturbed the presented and forecast sections of the series. Webby et al. (2005) suggest that more events increase forecasters' cognitive load and this leaves them more susceptible to cognitive biases, a notion that appears consistent with Kahneman's (2013) two-system theory of cognition.

3.3 *Judges' Use of Analogical Strategies to Make Forecasts When Series Are Disrupted by External Events*

Analogical forecasting is effective in other types of forecasting tasks (e.g., Green & Armstrong, 2007; Litsiou et al., 2019). The planned event is compared with events in the past and the outcome associated with the most *similar* past event is taken as the forecast for what will happen when the planned event occurs.

Would this approach be useful when people use their judgment to make forecasts from time series that are disrupted by external events? In Goodwin and Fildes' (1999) task, people were shown a graph of a sales series and, underneath it, bars were displayed on certain periods, the height of which indicated the size of the promotions on those periods. Judges could thereby see the effects of past promotions on sales. Goodwin and Fildes (1999) found that forecasts made for periods with planned promotions were correlated with sales on the previous periods that had the most similar promotions to those planned. They suggested that this indicated a pattern-matching or analogical strategy: forecasters searched for the past promotion expenditure that was most similar to the one for the forecast period and used the actual sales for that past period as a basis for their forecast.

Cognitive psychologists refer to this approach as an instance-based or exemplar-based inference strategy. These strategies are contrasted with rule-based strategies (e.g., Nosofsky et al., 1989; Ward & Churchill, 1998). In rule-based inference, people extract an internal representation of the relation between two variables from the information that they are given. The process is akin to a mental version of regression. It has the advantage of parsimony: very little information has to be stored in memory once the rule has been extracted: the original information can be forgotten and the rule can be updated if new information arrives. Furthermore, the rule can easily be used to make an inference about the value of one variable from a value of the other variable that has never been encountered before. With exemplar-based inference, more information must be held either in memory or in external media. Most of the cognitive effort required to make an inference must occur after the inference task has been specified (whereas, with rule-based inference, much of the work, such as rule extraction, can be done beforehand). Furthermore, making an inference from the value of a variable that has never been encountered before is more complex with instance-based inference: it involves identifying the previous values of that variable that are closest to the one on which the inference must be based, identifying their effects on the other variable, and then interpolating between those effects to produce the inference.

Goodwin and Fildes (1999) findings are consistent with a rule-based inference process as well as with the exemplar-based process that they discuss in their paper. Is it possible to distinguish between these possibilities empirically? Generally, rules might be expected to produce better results than exemplars when inferences must be based on values outside the range of those previously encountered. Exemplar-based inference would then need to be based on extrapolation rather than interpolation. In fact, it is far from easy to determine whether people's inferences are based on rules or instances: both approaches can be easily elaborated to explain data that initially appears to favor one strategy over the other (e.g., Rodrigues & Murre, 2007).

There are also practical difficulties. Past events are not always displayed for forecasters as they were in Goodwin and Fildes' (1999) study. Forecasters may have to recall them from memory. Such recall may be subject to various types of bias. For example, forecasters may be more likely to recall events for which their forecasts were accurate or successful than those for which their forecasts were inaccurate or unsuccessful. Also, in Goodwin and Fildes' (1999) study, events varied along just a single dimension (promotion size) which meant that the task of comparing those events to the planned one was relatively easy. In practice, however, promotions typically differ along a number of other dimensions, such as their type ("Buy two, get one free" versus a 25% discount), the way they are advertised (coupons versus use of various media), and their timing (e.g., whether they coincide with a competitor's action or with a seasonal event, such as Christmas sales). The multi-dimensional nature of disruptive events makes it much more difficult to assess their similarity. Should the metric used in this assessment be Euclidean, City-block, or something else (Tversky, 1977)? Even once similarity has been assessed, the problem of how much adjustment to make to allow for the differences between the retrieved analogies and the planned action remains.

To investigate some of these issues, Lee et al. (2007) examined the effects of providing judgmental forecasters with various types of support for their use of analogical reasoning when estimating promotional effects. They argued that using analogies involves three stages: recall of past promotions (i.e., candidate analogies); assessment of similarity between each of those past promotions and the planned promotion to select those that are most appropriate as analogies; adjustment of the effects of those selected past promotions to allow for differences between them and the planned promotion. Lee et al. (2007) provided support for memory recall by displaying a database of past cases. They provided support for similarity assessments by automatically highlighting cases in the database that were similar to the planned promotion. They provided adjustment support by indicating the relative size of promotional effects when promotions differed in each of three ways (duration, type, and store); for example, they could be informed that two-week promotions had an effect 1.5 times greater than one-week promotions, information that would be useful when all retrieved analogies were one-week promotions whereas the planned promotion was for two weeks. They found that increasing support (memory versus memory + similarity versus memory + similarity + adjustment) produced increased forecast accuracy.

4 Statistics to the Rescue?

Forecasting Support Systems (FSS) have been developed to forecast effects of promotions and other factors that disrupt sales time series. Typically, they are based on models that take brand, store, and week effects into account and that include indicator variables that allow specification of factors that may temporarily influence sales for a given brand, store and week (e.g., advertising, aisle display). They include SCAN*PRO (Andrews et al., 2008) and CHAN4CAST (Divakar et al., 2005). These systems are based on particular statistical models. Other models that could be used instead might produce better forecasts. In what follows, we will briefly consider some of the models that researchers have developed and tested against data. Broadly speaking, models can be divided between those that are transparent and those that are not. We make this distinction because the transparency of the model is likely to impact the degree to which it is acceptable to forecasters and users of forecasts.

4.1 Non-transparent Models

Non-transparent models include those based on machine learning, such as approaches employing neural networks of various types (e.g., feedforward neural networks, recurrent neural networks). By non-transparent, we mean that it is not made explicit what features of the data the model has used as a basis for its forecasts and how it has analyzed them.

The use of neural networks to make inferences increased when it was found that adding a layer of hidden units between input and output units made them much more powerful (Rumelhart, McClelland, & the PDP research Group, 1986). To add to their flexibility, network designs later became even more elaborate. Typically, networks are trained on a large set of data and their performance is then tested on a new set of data of the same type. Various studies have compared their forecasting accuracy with conventional statistical forecasting techniques. Conclusions vary: some have shown little difference between the two approaches (Ahmed et al., 2010; Crone et al., 2011) whereas others indicate that statistical techniques are to be preferred (Makridakis et al., 2018).

Recent studies have examined this issue in the context of forecasting from time series disrupted by external events. Nikolopoulos (2010) compared performance of a neural network with that of multiple linear regression and with that of an automated version of the analogical approach when forecasting time series disrupted by promotional events. He found that multiple linear regression was best at forecasting from linear series and that the neural network was best at forecasting from non-linear ones. He therefore suggested that a hybrid approach be used: for example, if the variance explained by the regression (R^2) is greater than 95% (indicating linearity) or, alternatively, if it is greater than that of the neural network, regression should be favored; otherwise, a neural network should be employed.

Huber and Stuckenschmidt (2020) compared statistical models (based on exponential smoothing and on multiple regression with binary dummy variables) and machine learning models (e.g., recurrent and feedforward neural networks) at forecasting time series of industrial bakery sales disrupted by various public holidays. Series were split into training and test periods. The training period was used to train the networks, to identify adjustments relative to baseline needed for public holidays in the exponential smoothing models, and to specify coefficients for the dummy variables in the regression model. Results indicated that machine learning approaches, particularly those using recurrent neural networks, produced better forecasts than the statistical approaches.

Huang, Fildes and Soopramanien (2014, p. 746) argue that machine learning models have several limitations. First, "they ignore the carry-over effect of promotions and/or overlook the effect of competitive information". These two factors are unlikely to have been present in the dataset that Huber and Stuckenschmidt (2020) used to make comparisons between models; hence, it would be unwise to generalize their conclusions to situations in which those factors are important. Second, machine learning "models are also complex and difficult to interpret. They rely on expertise that may well not be available and the company instead substitutes judgment for more formal modelling efforts".

The non-transparency of machine learning models is the primary reason that they are difficult to interpret. Even in the simplest neural network models that include just a single hidden layer, it can be difficult to determine what the units in that layer are doing. What features are they extracting from the data? However, in many application areas, consumers of forecasts want to know why forecasters have made the forecasts that they have given to them. They need reasons for why the forecasts are as they are. People are more likely to accept the views of others if a rational explanation of those views is provided (Heath & Gonzalez, 1995). Hence, being able to provide such an explanation for a set of forecasts may help forecasters to persuade others that they should accept those forecasts.

Recently, there has been some success in developing ways of explaining why a machine learning system comes to the conclusions that it does. A second system is trained to reproduce the performance of the primary system. This second model can be much simpler than the original one and it can be optimized to ensure the explanations that it provides are useful. However, this work is still in its initial stages (Edwards & Veale, 2017).

4.2 Transparent Models

When models are transparent, the way that they are formulated makes explicit both the features of the data that are used as a basis for forecasts and how those features are analyzed. Models of this type are typically based on exponential smoothing or multiple regression with exogenous variables to account for various types of promotion, calendar events, competitor influences, etc. Modelling often involves two stages. The first one is designed to reduce the number of explanatory variables by various means, such as factor analysis (Huang et al., 2014) or principal components analysis (Kourentzes & Petropoulos, 2016; Trapero et al., 2015). In the second stage, the selected or transformed explanatory variables may be incorporated into an exponential smoothing algorithm (Kourentzes & Petropoulos, 2016) or entered into a regression to predict future sales of the focal product (Huang et al., 2014; Trapero et al., 2015). Predictors in the regression might include values of previous sales, previous prices, previous prices of competing products, promotional index for the focal product, promotional indices for competing products, and dummy variables for calendar events. In some cases, the fitted model is then simplified (e.g., by removing variables with very small parameter coefficient) to achieve greater parsimony (Huang et al., 2014).

Huang et al. (2014, p. 741) point out with justification that their model 'has good interpretability compared to "black box" machine learning approaches which can hardly be understood by brand/category managers'. However, Abolghasemi et al. (2020, pp. 3–4) argue, with reference to all the types of model that we have discussed in this section: "such methods are highly complex, have demanding data requirements, and are difficult to interpret in terms of distinguishing the impact of individual promotional variables. . . . evidence indicates that lack of resources, expertise, and high costs hinder the widespread implementation of such methods and support systems in practice (Hughes, 2001)". They go on to say that: "We aim to tackle this issue by introducing an easy-to-implement and practical model that can be used to incorporate the impact of systematic promotions into the statistical models".

Abolghasemi et al.'s (2020) own approach also involved two stages. First, potential systematic events identified by a company's experts were classified in various ways (e.g., promotion type; display type; advertisement type) and analysis of variance was then used to determine whether each combination of these variables (each 'state') significantly raised sales above baseline. Second, an ARIMA model with an exogenous variable for each significant state was fitted to the data. Abolghasemi et al.'s (2020) emphasize that their approach is different from a conventional ARIMA model with exogenous regressors (ARIMAX) because they use their first-stage algorithm to identify and embed the regressors whereas ARIMAX is a single stage approach that concurrently fits the entire model to the dataset.

To examine the performance of their modelling approach, Abolghasemi et al.'s (2020) used weekly sales data (actual sales, baseline statistical forecasts, final forecasts, and promotional mechanics) for 253 products obtained from two food and beverage companies. Baseline statistical forecasts were obtained by exponential smoothing and final forecasts were produced by adjusting the baseline forecasts. They compared the accuracy of their approach with that of judgmental adjustment, ARIMA, ARIMAX, and two machine learning approaches (Support Vector Regression; Regression Trees) by examining the means and medians of the Mean Absolute Scaled Error (MASE). On the mean of MASE, their approach outperformed all other algorithmic approaches and tied in accuracy with judgmental adjustment. On the median of MASE, it outperformed all other approaches. Differences were significant except with respect to the two machine learning approaches. These differences arose primarily because Abolghasemi et al.'s (2020) model easily outperformed the others on promotional periods.

It seems fair to say that Abolghasemi et al. (2020, p. 8) have advanced towards their aim of providing "a robust, yet simple and practical, method to effectively model systematic events and produce reliable forecasts". However, we should bear in mind that they tested their model on a dataset where the external events were limited to different types of sales promotion. It would be interesting to see how well it generalizes to situations where time series could also be disrupted by activities of competitors, calendar events, and other such factors.

The approaches that we have discussed here are generally capable of using the event characteristics that we outlined in Sect. 2.1 as exogenous predictor variables and the event impacts that we discussed in Sect. 2.2 as dependent variables to be forecast. However, there does appear to be one exception. In Sect. 2.1.2, we discussed predictability as an event characteristic that often needs to be taken into account.

Managers do not always have full control over factors that affect sales. They can decide the timing and nature of their own company's promotional efforts. However, there are other factors over which they have little control; if these factors occur, they are likely to affect sales—but they may not, in fact, occur. For example, they may judge that there is an 80% chance that their main competitor will carry out a major promotional campaign for certain products in June. However, they judge that there is a 10% chance that the campaign will be delayed until July and a 10% chance that it will not occur at all this year. By analyzing past data, forecasters may be able to estimate how much a competitor's campaigns depress their own sales when those campaigns occur. However, in order to forecast their own sales for June and July, they also need to factor in their likelihood of occurring in the first place. In principle, there appears to be no reason why this cannot be done in the manner that we described in Sect. 2.1.3: point forecasts could be adjusted to take the uncertainty of the event into account and prediction intervals widened appropriately for the periods affected. Other types of uncertainty would have a more permanent effect. For example, based on hearsay and media reports, managers may judge that there is a 30% chance of their main competitor being taken over by a rival or going into liquidation next month. Such developments are likely to elevate sales for some time into the future. Forecasters would need to take account of this possibility of a regime change.

5 Summary

In this chapter, we first provided an overview of the dimensions along which external events vary (magnitude, duration, regularity, frequency, predictability) and along which their impacts vary (magnitude, direction, duration, type). Next we discussed various ways in which judgment is often engaged when forecasts are made from time series disrupted by external events: adjustment from statistical baseline forecasts; selecting and cleaning data to produce baseline forecasts; using of judgmentally based analogical strategies to make forecasts. Then we discussed recent statistical models that have been developed to make forecasts from time series disrupted by external events. First, we covered 'black-box' machine learning approaches in which forecasters and users of forecasts are provided with no insight into how the algorithm processes the data. We suggested that users are likely to prefer more transparent statistical techniques in which the way in which data is processed to produce forecasts is made explicit. So we then discussed models of this type. We suggested that transparency may not be enough. The approaches that users are likely to prefer are those which are simple and easily understandable. We outlined what currently seems to be the simplest, most understandable statistical model and summarized research into its performance. This model relies on judges to identify candidate external events that they think should be taken into account in the statistical model.

Currently, all methods seem to have some shortcomings. It is perhaps a utopia, but perhaps the ideal way of forecasting external events lies in improvements in the way in which we combine judgment and statistics. Statistics should aim to support judgment and vice versa: judgmental biases should be identified and moderated by statistical data analysis; statistical data analysis should be enriched by knowledge and judgment provided by the forecaster. Only then will we truly maximize our forecasting potential where external events are concerned.

References

Abolghasemi, M., Hurley, J., Eshragh, A., & Fahimnia, B. (2020). Demand forecasting in the presence of systematic events: Cases in capturing sales promotions. *International Journal of Production Economics, 230*, 107892. https://doi.org/10.1016/j.ijpe.2020.107892

Ahmed, N. K., Atiya, A. F., Gayar, N. E., & El-Shishiny, H. (2010). An empirical comparison of machine learning for time series forecasting. *Economic Review, 29*(5–6), 594–621. https://doi.org/10.1080/07474938.2010.481556

Andrews, R. L., Currim, I. S., Leeflang, P., & Lim, J. (2008). Estimating the SCAN*PRO model of store sales: HB, FM, or just OLS? *International Journal of Research in Marketing, 22*(1), 22–33. https://doi.org/10.1016/j.ijresmar.2007.10.001

Broad, W. J. (2006). *The oracle: The lost secrets and hidden message of ancient Delphi*. Penguin Press.

Clements, C. F., McCarthy, M. A., & Blanchard, J. L. (2019). Early warning signals of recovery in complex systems. *Nature Communications, 10*, 1681. https://doi.org/10.1038/s41467-019-09684-y

Cooksey, R. W. (1996). *Judgment analysis: Theory, methods, and applications*. Academic.

Crone, S. F., Hibon, M., & Nikolopoulos, K. (2011). Advances in forecasting with neural networks? Empirical evidence from the NN3 competition on time series prediction. *International Journal of Forecasting, 27*(3), 635–660. https://doi.org/10.1016/j.ijforecast.2011.04.001

De Baets, S., & Harvey, N. (2018). Forecasting from time series subject to sporadic perturbations: Effectiveness of different types of forecasting support. *International Journal of Forecasting, 34*(2), 163–180. https://doi.org/10.1016/j.ijforecast.2017.09.007

De Boer, J. Z., & Hale, J. R. (2000). The geological origins of the oracle at Delphi, Greece. *Geological Society, London, Special Publications, 171*(1), 399–412. https://doi.org/10.1144/GSL.SP.2000.171.01.29

Divakar, S., Ratchford, B. T., & Shankar, V. (2005). CHAN4CAST: A multichannel, multiregion sales forecasting model and decision support system for consumer packaged goods. *Marketing Science, 24*(3), 334–350. https://doi.org/10.1287/mksc.1050.0135

Edwards, L., & Veale, M. (2017). Slave to the algorithm? Why a "right to an explanation" is probably not the remedy you are looking for. *Duke Law and Technology Review, 16*, 18–84. https://doi.org/10.31228/osf.io/97upg

Eggleton, I. R. (1982). Intuitive time-series extrapolation. *Journal of Accounting Research, 20*(1), 68–102.

Fildes, R., & Petropoulos, F. (2015). Improving forecast quality in practice. *Foresight: The International Journal of Applied Forecasting, 36*, 5–12.

Fildes, R., Goodwin, P., & Önkal, D. (2019). Use and misuse of information in supply chain forecasting of promotion effects. *International Journal of Forecasting, 35*(1), 144–156. https://doi.org/10.1016/j.ijforecast.2017.12.006

Fischhoff, B. (1975). Hindsight is not equal to foresight: The effect of outcome knowledge on judgment under uncertainty. *Journal of Experimental Psychology: Human Perception and Performance, 1*(3), 288–299. https://doi.org/10.1037/0096-1523.1.3.288

Gerden, E. (2021). Russian IT market growing steadily after pandemic. *Computer Weekly*. Retrieved from https://www.computerweekly.com/news/252508694/Russian-IT-market-growing-steadily-after-pandemic

Goodwin, P., & Fildes, R. (1999). Judgmental forecasts of time series affected by special events: Does providing a statistical forecast improve accuracy? *Journal of Behavioral Decision Making, 12*(1), 37–53. https://doi.org/10.1002/(SICI)1099-0771(199903)12:1%3C37::AID-BDM319%3E3.0.CO;2-8

Goodwin, P., & Fildes, R. (2011). Forecasting in supply chain companies: Should you trust your judgment? *OR Insight, 24*(3), 159–167. https://doi.org/10.1057/ori.2011.5

Goodwin, P., Fildes, R., Lawrence, M., & Stephens, G. (2011). Restrictiveness and guidance in support systems. *Omega, 39*(3), 242–253. https://doi.org/10.1016/j.omega.2010.07.001

Gorodnichenko, Y., Pham, T., & Talavera, O. (2021). The voice of monetary policy. *VOX EU*. Retrieved from https://voxeu.org/article/voice-monetary-policy

Green, K. C., & Armstrong, J. S. (2007). Structured analogies for forecasting. *International Journal of Forecasting, 23*(3), 365–376. https://doi.org/10.1016/j.ijforecast.2007.05.005

Harvey, N., & Bolger, F. (1996). Graphs versus tables: Effects of data presentation format on judgemental forecasting. *International Journal of Forecasting, 12*(1), 119–137. https://doi.org/10.1016/0169-2070(95)00634-6

Harvey, N., & Fischer, I. (1997). Taking advice: Accepting help, improving judgment, and sharing responsibility. *Organizational Behavior and Human Decision Processes, 70*(2), 117–133. https://doi.org/10.1006/obhd.1997.2697

Harvey, N., Twyman, M., & Speekenbrink, M. (2018). Asymmetric detection of changes in volatility: Implications for risk perception. In G. Gunzelmann, A. Howes, T. Tenbrink, & E. J. Davelaar (Eds.), *Proceedings of the 39th Annual Conference of the Cognitive Science Society* (pp. 2162–2167). Cognitive Science Society.

Heath, C., & Gonzalez, R. (1995). Interaction with others increases decision confidence but not decision quality: Evidence against information collection views of interactive decision making. *Organizational Behavior and Human Decision Processes, 61*(3), 305–326. https://doi.org/10.1006/obhd.1995.1024

Hewage, H. C., Perera, H. N., & De Baets, S. (2022). Forecast adjustments during post-promotional periods. *European Journal of Operational Research, 300*(2), 461–472. https://doi.org/10.1016/j.ejor.2021.07.057

Huang, T., Fildes, R., & Soopramanien, D. (2014). The value of competitive information in forecasting FMCG retail product sales and the variable selection problem. *European Journal of Operational Research, 237*(2), 738–748. https://doi.org/10.1016/j.ejor.2014.02.022

Huber, J., & Stuckenschmidt, H. (2020). Daily retail demand forecasting using machine learning with emphasis on calendric special days. *International Journal of Forecasting, 36*(4), 1420–1438. https://doi.org/10.1016/j.ijforecast.2020.02.005

Hughes, M. C. (2001). Forecasting practice: Organizational issues. *Journal of the Operational Research Society, 52*(2), 143–149. https://doi.org/10.1057/palgrave.jors.2601066

Kahneman, D. (2013). *The marvels and the flaws of intuitive thinking. The new science of decision-making, Problem-Solving, and Prediction*. Harper Collins.

Kahneman, D., Sibony, O., & Sunstein, C. R. (2021). *Noise: A flaw in human judgment*. William Collins.

Karvetski, C. W., Meinel, C., Maxwell, D. T., Lu, Y., Mellers, B. A., & Tetlock, P. E. (2021). What do forecasting rationales reveal about thinking patterns of top geopolitical forecasters? *International Journal of Forecasting, 38*, 688–704. https://doi.org/10.1016/j.ijforecast.2021.09.003

Kleinmuntz, B. (1990). Why we still use our heads instead of formulas: Toward an integrative approach. *Psychological Bulletin, 107*(3), 296–310. https://doi.org/10.1037/0033-2909.107.3.296

Kourentzes, N., & Petropoulos, F. (2016). Forecasting with multivariate temporal aggregation: The case of promotional modelling. *International Journal of Production Economics, 181*(A), 145–153. https://doi.org/10.1016/j.ijpe.2015.09.011

Lawrence, M., & Makridakis, S. (1989). Factors affecting judgmental forecasts and confidence intervals. *Organizational Behavior and Human Decision Processes, 43*(2), 172–187. https://doi.org/10.1016/0749-5978(89)90049-6

Lee, W. Y., Goodwin, P., Fildes, R., Nikolopoulos, K., & Lawrence, M. (2007). Providing support for the use of analogies in demand forecasting tasks. *International Journal of Forecasting, 23*(3), 377–390. https://doi.org/10.1016/j.ijforecast.2007.02.006

Lim, J. S., & O'Connor, M. (1996). Judgmental forecasting with time series and causal information. *International Journal of Forecasting, 12*(1), 139–153. https://doi.org/10.1016/0169-2070(95)00635-4

Litsiou, K., Polychronakis, Y., Karami, A., & Nikolopoulos, K. (2019). Relative performance of judgmental methods for forecasting the success of megaprojects. *International Journal of Forecasting, 38*, 1185–1196. https://doi.org/10.1016/j.ijforecast.2019.05.018

Makridakis, S., Spiliotis, E., & Assimakopoulos, V. (2018). The M4 competition: Results, findings, conclusion and way forward. *International Journal of Forecasting, 34*(4), 802–808. https://doi.org/10.1016/j.ijforecast.2018.06.001

Matyas, T. A., & Greenwood, K. M. (1990). Visual analysis of single-case time series: Effects of variability, serial dependence, and magnitude of intervention effects. *Journal of Applied Behavior Analysis, 23*(3), 341–351. https://doi.org/10.1901/jaba.1990.23-341

Mellers, B., Stone, E., Murray, T., Minster, A., Rohrbaugh, N., Bishop, M., Chen, E., Baker, J., Hou, Y., Horowitz, M., Ungar, L., & Tetlock, P. E. (2015). Identifying and cultivating superforecasters as a method of improving probabilistic predictions. *Perspectives on Psychological Science, 10*(3), 267–281.

Nikolopoulos, K. (2010). Forecasting with quantitative methods: The impact of special events in time series. *Applied Economics, 42*(8), 947–955. https://doi.org/10.1080/00036840701721042

Nosofsky, R. M., Clark, S. E., & Shin, H. J. (1989). Rules and exemplars in categorization, identification, and recognition. *Journal of Experimental Psychology: Learning, Memory, and Cognition, 15*(2), 282–304. https://doi.org/10.1037/0278-7393.15.2.282

Önkal, D., & Gönül, M. S. (2005). Judgmental adjustment: A challenge for providers and users of forecasts. *Foresight: The International Journal of Applied Forecasting, 1*(1), 13–17.

Reuters. (2021). *Reuters*. Retrieved from https://www.reuters.com/business/retail-consumer/jeans-maker-levi-strauss-beats-quarterly-revenue-estimates-2021-07-08/

Rodrigues, P. M., & Murre, J. M. (2007). Rules-plus-exception tasks: A problem for exemplar models? *Psychonomic Bulletin & Review, 14*(4), 640–646. https://doi.org/10.3758/bf03196814

Rubio-Licht, N., Eichenstein, A., Roach, S., & Irwin, V. (2022). *The war in Ukraine is putting tech—From companies to governments—To the test*. PRO. Retrieved from https://www.protocol.com/policy/russia-ukraine-war-tech

Rumelhart, D. E., McClelland, J. L., & The PDP Research Group. (1986). *Parallel distributed processing: Explorations in the microstructure of cognition*. The MIT Press.

Speekenbrink, M., Twyman, M. A., & Harvey, N. (2012). Change detection under autocorrelation. In N. Miyake, D. Peebles, & R. P. Cooper (Eds.), *Proceedings of the 34th Annual Conference of the Cognitive Science Society* (pp. 1001–1006). Cognitive Science Society.

Tetlock, P. E., & Gardner, D. (2015). *Superforecasting: The art and science of prediction*. Crown Publishers.

Trapero, J. R., Pedregal, D. J., Fildes, R., & Kourentzes, N. (2013). Analysis of judgmental adjustments in the presence of promotions. *International Journal of Forecasting, 29*(2), 234–243. https://doi.org/10.1016/j.ijforecast.2012.10.002

Trapero, J. R., Kourentzes, N., & Fildes, R. (2015). On the identification of sales forecasting models in the presence of promotions. *Journal of the Operational Research Society, 66*(2), 299–307. https://doi.org/10.1057/jors.2013.174

Tversky, A. (1977). Features of similarity. *Psychological Review, 84*(4), 327–352. https://doi.org/10.1037/0033-295x.84.4.327

Tversky, A., & Kahneman, D. (1974). Judgment under uncertainty: Heuristics and biases: Biases in judgments reveal some heuristics of thinking under uncertainty. *Science, 185*(4157), 1124–1131. https://doi.org/10.1126/science.185.4157.1124

VIX. (2022). *Wikipedia*. Retrieved from https://en.wikipedia.org/wiki/VIX

Ward, G., & Churchill, E. F. (1998). Two tests of instance-based and abstract rule-based accounts of invariant learning. *Acta Psychologica, 99*(3), 235–253. https://doi.org/10.1016/s0001-6918(98)00014-6

Webby, R., O'Connor, M., & Edmundson, B. (2005). Forecasting support systems for the incorporation of event information: An empirical investigation. *International Journal of Forecasting, 21*(3), 411–423. https://doi.org/10.1016/j.ijforecast.2004.10.005

Chapter 11
Forecasting in Organizations: Reinterpreting Collective Judgment Through Mindful Organizing

Efrain Rosemberg Montes

Keywords Group judgment · Mindful organizing · Debiasing · Group forecasting

1 Introduction: Slow Progress Behind Paradigmatic Blinkers?

Several forecasting domains have experienced poor predictive accuracy beyond the expected variation driven by uncertainty (Kahneman & Lovallo, 1993; Makridakis et al., 2010; Makridakis & Taleb, 2009). Outside the frequently reviewed challenges of demand forecast accuracy (Goodwin et al., 2018; Lawrence et al., 2006; Stewart, 2001), practitioners and academics have reported similar outcomes in other domains. For instance, a large-scale study on 5400 IT projects with budgets above $15 M concluded that 45% of projects exceed their budgets and achieve 56% less value than forecasted (Bloch et al., 2012). Consistent with this finding, a 2015 report from the U.S. Government Accountability Office (U.S. GAO) offered support for those conclusions:

> Federal investments in information technology (I.T.) have often resulted in multimillion-dollar cost overruns and years-long schedule delays, with questionable mission-related achievements (GAO-15-675T, 2015, p. 2).

In a different industry, Nielsen's 2015 Breakthrough Innovation Report (Europe Edition, n.d.) has outlined success rates of 0.2% (18 out of 8650) in packaged goods launches. Remarkably, the lack of improvement over time persists, although evidence-based recommendations have been suggested repeatedly. Perhaps one the most salient example of forecast inaccuracy recidivism comes from research on Mega Projects (budgets over 1$bn), where nine out of ten projects exceed budgets with deviations of fifty percent being frequent (Flyvbjerg, 2014). The data reveals

E. R. Montes (✉)
Microsoft Spain & IE Business School – IE University, Madrid, Spain
e-mail: efrain.rosemberg@student.ie.edu

M. Seifert (ed.), *Judgment in Predictive Analytics*, International Series in Operations Research & Management Science 343,
https://doi.org/10.1007/978-3-031-30085-1_11

that these budgets overrun rates have remained constant for the last 70 years regardless of sectors and geographies.

In response to the unadaptive observations in many forecasting domains, some of the field's literature reviews in the last four decades have pointed in the direction of organizational challenges. For instance, early in this journey, several reviews highlighted the need for more attention to the organizational and behavioral aspects of the forecasting process, highlighting the importance of the relationship between forecast preparers and users to make improvements in the field (Mahmoud et al., 1992; Makridakis & Wheelwright, 1977; Scott Armstrong, 1988; Winklhofer et al., 1996). Around the same time but from a different angle, Hogarth and Makridakis (1981) suggested a reconceptualization of forecasting and planning (F&P) by adopting a multi-attribute analysis approach that evaluates F&P on organizational aspects (flexibility, motivation, control). However, despite these early calls, years later Lawrence et al. (2006) asserted that forecasts produced by groups had been largely neglected in their review of 25 years of judgmental forecasting; whereas Fildes (2006) identified a gap in the contribution of forecasting journals to the understanding of organizational issues, which had resulted in a persistent gap between theory and forecasting practice. From a research methods angle, Fildes et al. (2009) made an explicit call for revisiting the methods used to study organizational aspects in their review of the effectiveness of judgmental adjustments of demand forecasts:

> This indicates the need for organization-based studies that use interpretive research methods to establish, at a deep level, the beliefs and values of managers engaged in forecasting (p. 18).

More recently, Goodwin et al. (2018) have highlighted a lack of theoretical foundation on forecasting's organizational aspects in their forecast decisions review.

Notably, some researchers have made crucial attempts to respond to the call of the fields' reviews. For instance, Fildes and Hastings (1994) identified a lack of awareness of statistical techniques and organizational design as impediments to forecasting activity in a multi-divisional organization. Other relevant studies focused on the impact of the political dimension in the forecasting process, opening a stream of research that portrays more "naturalistic organizations" with power plays and attempts at influencing the forecast to accommodate individual incentives (Bretschneider et al., 1989; Deschamps, 2004; Galbraith & Merrill, 1996; Lawrence et al., 2000). More recent contributions have revolved around appropriate organizational structure design, i.e., who owns the forecasting process (Protzner & van de Velde, 2015), and the use of incentives such as accuracy penalties to improve accuracy via behavioral alignment (Scheele et al., 2018). Within the context of organizational forecasting literature, the few studies focused on processes deserve special attention. Davis and Mentzer (2007) have adopted a process approach conceptualizing forecasting as a dynamic capability embedded in a particular organizational climate and adjusted by feedback loops. Similarly, Oliva and Watson (2009, 2011) have used case studies to demonstrate that process design can mitigate functional inclinations' effects. The latter line of inquiry makes a more explicit reference to the interactions between team members and consensus reaching; however, the micro-processes that unveil the mechanisms at play are hinted at but not fully addressed.

Despite these efforts, the perception of a significant gap in the understanding of forecasting from an organizational perspective remains with emphasis on the lack of theoretical foundations (Goodwin et al., 2018). This persistence over time might indicate an underlying—and largely inadvertent—paradigmatic entrenchment, partially engendered by the kind of training and background of the field's researchers, overweighted in the objectivistic-functionalist approaches intrinsic to fields like statistics, operations, and econometrics (Fildes et al., 2003). The latter is relevant because the organizational aspects of forecasting cannot be studied exhaustively without relaxing the objectivistic presumption of the existence of a concrete reality "out there" that forecasters can accurately predict, neglecting the inherent subjectivity of some aspects of the forecasting process (Hogarth & Makridakis, 1981). Arguably, a relaxation of the concrete world assumption would increase the infusion of relevant organizational behavior and decision-making theory in the forecasting field, which could move the conversation into a "transition zone" between the dominant functionalist paradigm and interpretivist approaches (Gioia & Pitre, 1990), where reality is subjective i.e., observed through the eyes of the participants (Burrell & Morgan, 1979).

Dealing with subjectivity is not desirable but necessary when organizational dynamics such as informational ambiguity, longitudinal contexts, behavior-bending incentives, and the existence of hierarchical structures are internalized as inherent characteristics of any social interchange (Shapira, 2008). Importantly, those organizational factors will render accuracy as a partial goal, especially when predictions anticipate a non-desirable outcome and require decisive intervention. In line with this reasoning, Einhorn and Hogarth (1981) have stressed that judgment accuracy is conditional on contextual assumptions and a specific time frame, i.e., when the context creates inevitable tension between goals—accuracy, motivation, performance, long-term viability—, the only thing a decision-maker can do is balance trade-offs based on subjective preferences (Hogarth & Makridakis, 1981).

Occasionally, some forecasting researchers have recognized the multi-goal nature of the forecasting activity beyond accuracy. For instance, some researchers have suggested that organizations should judge the utility of sales forecasting on the extent it supports improved business performance metrics such as inventory levels, profitability, supply chain costs, and customer service (Mahmoud et al., 1992; Mentzer, 1999). Whereas McCarthy et al. reported that firms associated sales forecasting performance with business performance metrics like inventory levels (48%), customer service (30%), and supply chain costs (19%). However, these observations did not engender theoretical developments. Probably, at least in part, because conceptualizing the forecasting process as an organizational instrument to influence outcomes rather than merely predict the future is not appropriate in a functionalist-objectivistic frame but is only possible in a world in which beliefs could shape reality; making forecasting, decision-making, and enactment blend under the intentionality of actors through self-fulfilling prophecies (Bolger & Wright, 1994; Einhorn & Hogarth, 1981; Henshel, 1993).

These ideas have some empirical support; for instance, Sutcliffe and Weber (2003) have found a U-shaped curve relation between perceptual accuracy and firm performance. Unexpectedly, "humbly optimistic" managers—rather than accurate ones—improved performance using interpretation frameworks that mobilized action. In this case, inaccuracy propelled organizations to pursue goals that otherwise would be deemed unattainable:

> Misperceptions may be beneficial if they enable managers to overcome inertial tendencies and propel them to pursue goals that might look unattainable in environments assessed in utter objectivity. Because environments aren't seen accurately, managers may undertake potentially difficult courses of action with the enthusiasm, effort, and self-confidence necessary to bring about success. Having an accurate environmental map maybe less important than having some map that brings order to the world and prompts action. (Sutcliffe, 1994; p. 1374)

Organizational theory in the forecasting field has been deemed "trendy, theoretical and complex" (Fildes et al., 2003). Alas, this approach—surreptitiously enacted by the functionalistic assumptions—might have attenuated the potential of some streams of organizational research in the field. Describing these dynamics in detail might show the way froward.

2 Showcasing the Effects of Functionalism in Forecasting Research

> Your horse Is a zebra, and zebras can't be tamed (McGahan, 2007)

2.1 Extracting Forecasts from Groups

Research in extracting forecasts from groups has mainly focused on the mechanics—and risks—of forecast aggregation but has failed to describe the micro-processes that generate these outcomes and hence limited its ability to make prescriptions. For instance, the inclusion of the biases and heuristics (H&B) tradition in the forecasting field has almost exclusively highlighted individual's judgment flaws that hamper accuracy (Eroglu & Croxton, 2010; Flyvbjerg, 2008; Kahneman & Lovallo, 1993; Lawrence & O'Connor, 1995; Lee & Siemsen, 2016), where the collective aspect has emphasized the dangers of groupthink and bandwagon effects (Janis, 1982; Mellers et al., 2014; Sniezek, 1989; Surowiecki, 2005), and the potential bias accentuation effect of the group (Buehler et al., 2005). Consequently, the arsenal of the heuristics and biases tradition has cast a shadow of doubt on human judgment, especially when compared to algorithms (Fildes et al., 2009; Lawrence et al., 2006; Stewart, 2001).

However, this approach overlooks that the H&B program's objective was to generate a reliable descriptive model of human behavior so that researchers could design prescriptive models to improve judgment, and *"not to come up with a list of biases"* (Shapira, 2008, p. 4). Additionally, the literature often misses that the H&B program's hypotheses have been tested mainly in laboratory studies focused on individual's judgment against the outcome of normative models (Meehl, 1954; Tversky & Kahneman, 1974); hence, the validity of these findings in naturalistic settings is still undergoing (Klein, 2008; Mosier et al., 2018). Furthermore, the forecasting literature has devoted insufficient attention to the initial observation that superficial information search and processing are the main drivers of error in human judgment (Slovic et al., 1977; Slovic & Lichtenstein, 1971); and that those inclinations could be mitigated and manipulated by specific dynamics at the individual and group levels: sampling information from many sources, looking for disconfirming information and using roles as "devil's advocates" (Hogarth & Makridakis, 1981; Mellers et al., 2014; Sunstein, 2015; Tetlock & Gardner, 2015). The latter is especially relevant in teams that share mental models and are cohesive (Kerr & Tindale, 2004); hence, posing team interactions as the central dynamic in judgment within groups.

In one notable exception to the literature's neglect of team dynamics, Oliva and Watson (2009, 2011) elevate the importance of teams' interactions within ritualized forums, e.g., "forecast consensus meetings." These forums aim to forecast alignment via information sharing, anchoring discussions on an initial estimate, and open feedback, elements that could serve as group debiasing mechanisms and circumvent groupthink risks. Crucially, these studies circumvented some of the limitations of pure functionalistic approaches by adopting qualitative methods and addressing aspects such as organizational culture.

2.2 *Learning from Feedback*

Arguably, at least part of the observed repetitive forecasting gaps could be justified by the intrinsic difficulty of predicting the future. For instance, researchers have asserted that uncertainty levels determine an event's predictability (Makridakis et al., 2010), while other views have suggested the existence of predictability limits related to the time horizon of the forecasts (Hogarth & Makridakis, 1981), or the intrinsic characteristics of the tasks, e.g., "task learnability" (Bolger & Wright, 1994). However, forecasting deviations' recurrence highlights organizations' difficulties in consistently learning from mistakes (Edmondson, 1996; Finkelstein, 2003; Madsen & Desai, 2010). Cannon and Edmondson (2005) have captured the latter concisely:

> The idea that people and the organizations in which they work should learn from failure has considerable popular support—and even seems obvious—yet organizations that systematically learn from failure are rare. (p. 299)

Failing to learn from wrong forecasts has profound implications for the organizations' ability to plan and make decisions. Flyvbjerg (2014) refers to these implications in Mega Projects:

> With errors and biases of such magnitude in the forecasts that form the basis for business cases, cost-benefit analyses, and social and environmental impact assessments, such analyses will also, with a high degree of certainty, be strongly misleading. (p. 10).

Within this context, the scarce forecasting literature on feedback and learning has focused on the relative efficacy of different forms of feedback presentation to individuals: e.g., most recent error, feedback on biases, or forecast calibration (Bolger & Önkal-Atay, 2004; Goodwin et al., 2018; Lawrence et al., 2006). Whereas other studies have described feedback limitations: e.g., point forecast feedback with no rationale or guidance does not improve performance (Klayman, 1988), forecasters are not open to guidance as they overweight their own rationale (Goodwin et al., 2011), and individuals do not accept suggestions but prefer to arrive at their conclusions independently (Parikh et al., 2001). In one interesting approach, Legerstee and Franses (2014) presented a naturalistic study where forecasters received performance and cognitive process feedback, resulting in accuracy gains behind less frequent adjustments and making more downwards interventions. However, the study was silent about the reasons for the improvement and studied forecasters individually with no interactions with their teams or contexts.

Beyond cognitive mechanisms, the organizational reasons that impede learning and adjustment from feedback remain largely unexplored in the forecasting literature; however, a deeper understanding of these causes can unveil organizational interventions. For instance, understanding the unwillingness of individuals to discuss mistakes due to fear of penalties and individuals' instinctive tendency to deny or ignore errors to maintain self-esteem (Goleman, 1985; Sagan, 1995; Taylor, 1989) explains why singling out individuals diminishes the chances of reporting and learning from mistakes (Desai, 2014). These dynamics could also explain why the blame for inaccuracy is often placed on external factors, providing a plausible and no confrontational cause that preserves harmony but impedes learning (Baumard & Starbuck, 2005). In addition to understanding the individuals' psychological reactions to mistakes in organizational contexts, a focus on collective learning processes could help address agency issues such as forecasting game playing (Mello, 2009; Meyer & Zucker, 1989). Research in this direction could also address collective attention issues (Hoffman & Ocasio, 2001; Ocasio, 1997), where shared mental models might filter out some events or make them go unnoticed, effectively becoming organizational blind spots.

Inadvertently, the quest for forecasting accuracy by refining techniques dilutes the potential of the study of organizational processes where accuracy becomes a byproduct of effective learning mechanisms from error and contexts that incentivize performance improvement over time (Mezias & Starbuck, 2003, 2009). A potential new approach to learning could resemble the teachings in Zen in the Art of Archery, where looking at the aim is replaced by a focus on the process of aiming, resulting in more chances to hit the mark eventually (Herrigel & Hull, 1953).

3 Nuanced Organizational Aspects Towards a New Framework in Forecasting

God gave all the easy problems to the physicists. (James March)

This essay proposes that prospective theoretical frameworks in forecasting should acknowledge and integrate nuanced aspects of team interactions' form and content such as attitudes towards conversations about failure and success, team deliberation about performance, and the leader's role in groups. Importantly, organizational aspects in a forecasting context might refer to different levels of analysis ranging from culture to structure design to processes. However, if group interactions should be the primary consideration, focusing on small groups' dynamics might yield valuable insights since organizations have increasingly turned to work teams as a structural building block (Kozlowski & Ilgen, 2006).

3.1 Learning from Success Versus Failure

Intuitively, learning from best practices seems like a plausible strategy. However, in dynamic contexts, the tendency to focus on best practices could increase organizational inertia due to previous successful tasks' formalization (Baumard & Starbuck, 2005; Miller, 1994). More specifically, Miller (1994) associates an emphasis on learning from success with damaging outcomes: (a) inertial pressures in decision-making, (b) extreme procedural orientation to replicate previous successes, (c) reduced intelligence gathering, and (d) inadequate recognition of changes in the environment. Similarly, Baumard and Starbuck (2005) conclude that firms excessively focused on reproducing their successes create behavioral programs to make replication efficient, decreasing incentives to invest in information gathering and becoming *"less aware of events outside their immediate domains and less capable of diverse actions" (p. 283).*

Contrary to the focus on success, the organizational learning literature provides empirical support favoring learning from failure as the key for long term viability in different contexts such as the orbital vehicle launching industry (Madsen & Desai, 2010), cardiac surgery (Kc et al., 2013) and entrepreneurship (Eggers & Song, 2014). These authors find that experience with failures tends to yield the most significant gains in future performance, primarily when an error is acknowledged and processed.

In parallel, in the education field, comparative studies between Chinese and U.S. teaching techniques have found that the former students systematically outperform the latter in standardized math test scores. Stevenson and Stigler (1992) and Wang and Murphy (2004) explain this overperformance behind Chinese teachers' use of errors to prompt group discussion about mathematical concepts, promoting a classroom environment where students do not feel ashamed to make mistakes. Stevenson and Stigler (1992) linked these teaching practices to cultural beliefs:

> For Americans, errors tend to be interpreted as an indication of failure in learning the lesson. For Chinese and Japanese, they are an index of what still needs to be learned (p. 192).

In this context, teachers and students *"dwell on errors"*, correcting the error and then asking the students to explain the reasoning behind them. Schleppenbach et al. (2007) advance further in this direction in another comparative study and find that the normalization of error reinforces the shared belief that it is common to voice and discuss mistakes. They encountered that discussion of errors is so essential for Chinese teachers that sometimes they induce errors to start a class conversation.

These findings in the education field have certain parallelism with findings from a study on nursing teams that revealed that the best performing nursing teams reported more errors, were willing to discuss them, and received coaching from their superiors during these discussions (Edmondson, 1996). The study concluded that these teams had a better error climate, which resulted in increased error detection and correction.

3.2 Group Deliberation About Performance

The Delphi method is a collective forecasting method consisting of strictly separate forecasts from multiple individuals and then revised using anonymous feedback from other Delphi panel members (Linstone & Turoff, 1975). In a review of Delphi's empirical studies, Rowe and Wright (2001) found that Delphi groups outperformed 'statistical' groups (aggregations of non-interacting individuals) in twelve out of sixteen studies, and outperformed standard interacting groups in five out of seven studies. Even though the causes behind this observed superior performance are not fully understood, one critical insight is that Delphi improves judgmental accuracy when outcome feedback includes "reasons" and not only statistics (Rowe & Wright, 1996). Paradoxically, even though Delphi is primarily described as a method of forecast aggregation, the interchange of reasons hints at the efficacy of group deliberation since this process could be described as deliberation "by turns" in which forecasters can alert each other about the use of unfitting framings, cognitive biases, and inappropriate cause-effect relationships (Goodwin & Wright, 2010).

Research on debriefs and performance appraisals provides additional support for the effectiveness of deliberation and performance feedback across different fields ranging from medicine (Gaba et al., 2001) to organizational training (Garvin et al., 2008). One meta-analysis, including 46 samples from 31 studies, concluded that debriefs improved effectiveness over a control group by approximately 25% on average (Tannenbaum & Cerasoli, 2013). Additionally, the authors defined specific characteristics. Firstly, debriefs are active versus passive, which translates into participation and self-discovery rather than being told how to improve; secondly, the primary intent should be developmental and non-punitive. Third, debriefs are centered on specific episodes or performance events, and, finally, they include inputs from multiple team members and at least one external source.

One key observation is that team deliberation norms could have cultural implications on team interactions beyond effectiveness. For instance, the U.S. Army After Action Reviews (AAR)—formalized retrospective analyses—have been linked to the promotion of team cohesiveness via group interaction, consensus building, intragroup competition reduction, and preservation of a positive atmosphere (Morrison & Meliza, 1999).

3.3 Team Leaders as Facilitators

One seminal study on best-performing nursing teams highlighted the impact of leadership on teams' ability to learn from mistakes (Edmondson, 1996). The study concluded that a leadership model that establishes a context for openly handling errors induces a shared perception of error normalization. This normalization triggers openness to report errors and generates discussions about what went wrong, ultimately reducing errors in subsequent attempts. Conversely, the literature on group forecasting seldom integrates the fact that most forecasters operate in teams and small groups with a leader-staff configuration where voting systems are non-existent, but the leader executes some form of compilation after receiving cues from the different members; a model of interaction termed "hierarchical decision-making teams with distributed expertise." (Hollenbeck et al., 1995).

Team leader-staff dynamics have been widely explored in the U.S. Navy's Tactical Decision Making Under Stress program that spanned from 1990 to 1999 (TADMUS project). This project was one of the most extensive research efforts on team decision-making under stress which yielded essential findings regarding leadership's impact on teams' interactions and communication processes (Collyer & Malecki, 1998). The program's goal was to develop training plans to help leaders enhance team performance by setting team goals, managing conflict, ensuring participation, and observing group dynamics (Salas et al., 1998). Results showed that teams with leaders who had received formal leadership performed better in a series of tasks (Cannon-Bowers & Salas, 1998). Moreover, Tannenbaum et al. (1998) identified some effective leaders' behaviors and concluded they acted as facilitators, contrary to leaders' conception as supervisors. These leaders engaged in self-critique to signal the team members it was "safe" to make and admit errors, focused on feedback on tasks instead of people to avoid affective conflict (but foster cognitive conflict), provided task-specific suggestions, and ensured the discussion included teamwork feedback—team processes—instead of outcome feedback exclusively.

Kozlowski et al. attribute team effectiveness improvement behind facilitation to the triggering of collective metacognitive processing, especially when the "metacognitive musing" is verbalized as feedback about major task engagements and includes reflection upon performance and team processes. Given that discussions alone do not facilitate shared cognition (Mathieu et al., 2000), these facilitation strategies could create shared mental models, i.e., team members agree at least on the group's objectives and the basic models used to interpret reality. These mental models are one key element of deliberations' effectiveness since team members with dissimilar mental models hamper the team's performance (Mathieu et al., 2000; Smith-Jentsch et al., 2008).

4 Mindful Organizing: A Framework in the Interpretivist-Functionalist Transition Zone

There is nothing more practical than a good theory

Gioia and Pitre (1990) have stressed that the study of organizational dynamics such as power, conflict, and meaning negotiation cannot be analyzed thoroughly from a pure functionalist angle but need a bridge to an interpretivist view. As an illustration of an interpretivist-functionalist bridge, they put forward the idea of structuration, a circular process where agents' interactions enact structures that then influence agents' behavior. The focus on actors' interactions highlights the importance of communication within groups that constitutes "micro-level decision-making interaction patterns" that form the building blocks of teams' processes (McGrath et al., 2000). Hence, opposing the forecasting literature's approach where interactions are neglected or seen as potential sources of biases, relevant theoretical frameworks should focus primarily on social interaction dynamics, and their predictive power will depend on their grasp of these interactions' complexities (Poole et al., 1996).

Within this context, one relevant structurationistic frame is *sensemaking*. Sensemaking researchers have defined it as the perpetual process in which organizations try to interpret equivocal inputs and enact those interpretations back into the environment to make it more predictable (Starbuck, 2015; Weick, 1995; Weick et al., 2005). Within that context, organizations engaging in sensemaking notice or perceive cues, create negotiated interpretations and take deliberate action (Maitlis & Christianson, 2014). Crucially, these "sensemaking moves" are inherently social, in contrast to the mere exploitation of concrete cues in the individual decision-making models that collapse at the collective level when the "politics of meaning" are present (Patriotta, 2003).

One specific mode of implementing sensemaking is mindful organizing (Weick et al., 1999; Weick & Sutcliffe, 2007), a collective process consisting of learning mechanisms that focus attention, manage errors, and maintain alertness (Ocasio, 1997; Vogus & Sutcliffe, 2012; Weick et al., 1999); intrinsically ingrained in the interactions of groups (Vogus & Sutcliffe, 2012). Notably, one of the framework's theoretical accomplishments is describing five processes that create and sustain collective mindfulness (Weick et al., 1999; Weick & Sutcliffe, 2007). Preoccupation with error manifests by actively reporting errors, treating all failures as opportunities to learn, and by understanding the liabilities of success in the form of complacency.

Reluctance to simplify interpretations manifests as a continuous search for different perspectives to avoid the temptation of accepting one unique interpretation at face value (e.g., adversarial reviews, bringing employees with non-typical experience). Sensitivity to operations is characterized by a high level of situational awareness, i.e., a "struggle for alertness" to catch errors in the moment. Commitment to resilience is the belief in the fallibility of existing knowledge and the ability to recover from errors. Finally, under-specification of structures is characterized by fluid decision-making in fast-paced situations, where team hierarchy can be dissolved so that decision-makers change depending on the circumstances.

Firstly, the five effortful processes of mindful organizing could illuminate the types of interaction that mitigate propensities towards biased and auto-pilot cognitive modes that hinder judgment and learning in forecasting contexts (Kahneman, 2011; Kahneman & Lovallo, 1993; Lovallo & Kahneman, 2003). Secondly, some of the mindful organizing processes, such as preoccupation with failure and reluctance to simplify interpretations, direct attention towards concrete social and cognitive mechanisms to process error as an essential antecedent to learning (Sitkin, 1990), and scrutinize hypothesis to redefine previous categories (Langer, 2016). Finally, and central to advancing forecasting from an organizational perspective, the thoroughly social aspect of mindful organizing leaves little room for heroic individuals, omnipresent leaders, or experts but requires "dense interrelations" as an antecedent to learning and adaptation (Weick & Roberts, 1993).

5 Inducing Mindful Organizing to Debias Group Judgment

> A tendency toward mindlessness is characterized by a style of mental functioning in which people follow recipes, impose old categories to classify what they see, act with some rigidity, operate on automatic pilot, and mislabel unfamiliar new contexts as familiar old ones. (Weick & Sutcliffe, 2007)

As stated previously, the forecasting literature has emphasized the role of cognitive biases in diminishing the accuracy of judgment (Eroglu & Croxton, 2010; Flyvbjerg, 2008; Kahneman & Lovallo, 1993; Lawrence & O'Connor, 1995; Lee & Siemsen, 2016), where the collective aspect has emphasized the dangers of groupthink (Janis, 1982; Mellers et al., 2014; Sniezek, 1989; Surowiecki, 2005) and the potential bias accentuation effect of the group (Buehler et al., 2005).

In parallel, the group decision-making literature has uncovered that judgment quality may suffer during group discussions. For instance, research on brainstorming has highlighted that group discussion tends to interfere with people's ability to immerse in a productive train of thought (Kerr & Tindale, 2004; Nijstad & Stroebe, 2006). This line of research has also highlighted other factors that yield "process losses" during group deliberation techniques such as brainstorming: an unwillingness to contribute ideas because of evaluation apprehension, and convergence via social comparison on a relatively low standard of performance (Larey & Paulus, 1999; Mullen et al., 1991). The "cognitive centrality" effect could explain this

Table 11.1 Mindful organizing processes map to group judgment propensities

Mindful organizing processes	Group judgment propensities in forecasting and decision-making literature
Preoccupation with error	Complacency (Sitkin, 1990), extrapolation of past success (Baumard & Starbuck, 2005; Kc et al., 2013; Madsen & Desai, 2010), explaining away mistakes (Sagan, 1995), stigmatization of error (Edmondson, 1996, 1999), inattention to weak cues (Miller, 1994)
Reluctance to simplify interpretations	Confirmation biases and anchoring (Eroglu & Croxton, 2010; Kahneman & Lovallo, 1993; Lovallo et al., 2012), overreliance on experiential learning, group cognitive centrality (Kameda et al., 1997), overreliance on old categories, groupthink and bandwagon effects (Buehler et al., 2005; Janis, 1982)
Sensitivity to operations	Overgeneralization of previous experiences (Lovallo & Kahneman, 2003), superficial information search & processing (Slovic et al., 1977; Slovic & Lichtenstein, 1971)
Commitment to resilience	Failure as a definitive state, overreliance on old categories
Underspecification of structures	"Deans decease" effect (Bedeian, 2002), hierarchical rigidity

convergence, i.e., groups are more likely to choose ideas with the most significant overlap (Kameda et al., 1997); creating a convergence towards what is already known or believed, effectively working as an instance of collective confirmation bias (van den Bossche et al., 2011). For instance, Tindale (1993) analyzed videotaped discussions of conjunction problems and found groups typically exchange individual judgments and more than 60% of the time select an undefended judgment from one of the members without debate, if the individual preference is plausible within a shared representation.

Hence, it is clear that not all interactions are adaptive; alas, their content and form are key mechanisms influencing outcomes such as reflective reframing, rigorous discussions of errors, or hypothesis questioning (Sutcliffe et al., 2016). Within this context, mindful organizing processes offer a refreshing contrast to some of the most frequently cited challenges of group decision-making and judgment in forecasting and other adjacent domains (Table 11.1). Consequently, if mindlessness seems to be the default state of group interactions—leading to biased judgment—and mindful organizing seems to counteract those propensities, one central question would be how forecasting teams could reliably activate it in their interactions.

5.1 *Focus on Episodic, Dramatic Error*

Christianson et al. (2009) have conceived organizational learning as the revision of response repertoires triggered by errors that serve as audits—sometimes brutal—of existent repertoires. From a mindful organizing perspective, interactions need to be episodic, focus on dramatic events, and some degree of conflict—sustained for enough time—needs to be present.

Within this context, errors should generate significant interruptions i.e., exaggerated versions of the day-to-day stimulus, to justify the investment in effortful mindful engagements. For instance, Vogus and Colville (2017) have reported the beneficial effects of collective mindfulness only when a hospital nursing unit had a negative performance history of adverse events like harmful medication errors. Under those conditions, collective mindfulness acted as a problem-solving and emotion regulation resource. In contrast, on units without a history of adverse events, collective mindfulness was associated with higher emotional exhaustion levels, depleting personal resources without tangible benefits. Similarly, Ray et al. (2011) have stated that mindful organizing costs will not be questioned in high-reliability situations, such as nuclear power plants or aircraft carriers, because of the potentially fatal consequences. However, in other organizations, the constant pressures for efficiency could make it difficult to establish a relationship between organizational mindfulness and conventional financial performance measures unless managers recognize issues as critical, something more likely to happen if issues are dramatic or after major engagements incidents.

Another aspect of mindful encounters is the existence of conflict (Bossche et al., 2011). However, conflict only enhances team outcomes when it leads to deep-level processing, which allows a complete awareness of the complexity of the problems. Within this context, conflict—viewed as the adversarial review of information—should trigger "mindful scanning" (Vogus & Sutcliffe, 2012), which entails the exploration of the fringes of the current task (Fiol & O'Connor, 2003). Such practice could mitigate the temptation to fall into bandwagon effects and avoid the risks inherent to undisputed leaders' opinions, e.g., "deans decease" (Bedeian, 2002).

Finally, these team interactions need to allow for sufficient time since new cues seem to impact decision-making when groups use enough time to reach consensus, avoiding premature closure. Dwelling on some aspects of the discussions is especially relevant when most members already share the same preference since people are perceived as more competent and credible when they share information that others already know (Kerr & Tindale, 2004). For example, in an experiment involving undergraduate students, anchoring bias was only overcome when individuals were sufficiently motivated and able to think carefully about their answers, so satisficing with the initial estimates was avoided (Epley & Gilovich, 2006).

5.2 *Use of Analogical Reasoning and Reference Classes*

Gick and Holyoak (1980) have demonstrated that analogies are potent devices to encapsulate and transfer knowledge from different domains. Their seminal study provided an experimental demonstration that a target problem can be solved using a similar problem from an unrelated domain when the relations between the analogy story and the target problem are mapped effectively. Moreover, Gick and Holyoak

(1983) advanced this reasoning by empirically demonstrating that deriving a "problem schema"—a representation of the structural relationship from more than one analogy—enhanced knowledge transfer significantly; since the use of several analogies can reduce the risk of "superficial mappings" between the source and the target.

This approach finds an equivalent from the mindfulness perspective, Langer asserts that people can overcome biased judgment through the continuous creation and refinement of new categories and an implicit awareness of more than one perspective. Moreover, evidence exists that analogical thinking—via observation of analogous artifacts—can induce these mindfulness processes allowing for deconstruction, symbolization, parody, and allegory (Barry & Meisiek, 2010; Davies, 2006).

The literature on analogies in management contexts is scarce but provides insights into how analogies can enhance judgment. Extracting knowledge from multiple analogies is embodied in the concept of "outside view", a deliberate effort to make use of analogical reasoning based on historical data to counteract the natural tendency of decision-makers to use an "inside view" (Kahneman & Lovallo, 1993):

> Decision makers have a strong tendency to consider problems as unique. They isolate the current choice from future opportunities and neglect the statistics of the past in evaluating current plans.

Similarly, Goodwin and Wright (2010) have reasoned that a critical aspect of taking an outside view—through a technique termed reference class—will be the forecasters' ability to use the correct analogies to filter out superficial similarities and retain a smaller number of cases with the most profound structural similarities. The effectiveness of taking an "outside view" is supported empirically in different domains: effective competitive positions discovery (Gavetti et al., 2005), adjustments of stock valuation forecasts (Lovallo et al., 2012), geopolitical forecasts (Mellers et al., 2014), and making estimates about the costs of megaprojects (Flyvbjerg, 2008).

6 Conclusion

The incorporation of mindful organizing in group forecasting contexts can contribute to the forecasting field from three angles: (a) substantiates the inherent subjectivity of the forecast process where actors can influence prediction outcomes by enacting shared interpretations, (b) offers a representation of collective judgment debiasing mechanisms, and (c) emphasizes the process of collective learning via error deliberation. These elements allow for the integration of other de-facto organizational goals in forecasting such as improved performance, team motivation, intentions signaling, or mobilization of resources. Under this approach, achieving forecast accuracy is less critical than unveiling collective learning mechanisms, which will eventually yield higher forecast adaptation levels in the long run.

Reinventing performance debriefs (e.g., after-action reviews, postmortems) could be an effective way to induce collective mindfulness, resulting in collective judgment debiasing and improving forecast accuracy as a byproduct. Notably, recognizing that these interactions need to be episodic, structured around errors, possess an adversarial tone, and need to be facilitated by the leader provides an actionable template for practitioners. Consistent with the interpretivist approach, forecasting teams will deliver accuracy gains by enhancing their collective judgment and partially enacting shared beliefs.

The forecasting process is vastly present in firms through structured multidisciplinary interactions such as Sales and Operations meetings (S&OP); however, their influence on broader organizational dynamics has been seldom studied. Making the forecasting function a potential source of organizational mindfulness and consensus-building could amplify its role, becoming a bottom-up organizational force from which organizational traits such as psychological safety can emerge (Edmondson, 1999). Paraphrasing Donald Hebb's quote, "Neurons that fire together, wire together", a team that forecasts together learns together.

References

2015 Breakthrough Innovation Report, Europe Edition. (n.d.). Retrieved May 31, 2020, from http://innovation.nielsen.com/breakthrough2015EU

Barry, D., & Meisiek, S. (2010). Seeing more and seeing differently: Sensemaking, mindfulness, and the workarts. *Organization Studies, 31*(11), 1505–1530. https://doi.org/10.1177/0170840610380802

Baumard, P., & Starbuck, W. H. (2005). Learning from failures: Why it may not happen. *Long Range Planning, 38*(3), 281–298. https://doi.org/10.1016/j.lrp.2005.03.004

Bedeian, A. G. (2002). The Dean's disease: How the darker side of power manifests itself in the Office of Dean. *Academy of Management Learning & Education, 1*(2), 164–173. https://doi.org/10.5465/amle.2002.8509359

Bloch, M., Blumberg, S., & Laartz, J. (2012). *Delivering large-scale IT projects on time, on budget, and on value* (Vol. 27, p. 6). Springer.

Bolger, F., & Önkal-Atay, D. (2004). The effects of feedback on judgmental interval predictions. *International Journal of Forecasting, 20*(1), 29–39. https://doi.org/10.1016/S0169-2070(03)00009-8

Bolger, F., & Wright, G. (1994). Assessing the quality of expert judgment: Issues and analysis. *Decision Support Systems, 11*(1), 1–24. https://doi.org/10.1016/0167-9236(94)90061-2

Bretschneider, S., Gorr, W. L., Grizzle, G., & Klay, E. (1989). Political and organizational influences on the accuracy of forecasting state government revenues. *International Journal of Forecasting, 5*(3), 307–319.

Buehler, R., Messervey, D., & Griffin, D. (2005). Collaborative planning and prediction: Does group discussion affect optimistic biases in time estimation? *Organizational Behavior and Human Decision Processes, 97*(1), 47–63. https://doi.org/10.1016/j.obhdp.2005.02.004

Burrell, G., & Morgan, G. (1979). *Sociological paradigms and organisational analysis: Elements of the sociology of corporate life*. Heinemann.

Cannon, M. D., & Edmondson, A. C. (2005). Failing to learn and learning to fail (intelligently): How great organizations put failure to work to innovate and improve. *Long Range Planning, 38*(3), 299–319. https://doi.org/10.1016/j.lrp.2005.04.005

Cannon-Bowers, J. A., & Salas, E. (1998). Team performance and training in complex environments: Recent findings from applied research. *Current Directions in Psychological Science, 7*(3), 83–87. https://doi.org/10.1111/1467-8721.ep10773005
Christianson, M. K., Farkas, M. T., Sutcliffe, K. M., & Weick, K. E. (2009). Learning through rare events: Significant interruptions at the Baltimore & Ohio Railroad Museum. *Organization Science, 20*(5), 846–860. https://doi.org/10.1287/orsc.1080.0389
Collyer, S. C., & Malecki, G. S. (1998). Tactical decision making under stress: History and overview. In *Making decisions under stress: Implications for individual and team training* (pp. 3–15). American Psychological Association. https://doi.org/10.1037/10278-016
Davies, S. (2006). Aesthetic judgements, artworks and functional beauty. *The Philosophical Quarterly, 56*(223), 224–241. https://doi.org/10.1111/j.1467-9213.2006.00439.x
Davis, D. F., & Mentzer, J. T. (2007). Organizational factors in sales forecasting management. *International Journal of Forecasting, 23*(3), 475–495.
Desai, V. (2014). Learning through the distribution of failures within an organization: Evidence from heart bypass surgery performance. *Academy of Management Journal, 58*(4), 1032–1050. https://doi.org/10.5465/amj.2013.0949
Deschamps, E. (2004). The impact of institutional change on forecast accuracy: A case study of budget forecasting in Washington state. *International Journal of Forecasting, 20*(4), 647–657. https://doi.org/10.1016/j.ijforecast.2003.11.009
Edmondson. (1996). Learning from mistakes is easier said than done: Group and organizational influences on the detection and correction of human error. *The Journal of Applied Behavioral Science, 32*(1), 5–28. https://doi.org/10.1177/0021886396321001
Edmondson. (1999). Psychological safety and learning behavior in work teams. *Administrative Science Quarterly, 44*(2), 350–383. https://doi.org/10.2307/2666999
Eggers, J. P., & Song, L. (2014). Dealing with failure: Serial entrepreneurs and the costs of changing industries between ventures. *Academy of Management Journal, 58*(6), 1785–1803. https://doi.org/10.5465/amj.2014.0050
Einhorn, H. J., & Hogarth, R. M. (1981). Behavioral decision theory: Processes of judgment and choice. *Journal of Accounting Research, 19*(1), 1–31. https://doi.org/10.2307/2490959
Epley, N., & Gilovich, T. (2006). The anchoring-and-adjustment heuristic: Why the adjustments are insufficient. *Psychological Science, 17*(4), 311–318. https://doi.org/10.1111/j.1467-9280.2006.01704.x
Eroglu, C., & Croxton, K. L. (2010). Biases in judgmental adjustments of statistical forecasts: The role of individual differences. *International Journal of Forecasting, 26*(1), 116–133.
Fildes, R. (2006). The forecasting journals and their contribution to forecasting research: Citation analysis and expert opinion. *International Journal of Forecasting, 3*(22), 415–432. https://doi.org/10.1016/j.ijforecast.2006.03.002
Fildes, R., & Hastings, R. (1994). The organization and improvement of market forecasting. *Journal of the Operational Research Society, 45*(1), 1–16. https://doi.org/10.1057/jors.1994.1
Fildes, R., Bretschneider, S., Collopy, F., Lawrence, M., Stewart, D., Winklhofer, H., Mentzer, J. T., & Moon, M. A. (2003). Researching sales forecasting practice commentaries and authors' response on "conducting a sales forecasting audit" by M.A. Moon, J.T. Mentzer & C.D. Smith. *International Journal of Forecasting, 19*(1), 27–42. https://doi.org/10.1016/S0169-2070(02)00033-X
Fildes, R., Goodwin, P., Lawrence, M., & Nikolopoulos, K. (2009). Effective forecasting and judgmental adjustments: An empirical evaluation and strategies for improvement in supply-chain planning. *International Journal of Forecasting, 25*(1), 3–23. https://doi.org/10.1016/j.ijforecast.2008.11.010
Finkelstein, S. (2003). *Why smart executives fail and what you can learn from their mistakes*. Portfolio.
Fiol, C. M., & O'Connor, E. J. (2003). Waking up! Mindfulness in the face of bandwagons. *Academy of Management Review, 28*(1), 54–70. https://doi.org/10.5465/amr.2003.8925227

Flyvbjerg, B. (2008). Curbing optimism bias and strategic misrepresentation in planning: Reference class forecasting in practice. *European Planning Studies, 16*(1), 3–21. https://doi.org/10.1080/09654310701747936

Flyvbjerg, B. (2014). What you should know about megaprojects and why: An overview. *Project Management Journal, 45*(2), 6–19. https://doi.org/10.1002/pmj.21409

Gaba, D. M., Howard, S. K., Fish, K. J., Smith, B. E., & Sowb, Y. A. (2001). Simulation-based training in anesthesia crisis resource management (ACRM): A decade of experience. *Simulation & Gaming*. https://doi.org/10.1177/104687810103200206

Galbraith, C. S., & Merrill, G. B. (1996). The politics of forecasting: Managing the truth. *California Management Review, 38*(2), 29–43. https://doi.org/10.2307/41165831

GAO-15-675T, U. S. G. A. (2015). *Information technology: Additional actions and oversight urgently needed to reduce waste and improve performance in acquisitions and operations. GAO-15-675T.* Retrieved from https://www.gao.gov/products/gao-15-675t

Garvin, D. A., Edmondson, A. C., & Gino, F. (2008, March 1). Is yours a learning organization? *Harvard Business Review*. Retrieved from https://hbr.org/2008/03/is-yours-a-learning-organization

Gavetti, G., Levinthal, D. A., & Rivkin, J. W. (2005). Strategy making in novel and complex worlds: The power of analogy. *Strategic Management Journal, 26*(8), 691–712. https://doi.org/10.1002/smj.475

Gick, M. L., & Holyoak, K. J. (1980). Analogical problem solving. *Cognitive Psychology, 12*(3), 306–355. https://doi.org/10.1016/0010-0285(80)90013-4

Gick, M. L., & Holyoak, K. J. (1983). Schema induction and analogical transfer. *Cognitive Psychology, 15*(1), 1–38. https://doi.org/10.1016/0010-0285(83)90002-6

Gioia, D. A., & Pitre, E. (1990). Multiparadigm perspectives on theory building. *Academy of Management Review, 15*(4), 584–602. https://doi.org/10.5465/amr.1990.4310758

Goleman, D. (1985). *Vital lies, simple truths: The psychology of self deception*. Simon & Schuster.

Goodwin, P., & Wright, G. (2010). The limits of forecasting methods in anticipating rare events. *Technological Forecasting and Social Change, 77*(3), 355–368. https://doi.org/10.1016/j.techfore.2009.10.008

Goodwin, P., Fildes, R., Lawrence, M., & Stephens, G. (2011). Restrictiveness and guidance in support systems. *Omega, 39*(3), 242–253. https://doi.org/10.1016/j.omega.2010.07.001

Goodwin, P., Moritz, B., & Siemsen, E. (2018). Forecast decisions. In *The handbook of behavioral operations* (pp. 433–458). Wiley. https://doi.org/10.1002/9781119138341.ch12

Henshel, R. L. (1993). Do self-fulfilling prophecies improve or degrade predictive accuracy? How sociology and economics can disagree and both be right. *The Journal of Socio-Economics, 22*(2), 85–104. https://doi.org/10.1016/1053-5357(93)90017-F

Herrigel, E., & Hull, R. F. C. (1953). *Zen in the art of archery*. Pantheon Books.

Hoffman, A. J., & Ocasio, W. (2001). Not all events are attended equally: Toward a middle-range theory of industry attention to external events. *Organization Science, 12*(4), 414–434. https://doi.org/10.1287/orsc.12.4.414.10639

Hogarth, R. M., & Makridakis, S. (1981). Forecasting and planning: An evaluation. *Management Science, 27*(2), 115–138. https://doi.org/10.1287/mnsc.27.2.115

Hollenbeck, J., Ilgen, D., Sego, D., Hedlund, J., Major, D., & Phillips, J. (1995). Multilevel theory of team decision making: Decision performance in teams incorporating distributed expertise. *Journal of Applied Psychology, 80*(2), 292–316. https://doi.org/10.1037/0021-9010.80.2.292

Janis, I. L. (1982). *Groupthink: Psychological studies of policy decisions and fiascoes*. Houghton Mifflin.

Kahneman, D. (2011). *Thinking, fast and slow*.

Kahneman, D., & Lovallo, D. (1993). Timid choices and bold forecasts: A cognitive perspective on risk taking. *Management Science, 39*(1), 17–31.

Kameda, T., Ohtsubo, Y., & Takezawa, M. (1997). Centrality in sociocognitive networks and social influence: An illustration in a group decision-making context. *Journal of Personality and Social Psychology, 73*(2), 296–309. https://doi.org/10.1037/0022-3514.73.2.296

Kc, D., Staats, B. R., & Gino, F. (2013). Learning from my success and from others' failure: Evidence from minimally invasive cardiac surgery. *Management Science, 59*(11), 2435–2449.
Kerr, N. L., & Tindale, R. S. (2004). Group performance and decision making. *Annual Review of Psychology, 55*(1), 623–655. https://doi.org/10.1146/annurev.psych.55.090902.142009
Klayman, J. (1988). On the how and why (not) of learning from outcomes. In *Human judgment: The SJT view* (pp. 115–162). North-Holland. https://doi.org/10.1016/S0166-4115(08)62172-X
Klein, G. (2008). Naturalistic decision making. *Human Factors, 50*(3), 456–460. https://doi.org/10.1518/001872008X288385
Kozlowski, S. W. J., & Ilgen, D. R. (2006). Enhancing the effectiveness of work groups and teams. *Psychological Science in the Public Interest, 7*(3), 77–124. https://doi.org/10.1111/j.1529-1006.2006.00030.x
Langer, E. J. (2016). *The power of mindful learning*. Hachette UK.
Larey, T. S., & Paulus, P. B. (1999). Group preference and convergent tendencies in small groups: A content analysis of group brainstorming performance. *Creativity Research Journal, 12*(3), 175–184. https://doi.org/10.1207/s15326934crj1203_2
Lawrence, M., & O'Connor, M. (1995). The anchor and adjustment heuristic in time-series forecasting. *Journal of Forecasting, 14*(5), 443–451. https://doi.org/10.1002/for.3980140504
Lawrence, M., O'Connor, M., & Edmundson, B. (2000). A field study of sales forecasting accuracy and processes. *European Journal of Operational Research, 122*(1), 151–160. https://doi.org/10.1016/S0377-2217(99)00085-5
Lawrence, M., Goodwin, P., O'Connor, M., & Önkal, D. (2006). Judgmental forecasting: A review of progress over the last 25 years. *International Journal of Forecasting, 22*(3), 493–518. https://doi.org/10.1016/j.ijforecast.2006.03.007
Lee, Y. S., & Siemsen, E. (2016). Task decomposition and newsvendor decision making. *Management Science, 63*(10), 3226–3245. https://doi.org/10.1287/mnsc.2016.2521
Legerstee, R., & Franses, P. H. (2014). Do experts' SKU forecasts improve after feedback? *Journal of Forecasting, 33*(1), 69–79. https://doi.org/10.1002/for.2274
Linstone, H. A., & Turoff, M. (1975). *The Delphi method: Techniques and applications*. Addison-Wesley.
Lovallo, D., & Kahneman, D. (2003). Delusions of success. How optimism undermines executives' decisions. *Harvard Business Review, 81*(7), 56–63, 117.
Lovallo, D., Clarke, C., & Camerer, C. (2012). Robust analogizing and the outside view: Two empirical tests of case-based decision making. *Strategic Management Journal, 33*(5), 496–512. https://doi.org/10.1002/smj.962
Madsen, P. M., & Desai, V. (2010). Failing to learn? The effects of failure and success on organizational learning in the global orbital launch vehicle industry. *Academy of Management Journal, 53*(3), 451–476. https://doi.org/10.5465/amj.2010.51467631
Mahmoud, E., DeRoeck, R., Brown, R., & Rice, G. (1992). Bridging the gap between theory and practice in forecasting. *International Journal of Forecasting, 8*(2), 251–267. https://doi.org/10.1016/0169-2070(92)90123-Q
Maitlis, S., & Christianson, M. (2014). Sensemaking in organizations: Taking stock and moving forward. *The Academy of Management Annals, 8*(1), 57–125. https://doi.org/10.1080/19416520.2014.873177
Makridakis, S., & Taleb, N. (2009). Decision making and planning under low levels of predictability. *International Journal of Forecasting, 25*(4), 716–733. https://doi.org/10.1016/j.ijforecast.2009.05.013
Makridakis, S., & Wheelwright, S. C. (1977). Forecasting: Issues & challenges for marketing management. *Journal of Marketing, 41*(4), 24–38. https://doi.org/10.1177/002224297704100403
Makridakis, S. G., Gaba, A., & Hogarth, R. (2010). *Dance with chance: Making luck work for you*. Oneworld Publications.

Mathieu, J. E., Heffner, T. S., Goodwin, G. F., Salas, E., & Cannon-Bowers, J. A. (2000). The influence of shared mental models on team process and performance. *Journal of Applied Psychology, 85*(2), 273–283. https://doi.org/10.1037/0021-9010.85.2.273

McGahan, A. M. (2007). Academic research that matters to managers: On zebras, dogs, lemmings, hammers, and turnips. *Academy of Management Journal, 50*(4), 748–753. https://doi.org/10.5465/amj.2007.26279166

McGrath, J. E., Arrow, H., & Berdahl, J. L. (2000). The study of groups: Past, present, and future. *Personality and Social Psychology Review, 4*(1), 95–105. https://doi.org/10.1207/S15327957PSPR0401_8

Meehl, P. E. (1954). *Clinical versus statistical prediction: A theoretical analysis and a review of the evidence* (p. x, 149). University of Minnesota Press. https://doi.org/10.1037/11281-000

Mellers, B., Ungar, L., Baron, J., Ramos, J., Gurcay, B., Fincher, K., Scott, S. E., Moore, D., Atanasov, P., Swift, S. A., Murray, T., Stone, E., & Tetlock, P. E. (2014). Psychological strategies for winning a geopolitical forecasting tournament. *Psychological Science, 25*(5), 1106–1115. https://doi.org/10.1177/0956797614524255

Mello, J. (2009). The impact of sales forecast game playing on supply chains. *Foresight: The International Journal of Applied Forecasting, 13*, 13–22.

Mentzer, J. T., Flint, D. J., & Kent, J. L. (1999). Developing a logistics service quality scale. *Journal of Business Logistics, 20*(1), 9.

Meyer, M. W., & Zucker, L. G. (1989). *Permanently failing organizations*. Sage.

Mezias, J. M., & Starbuck, W. H. (2003). Studying the accuracy of managers' perceptions: A research Odyssey. *British Journal of Management, 14*(1), 3–17. https://doi.org/10.1111/1467-8551.00259

Mezias, J., & Starbuck, W. H. (2009). *Decision making with inaccurate, unreliable data*. The Oxford Handbook of Organizational Decision Making. https://doi.org/10.1093/oxfordhb/9780199290468.003.0004

Miller, D. (1994). What happens after success: The perils of excellence*. *Journal of Management Studies, 31*(3), 325–358. https://doi.org/10.1111/j.1467-6486.1994.tb00621.x

Morrison, J. E., & Meliza, L. L. (1999). *Foundations of the after action review process (IDA/HQ-D2332)*. Institute for Defense Analyses. Retrieved from https://apps.dtic.mil/docs/citations/ADA368651.

Mosier, K., Fischer, U., Hoffman, R. R., & Klein, G. (2018). Expert professional judgments and "naturalistic decision making". In *The Cambridge handbook of expertise and expert performance* (2nd ed., pp. 453–475). Cambridge University Press. https://doi.org/10.1017/9781316480748.025

Mullen, B., Johnson, C., & Salas, E. (1991). Productivity loss in brainstorming groups: A meta-analytic integration. *Basic and Applied Social Psychology, 12*(1), 3–23. https://doi.org/10.1207/s15324834basp1201_1

Nijstad, B. A., & Stroebe, W. (2006). How the group affects the mind: A cognitive model of idea generation in groups. *Personality and Social Psychology Review, 10*(3), 186–213. https://doi.org/10.1207/s15327957pspr1003_1

Ocasio, W. (1997). Towards an attention-based view of the firm. *Strategic Management Journal, 18*(S1), 187–206. https://doi.org/10.1002/(SICI)1097-0266(199707)18:1+<187::AID-SMJ936>3.0.CO;2-K

Oliva, R., & Watson, N. (2009). Managing functional biases in organizational forecasts: A case study of consensus forecasting in supply chain planning. *Production and Operations Management, 18*(2), 138–151. https://doi.org/10.1111/j.1937-5956.2009.01003.x

Oliva, R., & Watson, N. (2011). Cross-functional alignment in supply chain planning: A case study of sales and operations planning. *Journal of Operations Management, 29*(5), 434–448. https://doi.org/10.1016/j.jom.2010.11.012

Parikh, M., Fazlollahi, B., & Verma, S. (2001). The effectiveness of decisional guidance: An empirical evaluation. *Decision Sciences, 32*(2), 303–332. https://doi.org/10.1111/j.1540-5915.2001.tb00962.x

Patriotta, G. (2003). Sensemaking on the shop floor: Narratives of knowledge in organizations*. *Journal of Management Studies, 40*(2), 349–375. https://doi.org/10.1111/1467-6486.00343

Poole, M. S., Seibold, D. R., & McPhee, R. D. (1996). The structuration of group decisions. In *Communication and group decision making* (2nd ed., pp. 114–146). Sage. https://doi.org/10.4135/9781452243764.n5

Protzner, S., & van de Velde, S. (2015). *Mind the gap between demand and supply: A behavioral perspective on demand forecasting*.

Ray, J. L., Baker, L. T., & Plowman, D. A. (2011). Organizational mindfulness in business schools. *Academy of Management Learning & Education, 10*(2), 188–203. https://doi.org/10.5465/amle.10.2.zqr188

Rowe, G., & Wright, G. (1996). The impact of task characteristics on the performance of structured group forecasting techniques. *International Journal of Forecasting, 12*(1), 73–89. https://doi.org/10.1016/0169-2070(95)00658-3

Rowe, G., & Wright, G. (2001). Expert opinions in forecasting: The role of the Delphi technique. In J. S. Armstrong (Ed.), *Principles of forecasting: A handbook for researchers and practitioners* (pp. 125–144). Springer US. https://doi.org/10.1007/978-0-306-47630-3_7

Sagan, S. D. (1995). *The limits of safety: Organizations, accidents, and nuclear weapons*. Princeton University Press.

Salas, E., Cannon-Bowers, J. A., & Johnston, J. H. (1998). Lessons learned from conducting the TADMUS program: Balancing science, practice, and more. In *Making decisions under stress: Implications for individual and team training* (pp. 409–413). American Psychological Association. https://doi.org/10.1037/10278-017

Scheele, L. M., Thonemann, U. W., & Slikker, M. (2018). Designing incentive systems for truthful forecast information sharing within a firm. *Management Science, 64*(8), 3690–3713. https://doi.org/10.1287/mnsc.2017.2805

Schleppenbach, M., Flevares, L. M., Sims, L. M., & Perry, M. (2007). Teachers' responses to student mistakes in Chinese and U.S. mathematics classrooms. *The Elementary School Journal, 108*(2), 131–147. https://doi.org/10.1086/525551

Scott Armstrong, J. (1988). Research needs in forecasting. *International Journal of Forecasting, 4*(3), 449–465. https://doi.org/10.1016/0169-2070(88)90111-2

Shapira, Z. (2008, March 6). On the implications of behavioral decision theory for managerial decision making: Contributions and challenges. The Oxford handbook of organizational decision making. Oxford University Press https://doi.org/10.1093/oxfordhb/9780199290468.003.0015.

Sitkin, S. B. (1990). *Learning through failure: The strategy of small losses*. Department of Management [University of Texas at Austin].

Slovic, P., & Lichtenstein, S. (1971). Comparison of Bayesian and regression approaches to the study of information processing in judgment. *Organizational Behavior and Human Performance, 6*(6), 649–744. https://doi.org/10.1016/0030-5073(71)90033-X

Slovic, P., Fischhoff, B., & Lichtenstein, S. (1977). Behavioral decision theory. *Annual Review of Psychology, 28*(1), 1–39. https://doi.org/10.1146/annurev.ps.28.020177.000245

Smith-Jentsch, K. A., Cannon-Bowers, J. A., Tannenbaum, S. I., & Salas, E. (2008). Guided team self-correction: Impacts on team mental models, processes, and effectiveness. *Small Group Research, 39*(3), 303–327. https://doi.org/10.1177/1046496408317794

Sniezek, J. A. (1989). An examination of group process in judgmental forecasting. *International Journal of Forecasting, 5*(2), 171–178. https://doi.org/10.1016/0169-2070(89)90084-8

Starbuck, W. H. (2015). Karl E. Weick and the dawning awareness of organized cognition. *Management Decision, 53*(6), 1287–1299. https://doi.org/10.1108/MD-04-2014-0183

Stevenson, H. W., & Stigler, J. W. (1992). *The learning gap: Why our schools are failing and what we can learn from Japanese and Chinese education*. Summit Books.

Stewart, T. R. (2001). Improving reliability of judgmental forecasts. In J. S. Armstrong (Ed.), *Principles of forecasting: A handbook for researchers and practitioners* (pp. 81–106). Springer US. https://doi.org/10.1007/978-0-306-47630-3_5

Sunstein, C. R. (2015). *Wiser: Getting beyond groupthink to make better decisions*. Harvard Business Review Press.

Surowiecki, J. (2005). *The wisdom of crowds*. Anchor.

Sutcliffe, K. M. (1994). What executives notice: Accurate perceptions in top management teams. *The Academy of Management Journal, 37*(5), 1360–1378. JSTOR. https://doi.org/10.2307/256677

Sutcliffe, K. M., & Weber, K. (2003). The high cost of accurate knowledge. *Harvard Business Review, 81*(5), 74–82.

Sutcliffe, K. M., Vogus, T. J., & Dane, E. (2016). Mindfulness in organizations: A cross-level review. *Annual Review of Organizational Psychology and Organizational Behavior, 3*(1), 55–81. https://doi.org/10.1146/annurev-orgpsych-041015-062531

Tannenbaum, S. I., & Cerasoli, C. P. (2013). Do team and individual debriefs enhance performance? A meta-analysis. *Human Factors, 55*(1), 231–245. https://doi.org/10.1177/0018720812448394

Tannenbaum, S. I., Smith-Jentsch, K. A., & Behson, S. J. (1998). Training team leaders to facilitate team learning and performance. In *Making decisions under stress: Implications for individual and team training* (pp. 247–270). American Psychological Association. https://doi.org/10.1037/10278-009

Taylor, S. E. (1989). *Positive illusions: Creative self-deception and the healthy mind*. Basic Books.

Tetlock, P. E., & Gardner, D. (2015). *Superforecasting: The art and science of prediction* (p. 340). Crown Publishers/Random House.

Tindale, R. S. (1993). Decision errors made by individuals and groups. In *Individual and group decision making: Current issues* (pp. 109–124). Lawrence Erlbaum Associates.

Tversky, A., & Kahneman, D. (1974). Judgment under uncertainty: Heuristics and biases. *Science, 185*(4157), 1124–1131. https://doi.org/10.1126/science.185.4157.1124

van den Bossche, P., Gijselaers, W., Segers, M., Woltjer, G. B., & Kirschner, P. (2011). Team learning: Building shared mental models. *Instructional Science, 39*(3), 283–301. https://doi.org/10.1007/s11251-010-9128-3

Vogus, T., & Colville, I. (2017). Sensemaking, simplexity and mindfulness. In *The Sage handbook of process organization studies* (pp. 340–355). Sage.

Vogus, T. J., & Sutcliffe, K. M. (2012). Organizational mindfulness and mindful organizing: A reconciliation and path forward. *Academy of Management Learning & Education, 11*(4), 722–735. https://doi.org/10.5465/amle.2011.0002c

Wang, T., & Murphy, J. (2004). An examination of coherence in a Chinese mathematics classroom. In *How Chinese learn mathematics* (Vol. 1, pp. 107–123). World Scientific. https://doi.org/10.1142/9789812562241_0004

Weick, K. E. (1995). *Sensemaking in organizations*. Sage.

Weick, K. E., & Roberts, K. H. (1993). Collective mind in organizations: Heedful interrelating on flight decks. *Administrative Science Quarterly, 38*(3), 357–381. https://doi.org/10.2307/2393372

Weick, K., & Sutcliffe, K. (2007). *Managing the unexpected: Resilient performance in an age of uncertainty* (2nd ed.). Wiley.

Weick, K. E., Sutcliffe, K. M., & Obstfeld, D. (1999). Organizing for high reliability: Processes of collective mindfulness. In *Research in organizational behavior* (Vol. 21, pp. 81–123). Elsevier Science/JAI Press.

Weick, K. E., Sutcliffe, K. M., & Obstfeld, D. (2005). Organizing and the process of Sensemaking. *Organization Science, 16*(4), 409–421. https://doi.org/10.1287/orsc.1050.0133

Winklhofer, H., Diamantopoulos, A., & Witt, S. F. (1996). Forecasting practice: A review of the empirical literature and an agenda for future research. *International Journal of Forecasting, 12*(2), 193–221. https://doi.org/10.1016/0169-2070(95)00647-8

Correction to: Performance-Weighted Aggregation: Ferreting Out Wisdom Within the Crowd

Robert N. Collins, David R. Mandel, and David V. Budescu

Correction to:
Chapter 7 in: M. Seifert (ed.), *Judgment in Predictive Analytics*, International Series in Operations Research & Management Science 343, https://doi.org/10.1007/978-3-031-30085-1_7

The copyright holder of this chapter has been retrospectively corrected. The correct copyright holder name is:

© His Majesty the King in Right of Canada as represented by Department of National Defence (2023)

The updated version of this chapter can be found at https://doi.org/10.1007/978-3-031-30085-1_7

M. Seifert (ed.), *Judgment in Predictive Analytics*, International Series in Operations Research & Management Science 343,
https://doi.org/10.1007/978-3-031-30085-1_12

Index

M. Seifert (ed.), *Judgment in Predictive Analytics*, International Series in Operations Research & Management Science 343,
https://doi.org/10.1007/978-3-031-30085-1

MIX
Papier aus verantwortungsvollen Quellen
Paper from responsible sources
FSC® C105338

If you have any concerns about our products,
you can contact us on
ProductSafety@springernature.com

In case Publisher is established outside the EU,
the EU authorized representative is:
Springer Nature Customer Service Center GmbH
Europaplatz 3, 69115 Heidelberg, Germany

Printed by Libri Plureos GmbH
in Hamburg, Germany